BASIC MATHEMATICS

BASIC MATHEMATICS

Howard A. Silver
Department of Mathematics
Chicago State University

PRENTICE-HALL, INC., ENGLEWOOD CLIFFS, NEW JERSEY 07632

Library of Congress Cataloging in Publication Data

Silver, Howard A., (date)
Basic mathematics.

Includes indexes.
1. Mathematics—1961–   . I. Title.
QA39.2.S518   513'.142   81-15745
ISBN 0-13-062802-6   AACR2

Editorial/production supervision: *Kathleen M. Lafferty*
Manufacturing buyer: *John B. Hall*

To my family

Printed in the United States of America

10 9 8 7 6 5 4

ISBN 0-13-062802-6

Prentice-Hall International, Inc., *London*
Prentice-Hall of Australia Pty. Limited, *Sydney*
Prentice-Hall of Canada, Ltd., *Toronto*
Prentice-Hall of India Private Limited, *New Delhi*
Prentice-Hall of Japan, Inc., *Tokyo*
Prentice-Hall of Southeast Asia Pte. Ltd., *Singapore*
Whitehall Books Limited, *Wellington, New Zealand*

CONTENTS

PREFACE

Basic Mathematics is written for college students who need to review the basic skills and ideas of mathematics. The text is designed to be covered in one semester, and as such it cannot be a complete discourse on the theory of arithmetic; rather, it is a skills-oriented approach to the most important and most needed skills in basic mathematics.

The text can be used in either a lecture or an individualized format. Below are the main features of the text that will help to provide mathematical competency for the student.

Section Objectives At the beginning of each section within the chapters, the goals or objectives for the students are spelled out as follows: "After completing this unit, you will be able to. . . ."

Typical Problems Before the actual instruction in each section, four to eight typical problems from the section are shown, with the answers given below. These problems serve to enhance the objectives and give the student a chance to preview the actual material to be covered.

Discussion Each section starts with a brief discussion of the concepts behind the skills. Every effort has been made to keep this text as simple and readable as possible. The procedures for the skills are then given and set off from the page with a box.

Examples Basically, students learn mathematics by examples and problems. This text is filled with carefully chosen and worked examples, with each step highlighted by a flow chart.

Why Does It Work? For many of the rules in the earlier chapters, a brief picture of why these procedures for adding, subtracting, multiplying, and dividing work is shown.

Problem Sets Students learn by doing. These "bite-sized" problem sets appear throughout the sections. The student should work as many of these problems as possible before going on to the next material. Also, the answers to these problems appear immediately after that problem set.

Miscellaneous Exercises At the end of every section, there is a collection of problems that review all the skills of that section. The problems are keyed to the examples. For instance, beside Problems 10 to 18 might appear a box that reads "See Examples 21 to 25." This directs the student back to the examples for help, when necessary.

Applications Within the examples and at the end of almost every miscellaneous exercise are various applications of the skills to fields such as business, health, chemistry, physical science, life science, electricity, and so on. These applications serve to reinforce the material as well as stress the relevance of the material.

Yes/No Boxes These boxes appear throughout the entire text and serve to warn the student of potential danger in a common mistake. The mathematical statement is made in two forms: correctly, under the YES column; and incorrectly, under the NO column with a big X through it.

Chapter Summary Every chapter has a summary that includes the important words and phrases of the chapter. Also included is a list of the key ideas and skills of the chapter.

Chapter Review Test The review test is a sample multiple-choice test that covers all the material of the chapter and that prepares the student for an in-class test.

Answers Answers to all the miscellaneous exercises and all the chapter review tests appear at the back of the book.

Pre-Tests and Post-Tests The perforated, tabbed pages at the back of the text are supplemental exercises. They include the following:

1. *Chapter Pre-Test* Pre-tests can be used as a diagnostic aid to place the student into the proper sections needed for review. The number beside a group of problems refers the section in the chapter where that material can be found.
2. *Section Post-Tests* For each section, there are two ten-questioned post-tests that serve as quizzes on that section. The numbers beside the answer blanks refer to examples in the section that show a particular skill.
3. *Chapter Post-Tests* Like the chapter pre-tests, these post-tests summarize the material of the chapter. They can be used to prepare the students for an in-class test.

I have worked very hard on this text, and I hope it proves helpful to the students using it. I am grateful for the very valuable comments of my colleagues: Jackie Blackmon, Sue Anderson, Bob Bernhardt, Dorothy Dierickx, Tim Carroll, Miguel Campos, Ramona Choos, Jeni Morison, and the mathematics tutors at Chicago State University.

I gratefully acknowledge the help of the following reviewers during the preparation of the manuscript: Mary Kay Beavers, San Francisco City College; Linda K. Hines, Tarrant County Junior College; Martin Johnson, Montgomery County Community College; Judith F. Kneen, Montgomery College; Janice Koop, University of San Diego; Donald Perry, Lee College; and Edward Wright, Oregon College of Education.

Finally, I would like to thank the great Prentice–Hall staff for all their help, encouragement, and affection throughout the duration of this project: my acquisitions editors, Harry Gaines and Bob Sickles; my production editor, Kathi Lafferty; a concatenation of field editors, Jo Slivinski, Beth Elderd, Cindy Marione Wallach, and Mike Ruhe; and all the copyeditors, artists, and so on whose names I never learned. Lastly, I would like to thank my typists, Karen Utes and Cecelia Roberts.

H. Silver

1

WHOLE NUMBERS

Before you begin this chapter, complete the Chapter 1 Pre-Test on page T3.

1.1 PLACE VALUE, NAMING WHOLE NUMBERS, AND ROUNDING

SECTION OBJECTIVES

After completing this unit, you will be able to:

1. Determine the place values of whole numbers.

2. Name a whole number in words.

3. Round a whole number to any accuracy.

TYPICAL PROBLEMS

You will be able to work problems like the ones below.

(a) Write 396 in expanded form.

(b) Write 21,098 in expanded form.

(c) In 407, what is the value of the 4?

(d) In 61,347, what is the value of the 1?

(e) In 32,906, what digit is in the tens' place?

(f) In 218,097, what digit is in the ten-thousands' place?

(g) Write 2798 in words.

(h) Write 45,897,410 in words.

(i) Round 713,380 to the nearest ten thousand.

ANSWERS

(a) 300 + 90 + 6 **(b)** 20,000 + 1000 + 90 + 8 **(c)** 400 **(d)** 1000 **(e)** 0 **(f)** 1
(g) Two thousand, seven hundred ninety-eight
(h) Forty-five million, eight hundred ninety-seven thousand, four hundred ten **(i)** 710,000

The number system we use for whole numbers and decimals is called the **base-10 place-value system.** This means that there are columns, and each column has a value. Each value is a power of 10: 1, 10, 100, 1000, and so on. The value of each column is 10 times the value of the column to the right of it.

. . .	10,000,000,000	1,000,000	100,000	10,000	1000	100	10	1
. . .	ten millions	millions	hundred thousands	ten thousands	thousands	hundreds	tens	ones

We can write any whole number in **expanded form,** in which we express a whole number as the sum of all its column values. For example, 67 means 6 *tens* and 7 *ones*, or $6 \times 10 + 7 \times 1$.

EXAMPLE 1 Write 217 in expanded form.

SOLUTION We write this in terms of powers of 10.

$$\begin{aligned} 217 &= 2 \times 100 + 1 \times 10 + 7 \times 1 \\ &= 200 + 10 + 7 \end{aligned}$$

EXAMPLE 2 Write 35,698 in expanded form.

SOLUTION We write this in terms of powers of 10.

$$\begin{aligned} 35{,}698 &= 3 \times 10{,}000 + 5 \times 1000 + 6 \times 100 + 9 \times 10 + 8 \times 1 \\ &= 30{,}000 + 5000 + 600 + 90 + 8 \end{aligned}$$

EXAMPLE 3 Write 5003 in expanded form.

SOLUTION The zeros are used here as **place holders.**

$$\begin{aligned} 5003 &= 5 \times 1000 + 0 \times 100 + 0 \times 10 + 3 \times 1 \\ &= 5000 + 3 \end{aligned}$$

YES	NO
8052 = 8000 + 50 + 2	~~8052 = 800 + 50 + 2~~
The zero is a place holder.	

PROBLEM SET

See Examples 1 to 3

1. Write 27 in expanded form.
2. Write 356 in expanded form.
3. Write 298 in expanded form.
4. Write 309 in expanded form.
5. Write 4579 in expanded form.
6. Write 5907 in expanded form.
7. Write 47,247 in expanded form.
8. Write 50,902 in expanded form.
9. Write 883,726 in expanded form.
10. Write 1,205,008 in expanded form.

ANSWERS 1. 20 + 7 2. 300 + 50 + 6 3. 200 + 90 + 8 4. 300 + 9
5. 4000 + 500 + 70 + 9 6. 5000 + 900 + 7
7. 40,000 + 7000 + 200 + 40 + 7 8. 50,000 + 900 + 2
9. 800,000 + 80,000 + 3000 + 700 + 20 + 6
10. 1,000,000 + 200,000 + 5000 + 8

> To find the value of a digit in a whole number:
> 1. Determine the column values.
> 2. Multiply the digit by the value of the column.

EXAMPLE 4 In 83,051, what is the value of the 3?

SOLUTION We first determine the column values.

10,000	1000	100	10	1
8	3	0	5	1

↑

value of 3 = 3 × 1000 = 3000

Determine column values → Multiply digit by column value

EXAMPLE 5 In 1,753,039, what is the value of the 7?

SOLUTION We first determine the column values.

1,000,000	100,000	10,000	1000	100	10	1
1	7	5	3	0	3	9

↑

value of 7 = 7 × 100,000 = 700,000

Determine column values → Multiply digit by column value

EXAMPLE 6 In 5085, what is the value of the 8?

SOLUTION

1000	100	10	1
5	0	8	5

↑

value of 8 = 8 X 10 = 80

Determine column values → Multiply digit by column value

PROBLEM SET

See Examples 4 to 6

1. In 257, what is the value of the 5?
2. In 35, what is the value of the 3?
3. In 604, what is the value of the 6?
4. In 4579, what is the value of the 7?
5. In 2089, what is the value of the 2?
6. In 12,459, what is the value of the 4?
7. In 204,897, what is the value of the 2?
8. In 23,560,872, what is the value of the 6?
9. In 123,496,593, what is the value of the 2?
10. In 200,400,000,000, what is the value of the 4?

ANSWERS 1. 50 2. 30 3. 600 4. 70 5. 2000 6. 400
7. 200,000 8. 60,000 9. 20,000,000 10. 400,000,000

EXAMPLE 7 In 4579, what digit is in the tens' column?

SOLUTION

1000	100	10	1
4	5	7	9

↑

digit in tens' column = 7

Determine column values

EXAMPLE 8 In 6053, what digit is in the hundreds' column?

SOLUTION

1000	100	10	1
6	0	5	3

↑

digit in hundreds' column = 0

Determine column values

EXAMPLE 9 In 8,943,682, what digit is in the hundred-thousands' column?

SOLUTION

1,000,000	100,000	10,000	1000	100	10	1
8	9	4	3	6	8	2

Determine column values

↑ digit in hundred-thousands' column = 9

PROBLEM SET

See Examples 7 to 9

1. In 285, what digit is in the tens' column?
2. In 1097, what digit is in the ones' column?
3. In 3497, what digit is in the thousands' column?
4. In 23,597, what digit is in the hundreds' column?
5. In 305,789, what digit is in the ten-thousands' column?
6. In 12,475,684, what digit is in the hundred-thousands' column?
7. In 3,356,893, what digit is in the thousands' column?
8. In 45,604,301, what digit is in the millions' column?
9. In 309,298,677, what digit is in the hundred-millions' column?
10. In 42,965,008, what digit is in the ten-thousands' column?

ANSWERS **1.** 8 **2.** 7 **3.** 3 **4.** 5 **5.** 0 **6.** 4 **7.** 6 **8.** 5 **9.** 3 **10.** 6

> To name a three-digit number in words:
>
> 1. Think of number expanded into hundreds, tens, and ones.
> 2. Name the hundreds, tens, and ones using Table 1.1.

Table 1.1

100 = one hundred	10 = ten	1 = one
200 = two hundred	20 = twenty	2 = two
300 = three hundred	30 = thirty	3 = three
400 = four hundred	40 = forty	4 = four
500 = five hundred	50 = fifty	5 = five
600 = six hundred	60 = sixty	6 = six
700 = seven hundred	70 = seventy	7 = seven
800 = eight hundred	80 = eighty	8 = eight
900 = nine hundred	90 = ninety	9 = nine

EXAMPLE 10 Write 652 in words.

SOLUTION

652 = 600 + 50 + 2

= six hundred fifty-two

Expand number → Name each number

EXAMPLE 11 Write 798 in words.

SOLUTION 798 = 700 + 90 + 8

= seven hundred ninety-eight

Expand number → Name each number

EXAMPLE 12 Write 319 in words.

SOLUTION 319 = 300 + 10 + 9

= three hundred nineteen

Expand number → Name each number

Notice that we write 10 + 9 as nineteen, and not as ten nine.

Notice that we hyphenate two-digit numbers, such as fifty-two and ninety-eight.

> To name any whole number:
>
> 1. Split the number into groups of three (starting from the right).
> 2. Name each group as a three-digit number.
> 3. Label each group as follows.
>
> | 100,000,000 | 10,000,000 | 1,000,000 | 100,000 | 10,000 | 1000 | 100 | 10 | 1 |
> |---|---|---|---|---|---|---|---|---|
> | -million | | | -thousand | | | | | |

EXAMPLE 13 Write 189,521 in words.

SOLUTION 189, 521

521 → five hundred twenty-one

189 → one hundred eighty-nine *thousand*,

Split into groups of three → Name each three-digit number

Thus, 189,521 is one hundred eighty-nine thousand, five hundred twenty-one.

EXAMPLE 14 Write 21,692,044 in words.

SOLUTION 21, 692, 044

21 → twenty-one *million*,

692 → six hundred ninety-two *thousand*,

044 → forty-four

Split into groups of three → Name each three-digit number

Thus, 21,692,044 is twenty-one million, six hundred ninety-two thousand, forty-four.

EXAMPLE 15 Write 614,267,831 in words.

SOLUTION

614, 267, 831

six hundred fourteen *million*, two hundred sixty-seven *thousand*, eight hundred thirty-one

Thus, 614,267,831 is six hundred fourteen million, two hundred sixty-seven thousand, eight hundred thirty-one.

PROBLEM SET

See Examples 10 to 15

1. Write 36 in words.
2. Write 278 in words.
3. Write 491 in words.
4. Write 906 in words.
5. Write 3468 in words.
6. Write 8043 in words.
7. Write 12,697 in words.
8. Write 39,091 in words.
9. Write 203,700 in words.
10. Write 3,098,678 in words.

ANSWERS

1. Thirty-six 2. Two hundred seventy-eight
3. Four hundred ninety-one 4. Nine hundred six
5. Three thousand, four hundred sixty-eight 6. Eight thousand, forty-three
7. Twelve thousand, six hundred ninety-seven
8. Thirty-nine thousand, ninety-one
9. Two hundred three thousand, seven hundred
10. Three million, ninety-eight thousand, six hundred seventy-eight

---

Often, we are given a number that has more digits than we are interested in. In these cases, we wish to simplify the number by **rounding off.**

> To round off whole-number places:
>
> 1. Determine the place to be rounded.
> 2. Go one extra place to the right.
>
> 3a. If this digit is 4 or less (0, 1, 2, 3, or 4), change this digit and all those to the right to 0. (This is called **rounding down.**)
>
> 3b. If this digit is 5 or more (5, 6, 7, 8, or 9), make rounded place one more and change all those to the right to 0. (This is called **rounding up**.)

EXAMPLE 16 Round 9468 to the nearest hundred.

SOLUTION We first find the hundreds' place; then we go one place right.

9468 ↑	Determine place to be rounded ↓
94[6]8 ↑	Go one place right ↓
9500	Round up (and change digits to 0) since digit is 6

EXAMPLE 17 Round 178,466,592 to the nearest million.

SOLUTION We start at the left and find the millions' column.

178,466,592 ↑	Determine place to be rounded ↓
178,[4]66,592 ↑	Go one place right ↓
178,000,000	Round down (and change digits to 0) since digit is 4

EXAMPLE 18 A certain city has a population of 1,681,453 people. Round this off to the nearest hundred thousand.

SOLUTION We solve this like any other round-off problem.

1,681,453 ↑	Determine place to be rounded ↓
1,6[8]1,453 ↑	Go one place right ↓
1,700,000	Round up (and change digits to 0) since digit is 8

PROBLEM SET

See Examples 16 to 18

1. Round 473 to the nearest ten.
2. Round 385,937 to the nearest hundred.
3. Round 36,962 to the nearest ten.
4. Round 45,937 to the nearest thousand.
5. Round 367,902 to the nearest ten thousand.

6. Round 29,950,735 to the nearest million.

7. Round 57,805 to the nearest ten.

8. Round 9873 to the nearest thousand.

9. Round 5826 to the nearest hundred.

10. Round 12,345,678,921 to the nearest hundred thousand.

ANSWERS 1. 470 2. 385,900 3. 36,960 4. 46,000 5. 370,000
6. 30,000,000 7. 57,810 8. 10,000 9. 5,800
10. 12,345,700,000

Miscellaneous Exercises 1.1

Write the following numbers in expanded form.

See Examples 1 to 3

1. 45
2. 69
3. 18
4. 579
5. 210
6. 307
7. 5396
8. 40,081
9. 805,703

Determine the column values of the following digits.

See Examples 4 to 6

10. In 34,971, what is the value of the 9?
11. In 12,086, what is the value of the 2?
12. In 304,981, what is the value of the 8?
13. In 2,903,812, what is the value of the 9?
14. In 35,812,042, what is the value of the 5?
15. In 30,062,002, what is the value of the 6?

Determine the digit in the indicated column.

See Examples 7 to 9

16. In 3982, what digit is in the tens' column?
17. In 34,981, what digit is in the hundreds' column?
18. In 42,097, what digit is in the ones' column?
19. In 403,985, what digit is in the thousands' column?
20. In 1,084,742, what digit is in the hundred-thousands' column?
21. In 32,083,173, what digit is in the ten-thousands' column?

Write the following numbers in words.

See Examples 10 to 12

22. 369
23. 612
24. 128
25. 903

See Examples 13 to 15

26. 3583
27. 61,346
28. 109,854
29. 23,547,901

Round the following numbers to the indicated accuracy.

See Examples 16 to 18

30. Round 4673 to the nearest hundred.
31. Round 23,694 to the nearest ten thousand.

32. Round 573,964 to the nearest thousand.

33. Round 485,963,476 to the nearest million.

Consumer application

34. When writing checks, we must write the amount of the check both as a numeral and in words. Fill in the words for the following imaginary checks. (Real checks also have a decimal part; we study these in Chapter 3.)

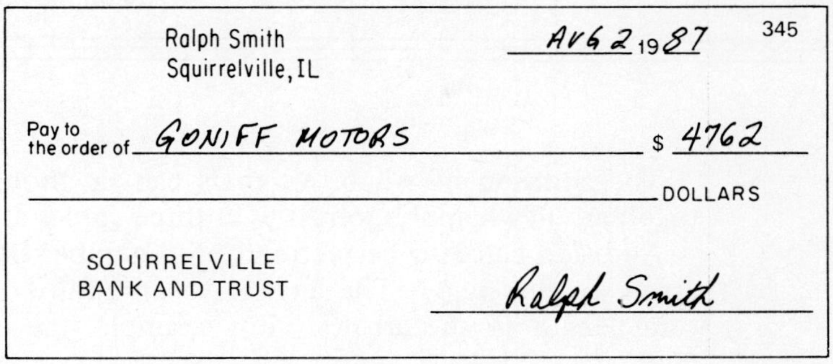
Ralph Smith
Squirrelville, IL
AUG 2 19 87
345
Pay to the order of GONIFF MOTORS $ 4762
DOLLARS
SQUIRRELVILLE
BANK AND TRUST
Ralph Smith

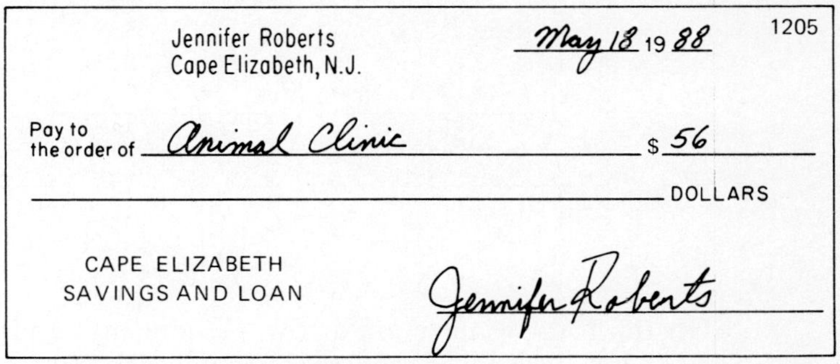
Jennifer Roberts
Cape Elizabeth, N.J.
May 18 19 88
1205
Pay to the order of Animal Clinic $ 56
DOLLARS
CAPE ELIZABETH
SAVINGS AND LOAN
Jennifer Roberts

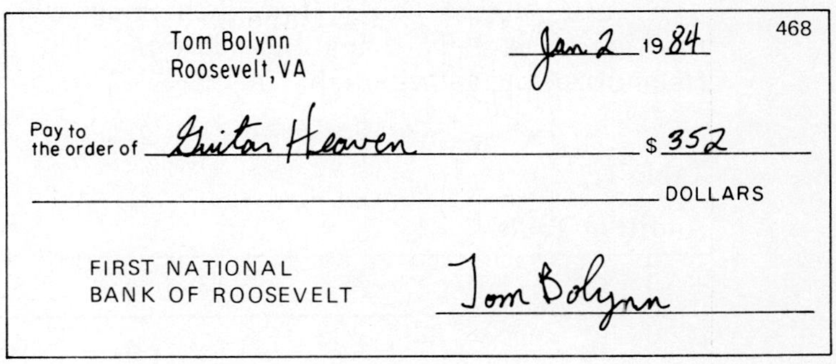
Tom Bolynn
Roosevelt, VA
Jan 2 19 84
468
Pay to the order of Guitar Heaven $ 352
DOLLARS
FIRST NATIONAL
BANK OF ROOSEVELT
Tom Bolynn

Now complete Post-Test 1 for Section 1.1 on page T5.

1.2 ADDITION OF WHOLE NUMBERS

SECTION OBJECTIVES After completing this unit, you will be able to add several whole numbers of all sizes.

TYPICAL PROBLEMS You will be able to work problems like the ones below.

(a) Add 21 + 148.

(b) Add 452 + 8036 + 98 + 5384.

(c)	79	(d)	893	(e)	26982
	152		18237		41908
	83		+ 6955		84563
	+ 21				+ 5287

ANSWERS **(a)** 169 **(b)** 13,970 **(c)** 335 **(d)** 26,085 **(e)** 158,740

The **addition** of whole numbers can be thought of as two sets being joined together. (Two apples joined with three apples are five apples.)

Addition can also be pictured on a **number line**. Each number is shown by an arrow to the right. The answer to an addition problem, called the **sum**, is the total length of the arrows. For example, the following number line shows the picture for $3 + 4 = 7$:

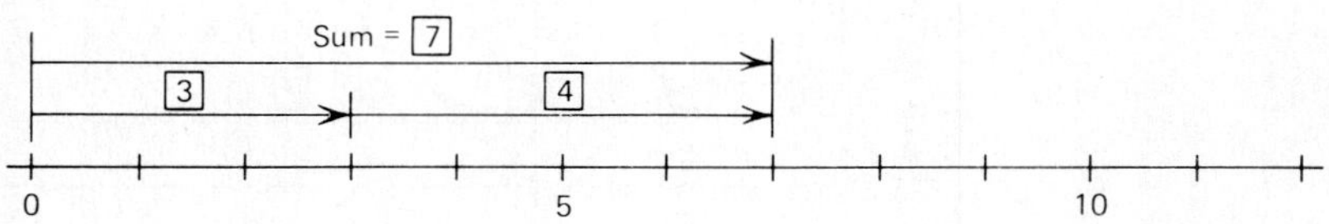

On the top, we have a 3-arrow. Where that ends, we start a 4-arrow. Together, they are the same as the 7-arrow below the line.

We assume that you pretty much know what addition is but that you need practice in actually doing the addition. The first thing you must do is learn (memorize) the addition table in Table 1.2.

Table 1.2

Addition Table

+	0	1	2	3	4	5	6	7	8	9
0	0	1	2	3	4	5	6	7	8	9
1	1	2	3	4	5	6	7	8	9	10
2	2	3	4	5	6	7	8	9	10	11
3	3	4	5	6	7	8	9	10	11	12
4	4	5	6	7	8	9	10	11	12	13
5	5	6	7	8	9	10	11	12	13	14
6	6	7	8	9	10	11	12	13	14	15
7	7	8	9	10	11	12	13	14	15	16
8	8	9	10	11	12	13	14	15	16	17
9	9	10	11	12	13	14	15	16	17	18

The numbers that we are adding together are called **addends.** To use the addition table, find one addend in the left-hand row and the other addend in the top column. Where they meet is the sum.

EXAMPLE 19 Add $4 + 8$.

SOLUTION

$$\begin{array}{rl} 4 & \text{addend} \\ +\ 8 & \text{addend} \\ \hline 12 & \text{sum} \end{array}$$

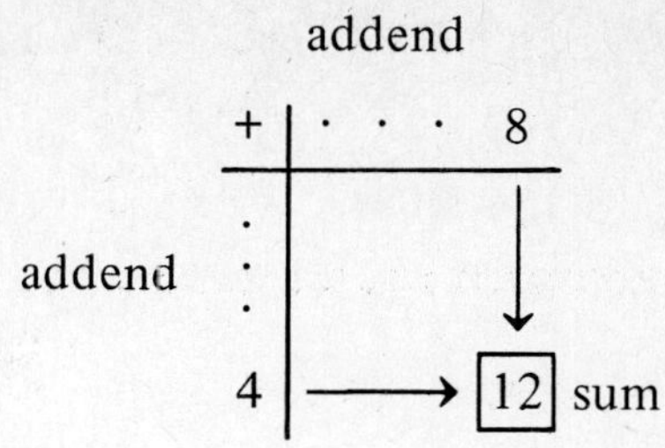

PROBLEM SET Add the following one-digit numbers.

See Example 19

1. $3 + 5$ 2. $7 + 5$ 3. $8 + 2$ 4. $6 + 6$ 5. $9 + 4$
6. $8 + 0$ 7. $9 + 2$ 8. $3 + 7$ 9. $0 + 7$ 10. $6 + 9$
11. $8 + 3$ 12. $9 + 7$ 13. $1 + 8$ 14. $2 + 7$ 15. $2 + 0$

16. $\begin{array}{r} 6 \\ +7 \\ \hline \end{array}$ 17. $\begin{array}{r} 8 \\ +5 \\ \hline \end{array}$ 18. $\begin{array}{r} 3 \\ +9 \\ \hline \end{array}$ 19. $\begin{array}{r} 9 \\ +9 \\ \hline \end{array}$ 20. $\begin{array}{r} 8 \\ +8 \\ \hline \end{array}$

21. $\begin{array}{r} 1 \\ +9 \\ \hline \end{array}$ 22. $\begin{array}{r} 8 \\ +7 \\ \hline \end{array}$ 23. $\begin{array}{r} 7 \\ +7 \\ \hline \end{array}$ 24. $\begin{array}{r} 0 \\ +7 \\ \hline \end{array}$ 25. $\begin{array}{r} 5 \\ +9 \\ \hline \end{array}$

ANSWERS **1.** 8 **2.** 12 **3.** 10 **4.** 12 **5.** 13 **6.** 8 **7.** 11 **8.** 10 **9.** 7 **10.** 15 **11.** 11 **12.** 16 **13.** 9 **14.** 9 **15.** 2 **16.** 13 **17.** 13 **18.** 12 **19.** 18 **20.** 16 **21.** 10 **22.** 15 **23.** 14 **24.** 7 **25.** 14

> To add larger whole numbers:
>
> 1. Line the numbers up on the right (ones' column).
> 2. Add the ones' column, add the tens' column, add the hundreds' column, and so on.

EXAMPLE 20 Add 83 + 15.

SOLUTION We line these up on the right and add each column. We show both the usual way that this is written as well as in "slow motion."

Usual

$$\begin{array}{r} 83 \\ +15 \\ \hline 98 \end{array}$$

Slow Motion

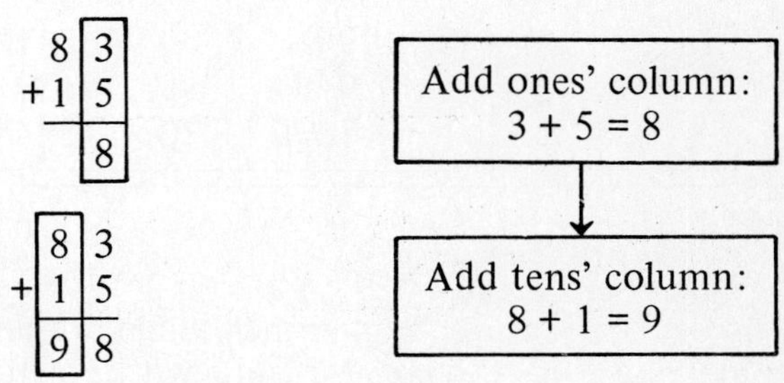

The sum is 98.

EXAMPLE 21 Add 305 + 121 + 232.

SOLUTION We again line these up on the right and add by columns.

Usual	Slow Motion	
$\begin{array}{r} 305 \\ 121 \\ +232 \\ \hline 658 \end{array}$	$\begin{array}{rr\|c\|} \cline{3-3} 3 & 0 & 5 \\ 1 & 2 & 1 \\ +2 & 3 & 2 \\ \hline & & 8 \\ \cline{3-3} \end{array}$	Add ones' column: $5 + 1 + 2 = 8$
	$\begin{array}{r\|c\|r} \cline{2-2} 3 & 0 & 5 \\ 1 & 2 & 1 \\ +2 & 3 & 2 \\ \hline & 5 & 8 \\ \cline{2-2} \end{array}$	Add tens' column: $0 + 2 + 3 = 5$
	$\begin{array}{r\|c\|rr} \cline{2-2} & 3 & 0 & 5 \\ & 1 & 2 & 1 \\ + & 2 & 3 & 2 \\ \hline & 6 & 5 & 8 \\ \cline{2-2} \end{array}$	Add hundreds' column: $3 + 1 + 2 = 6$

The sum is 658.

PROBLEM SET Add the following whole numbers.

See Examples 20 and 21

1. $\begin{array}{r} 34 \\ +51 \\ \hline \end{array}$	**2.** $\begin{array}{r} 62 \\ +17 \\ \hline \end{array}$	**3.** $\begin{array}{r} 73 \\ +23 \\ \hline \end{array}$	**4.** $\begin{array}{r} 90 \\ +\ 8 \\ \hline \end{array}$	**5.** $\begin{array}{r} 207 \\ +341 \\ \hline \end{array}$
6. $\begin{array}{r} 691 \\ +106 \\ \hline \end{array}$	**7.** $\begin{array}{r} 731 \\ +148 \\ \hline \end{array}$	**8.** $\begin{array}{r} 906 \\ +\ 93 \\ \hline \end{array}$	**9.** $\begin{array}{r} 480 \\ +514 \\ \hline \end{array}$	**10.** $\begin{array}{r} 6290 \\ +\ 305 \\ \hline \end{array}$
11. $\begin{array}{r} 2178 \\ +6320 \\ \hline \end{array}$	**12.** $\begin{array}{r} 2682 \\ +4115 \\ \hline \end{array}$	**13.** $\begin{array}{r} 5729 \\ +3240 \\ \hline \end{array}$	**14.** $\begin{array}{r} 8371 \\ +1126 \\ \hline \end{array}$	**15.** $\begin{array}{r} 6477 \\ +2212 \\ \hline \end{array}$
16. $\begin{array}{r} 72632 \\ +12345 \\ \hline \end{array}$	**17.** $\begin{array}{r} 267 \\ 212 \\ +210 \\ \hline \end{array}$	**18.** $\begin{array}{r} 363 \\ 214 \\ +312 \\ \hline \end{array}$	**19.** $\begin{array}{r} 6257 \\ 1522 \\ +\ 120 \\ \hline \end{array}$	**20.** $\begin{array}{r} 12345 \\ 6420 \\ +\ \ 213 \\ \hline \end{array}$

ANSWERS **1.** 85 **2.** 79 **3.** 96 **4.** 98 **5.** 548 **6.** 797 **7.** 879 **8.** 999 **9.** 994 **10.** 6595 **11.** 8498 **12.** 6797 **13.** 8969 **14.** 9497 **15.** 8689 **16.** 84,977 **17.** 689 **18.** 889 **19.** 7899 **20.** 18,978

Sometimes when we add, the digits in one column add up to a number more than 9. Here we have to **carry**. This means that we put the ones' digit of the column sum in the column and "carry" the tens' digit over to be added with the next column to the left.

In addition problems, if the digits in a column add to 10 or more, record the ones' digit of the sum and "carry" the tens' digit of the sum to the next column to the left.

EXAMPLE 22 Add 48 + 27.

SOLUTION We "carry" in this problem.

Usual

$$\begin{array}{r} {\scriptstyle 1} \\ 48 \\ +27 \\ \hline 75 \end{array}$$

Slow Motion

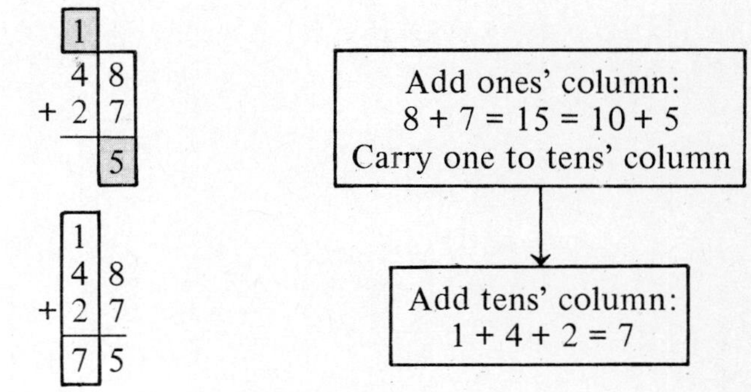

The sum is 75.

EXAMPLE 23 Add 3245 + 486.

SOLUTION We do this in the usual notation as well as in slow motion.

Usual

$$\begin{array}{r} {\scriptstyle 11} \\ 3245 \\ +\ 486 \\ \hline 3731 \end{array}$$

Slow Motion

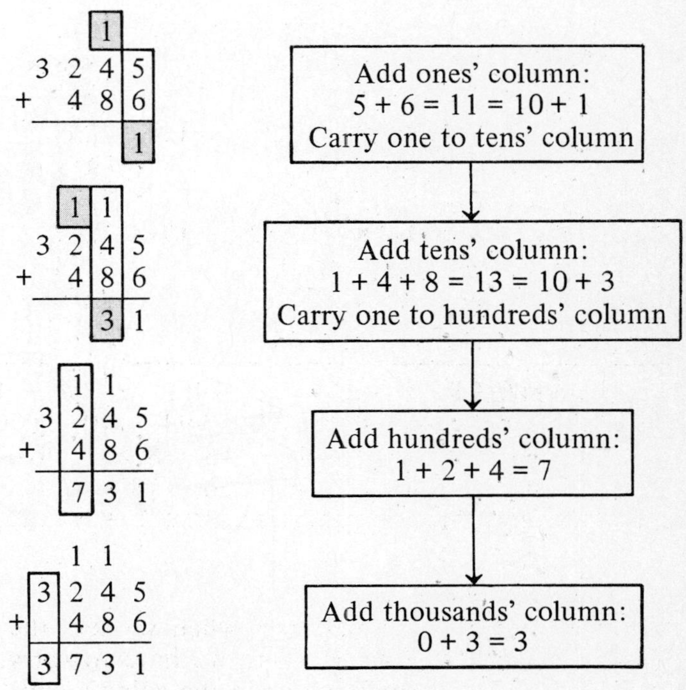

The sum is 3731.

EXAMPLE 24 Add 583 + 96 + 861.

SOLUTION We line these up on the right and add by columns.

Usual

```
  21
  583
   96
+ 861
 1540
```

Slow Motion

```
    1
  5 8 3
    9 6
 +8 6 1
      0
```

Add ones' column:
3 + 6 + 1 = 10
Carry one to tens' column

```
  2 1
  5 8 3
    9 6
 +8 6 1
    4 0
```

Add tens' column:
1 + 8 + 9 + 6 = 24 = 20 + 4
Carry two to hundreds' column

```
    2 1
    5 8 3
      9 6
 +  8 6 1
 1 5 4 0
```

Add hundreds' column:
2 + 5 + 0 + 8 = 15

The sum is 1540.

EXAMPLE 25 Charlie is itemizing his income tax deductions: medical, \$238; taxes, \$1986; interest, \$1834; contributions, \$751; losses, \$25; miscellaneous, \$296. What is his total deduction?

SOLUTION Whenever we see the word *total*, we should immediately think of addition. Here, we add all the deductions together.

Deduction	Amount
Medical	\$ 238
Taxes	1986
Interest	1834
Contributions	751
Losses	25
Miscellaneous	296
Total	\$5130

Thus Charlie's total deduction is \$5130.

EXAMPLE 26 Fran is on a diet and must total her calorie intake for the day: breakfast, 235; snack, 350; lunch, 550; snack, 450; dinner, 975; snack, 225. Find her total intake.

SOLUTION The word *total* means add.

Meal	Calories
Breakfast	235
Snack	350
Lunch	550
Snack	450
Dinner	975
Snack	225
Total	2785

Thus her calorie intake for the day is 2785.

Why Does It Work?

This rule works because we are adding common terms: ones, tens, hundreds, and so on.

$$\begin{array}{rcl} 48 & = & 4\ \textit{tens} + 8\ \textit{ones} \\ +\underline{27} & = & \underline{2\ \textit{tens} + 7\ \textit{ones}} \\ & & 6\ \textit{tens} + 15\ \textit{ones} \\ & = & \underbrace{6\ \textit{tens} + 1\ \textit{ten}} + 5\ \textit{ones} \\ 75 & = & 7\ \textit{tens} + 5\ \textit{ones} \end{array}$$

When we carry, we are rewriting 15 *ones* as 1 *ten* + 5 *ones*.

PROBLEM SET Add the following whole numbers.

See Examples 22 to 26

1. $\begin{array}{r} 14 \\ +\underline{23} \end{array}$ **2.** $\begin{array}{r} 58 \\ +\underline{29} \end{array}$ **3.** $\begin{array}{r} 19 \\ +\underline{54} \end{array}$ **4.** $\begin{array}{r} 203 \\ +\underline{159} \end{array}$

5. 92 + 19	**6.** 84 + 17	**7.** 79 + 8	**8.** 36 + 49
9. 795 + 328	**10.** 268 + 567	**11.** 479 + 578	**12.** 274 + 3969
13. 3841 + 76844	**14.** 3456 + 6871	**15.** 4857 + 3856	**16.** 2758 + 3869
17. 173 827 + 98	**18.** 402 73 + 1249	**19.** 823 449 + 747	**20.** 1292 6733 + 33841
21. 23 42 67 + 91	**22.** 241 73 822 + 446	**23.** 7822 22453 59167 + 82896	**24.** 16082 43381 52769 + 1323

25. Add 27 + 561.

26. Add 617 + 45 + 1801.

27. Add 19 + 206 + 442.

28. Add 1259 + 692 + 5811 + 26.

29. Add 607 + 916 + 4422 + 8.

30. Add 35,812 + 42,694 + 129,825.

31. Add 36 + 58.

32. Add 573 + 948.

33. Add 3967 + 4939.

34. Add 6893 + 39,599.

35. Add 483 + 583 + 485.

36. Add 573 + 697 + 294 + 496.

37. Add 582 + 58 + 3384 + 69 + 4966.

ANSWERS **1.** 37 **2.** 87 **3.** 73 **4.** 362 **5.** 111 **6.** 101 **7.** 87 **8.** 85 **9.** 1123 **10.** 835 **11.** 1057 **12.** 4243 **13.** 80,685 **14.** 10,327 **15.** 8713 **16.** 6627 **17.** 1098 **18.** 1724 **19.** 2019 **20.** 41,866 **21.** 223 **22.** 1582 **23.** 172,338 **24.** 113,555 **25.** 588 **26.** 2463 **27.** 667 **28.** 7788 **29.** 5953 **30.** 208,331 **31.** 94 **32.** 1521 **33.** 8906 **34.** 46,492 **35.** 1551 **36.** 2060 **37.** 9059

Miscellaneous Exercises 1.2

Add the following whole numbers.

See Example 19 **1.** 3 + 6 **2.** 6 + 9 **3.** 7 + 8

See Examples 20 and 21

4. 52 + 33	**5.** 231 342 + 13	**6.** 2136 1452 + 310

See Examples 22 to 24

7. $\begin{array}{r} 35 \\ +67 \\ \hline \end{array}$

8. $\begin{array}{r} 49 \\ +38 \\ \hline \end{array}$

9. $\begin{array}{r} 296 \\ 395 \\ +468 \\ \hline \end{array}$

10. $\begin{array}{r} 3758 \\ 4739 \\ +7563 \\ \hline \end{array}$

11. $\begin{array}{r} 4738 \\ 4781 \\ +8751 \\ \hline \end{array}$

12. $\begin{array}{r} 4736 \\ 9572 \\ 57839 \\ +37593 \\ \hline \end{array}$

13. 38 + 84

14. 68 + 483

15. 49 + 483 + 492

16. 96 + 488 + 579

17. 385 + 483 + 591

18. 693 + 4932 + 5084

Business application

19. The ABC Insurance Co. has four salespersons and sells three types of insurance. The sales for the preceding year were as follows:

Salesperson	Life	Automobile	Home
Arbuckle	$235,000	$53,000	$26,000
Brown	98,000	86,700	35,500
Munchkin	126,000	49,000	69,000
Zimkin	305,000	21,000	42,000

(a) How much insurance did Arbuckle sell? Brown? Munchkin? Zimkin? (Add the rows.)

(b) How much life insurance was sold? automobile insurance? home insurance? (Add the columns.)

Health application

20. A nurse is monitoring the caloric intake of three patients. He has the following data:

Patient	Breakfast	Lunch	Dinner
Bruce	450	585	1225
Susan	225	495	925
Nancy	335	615	845

(a) What was the daily calorie intake for Bruce? Susan? Nancy? (Add the rows.)

(b) How many total calories were consumed at breakfast? lunch? dinner? (Add the columns.)

Consumer application

21. Following are the itemized deductions from three families' income tax. Compute each of their total deductions.

Item	Smiths	O'Connors	LaRues
Medical	126	269	23
Taxes	1376	2482	921
Interest	1285	1985	1109
Contributions	369	482	123
Losses	137	56	0
Miscellaneous	428	239	386
Total	?	?	?

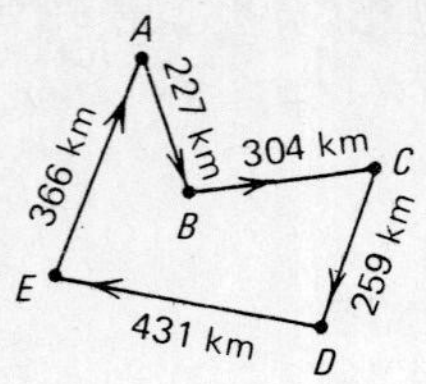

22. Roger is a traveling salesperson. At the left is a sketch of one of his business trips (starting and ending at town A). What is the total distance for this trip?

Now complete Post-Test 1 for Section 1.2 on page T7.

1.3 SUBTRACTION OF WHOLE NUMBERS

SECTION OBJECTIVE After completing this unit, you will be able to subtract two whole numbers of any size.

TYPICAL PROBLEMS You will be able to work problems like the ones below.

(a) Subtract 121 - 79.

(b) Subtract 1073 - 826.

(c) $\begin{array}{r} 653 \\ -492 \\ \hline \end{array}$ **(d)** $\begin{array}{r} 18258 \\ -\ 6929 \\ \hline \end{array}$ **(e)** Find n: $n + 10 = 21$.

ANSWERS **(a)** 42 **(b)** 247 **(c)** 161 **(d)** 11,329 **(e)** 11

The **subtraction** of two whole numbers can be thought of in two ways:

(a) *Take-away.* Here we have a set and wish to take a piece of it away. Consider the subtraction

$$8 - 3 = 5$$

The original set has eight objects. When we take three away, we are left with five objects.

(b) *Opposite of addition.* In addition to *take-away*, subtraction can also be thought of as the opposite of addition. The illustration pictures how the subtraction

$$10 - 4 = 6$$

answers the question: "What number added to 4 gives 10?"

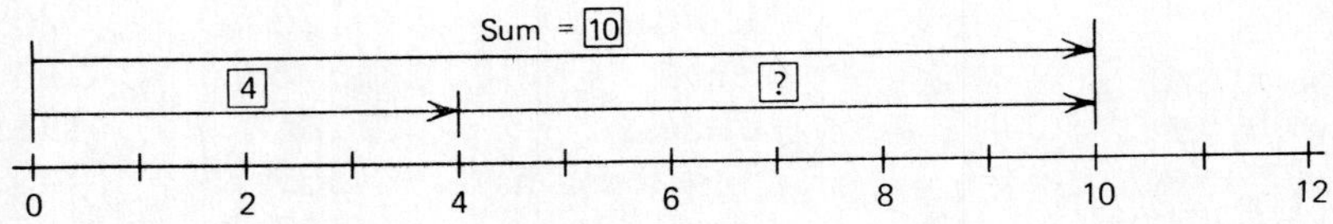

Later in this section and in the text we use the fact that subtraction is the opposite of addition. We use letters such as n or x instead of $\boxed{?}$. As examples,

$$4 + n = 10 \quad \text{means} \quad n = 10 - 4 = 6$$

$$n - 3 = 2 \quad \text{means} \quad n = 3 + 2 = 5$$

When subtracting, the first (or top) number is called the **minuend.** The second (or bottom) number is called the **subtrahend.** The result is called the **difference.** The subtraction facts, based on addition facts, are given in Table 1.3.

Table 1.3

Subtraction Facts

⊖	0	1	2	3	4	5	6	7	8	9
0	0									
1	1	0								
2	2	1	0							
3	3	2	1	0						
4	4	3	2	1	0					
5	5	4	3	2	1	0				
6	6	5	4	3	2	1	0			
7	7	6	5	4	3	2	1	0		
8	8	7	6	5	4	3	2	1	0	
9	9	8	7	6	5	4	3	2	1	0
10		9	8	7	6	5	4	3	2	1
11			9	8	7	6	5	4	3	2
12				9	8	7	6	5	4	3
13					9	8	7	6	5	4
14						9	8	7	6	5
15							9	8	7	6
16								9	8	7
17									9	8
18										9

EXAMPLE 27 Subtract 11 - 6.

We find the 11-row at the left and the 6-column at the top. They meet at 5. This is the answer.

$$\begin{array}{rl} 11 & \text{minuend} \\ -\ \ 6 & -\,\text{subtrahend} \\ \hline 5 & \text{difference} \end{array}$$

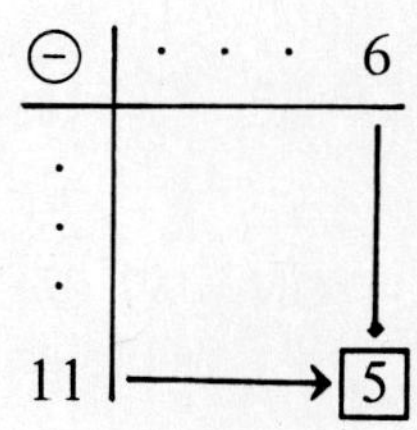

PROBLEM SET Subtract the following whole numbers.

See Example 27

1. 7 - 5	**2.** 6 - 2	**3.** 8 - 0	**4.** 9 - 8	**5.** 6 - 1
6. 10 - 3	**7.** 11 - 6	**8.** 8 - 5	**9.** 14 - 9	**10.** 16 - 8
11. 17 - 9	**12.** 15 - 6	**13.** 12 - 8	**14.** 13 - 7	**15.** 10 - 8

16. $\begin{array}{r} 9 \\ -8 \\ \hline \end{array}$	**17.** $\begin{array}{r} 13 \\ -\ 5 \\ \hline \end{array}$	**18.** $\begin{array}{r} 15 \\ -\ 8 \\ \hline \end{array}$	**19.** $\begin{array}{r} 16 \\ -\ 9 \\ \hline \end{array}$	**20.** $\begin{array}{r} 12 \\ -\ 7 \\ \hline \end{array}$
21. $\begin{array}{r} 9 \\ -0 \\ \hline \end{array}$	**22.** $\begin{array}{r} 14 \\ -\ 8 \\ \hline \end{array}$	**23.** $\begin{array}{r} 11 \\ -\ 9 \\ \hline \end{array}$	**24.** $\begin{array}{r} 10 \\ -\ 1 \\ \hline \end{array}$	**25.** $\begin{array}{r} 11 \\ -\ 4 \\ \hline \end{array}$

ANSWERS **1.** 2 **2.** 4 **3.** 8 **4.** 1 **5.** 5 **6.** 7 **7.** 5 **8.** 3 **9.** 5 **10.** 8 **11.** 8 **12.** 9 **13.** 4 **14.** 6 **15.** 2 **16.** 1 **17.** 8 **18.** 7 **19.** 7 **20.** 5 **21.** 9 **22.** 6 **23.** 2 **24.** 9 **25.** 7

> To subtract larger numbers:
>
> 1. Line the numbers up on the right (the ones' column).
> 2. Subtract the ones' column, the tens' column, the hundreds' column, and so on.

EXAMPLE 28 Subtract 69 - 42.

SOLUTION As with the addition examples, we do this in both the usual manner and in slow motion.

Usual

$$\begin{array}{r} 69 \\ -42 \\ \hline 27 \end{array}$$

Slow Motion

$$\begin{array}{rr} 6 & \boxed{9} \\ -4 & \boxed{2} \\ \hline & \boxed{7} \end{array}$$

Subtract ones' column: 9 - 2 = 7

↓

$$\begin{array}{rr} \boxed{6} & 9 \\ -\boxed{4} & 2 \\ \hline \boxed{2} & 7 \end{array}$$

Subtract tens' column: 6 - 4 = 2

The answer (difference) is 27.

EXAMPLE 29 Subtract 2674 - 510.

SOLUTION We subtract these column by column after we line the numbers up on the right (the ones' column).

Usual

$$\begin{array}{r} 2674 \\ -\ \ 510 \\ \hline 2164 \end{array}$$

Slow Motion

$$\begin{array}{rrrr} 2 & 6 & 7 & \boxed{4} \\ - & 5 & 1 & \boxed{0} \\ \hline & & & \boxed{4} \end{array}$$

Subtract ones' column: 4 - 0 = 4

↓

$$\begin{array}{rrrr} 2 & 6 & \boxed{7} & 4 \\ - & 5 & \boxed{1} & 0 \\ \hline & & \boxed{6} & 4 \end{array}$$

Subtract tens' column: 7 - 1 = 6

↓

$$\begin{array}{r} 2\,\boxed{6}\,7\,4 \\ -\ \ \boxed{5}\,1\,0 \\ \hline \boxed{1}\,6\,4 \end{array}$$

Subtract hundreds' column:
6 - 5 = 1

$$\begin{array}{r} \boxed{2}\,6\,7\,4 \\ -\ \ \ 5\,1\,0 \\ \hline \boxed{2}\,1\,6\,4 \end{array}$$

Subtract thousands' column:
2 - 0 = 2

The answer (difference) is 2164. Since subtraction is the opposite of addition, we can check this answer by adding the subtrahend to the difference.

$$\begin{array}{r} 2674 \\ -\ \boxed{510} \\ \hline +\ 2164 \\ \hline \boxed{2674} \end{array}$$

This is the original minuend, so the answer checks.

PROBLEM SET Subtract the following whole numbers.

See Examples 28 and 29

1. $\begin{array}{r} 45 \\ -31 \\ \hline \end{array}$ 2. $\begin{array}{r} 86 \\ -73 \\ \hline \end{array}$ 3. $\begin{array}{r} 49 \\ -31 \\ \hline \end{array}$ 4. $\begin{array}{r} 99 \\ -16 \\ \hline \end{array}$

5. $\begin{array}{r} 58 \\ -13 \\ \hline \end{array}$ 6. $\begin{array}{r} 78 \\ -20 \\ \hline \end{array}$ 7. $\begin{array}{r} 567 \\ -\ 42 \\ \hline \end{array}$ 8. $\begin{array}{r} 489 \\ -\ 33 \\ \hline \end{array}$

9. $\begin{array}{r} 685 \\ -\ 61 \\ \hline \end{array}$ 10. $\begin{array}{r} 478 \\ -137 \\ \hline \end{array}$ 11. $\begin{array}{r} 509 \\ -307 \\ \hline \end{array}$ 12. $\begin{array}{r} 593 \\ -342 \\ \hline \end{array}$

13. $\begin{array}{r} 3456 \\ -\ 321 \\ \hline \end{array}$ 14. $\begin{array}{r} 6782 \\ -\ 360 \\ \hline \end{array}$ 15. $\begin{array}{r} 5943 \\ -\ 933 \\ \hline \end{array}$ 16. $\begin{array}{r} 6709 \\ -4400 \\ \hline \end{array}$

17. $\begin{array}{r} 3859 \\ -1126 \\ \hline \end{array}$ 18. $\begin{array}{r} 48933 \\ -\ 3333 \\ \hline \end{array}$ 19. $\begin{array}{r} 12456 \\ -10000 \\ \hline \end{array}$ 20. $\begin{array}{r} 475893 \\ -\ 44752 \\ \hline \end{array}$

ANSWERS **1.** 14 **2.** 13 **3.** 18 **4.** 83 **5.** 45 **6.** 58 **7.** 525 **8.** 456 **9.** 624 **10.** 341 **11.** 202 **12.** 251 **13.** 3135 **14.** 6422 **15.** 5010 **16.** 2309 **17.** 2733 **18.** 45,600 **19.** 2456 **20.** 431,141

In subtraction problems, if the bottom number in a column is larger than the top number, we "borrow" 10 units from the column to the left.

EXAMPLE 30 Subtract 42 - 15.

SOLUTION This problem involves borrowing, and we do it in slow motion.

Usual

```
  3
  4̸¹2
- 1 5
-----
  2 7
```

Slow Motion

```
   4 [2]
 - 1 [5]
 -------
     [ ]
```

We cannot subtract ones' column: 2 - 5 = ?

↓

Borrow 10 from tens' column (rewrite 42 = 30 + 12)

```
  (3)
   4̸ [(12)]
 - 1 [ 5 ]
 ---------
     [ 7 ]
```

↓

Subtract ones' column: 12 - 5 = 7

```
 [(3)]
 [ 4̸ ] (12)
 -[ 1 ]  5
 ----------
 [ 2 ]   7
```

↓

Subtract tens' column: 3 - 1 = 2

We check the difference, 27, by adding:

```
   42 ←┐
 - 15  │
 + 27  │
   42 ←┘
```

EXAMPLE 31 Subtract 520 - 275.

SOLUTION In this problem, we have to borrow twice.

Usual

```
    11
  4 1̸
  5̸ 2̸¹0
- 2 7 5
-------
  2 4 5
```

Slow Motion

```
   5  2 [0]
 - 2  7 [5]
 ----------
        [ ]
```

We cannot subtract ones' column: 0 - 5 = ?

↓

Borrow 10 from tens' column (rewrite 20 = 10 + 10)

```
     (1)
   5  2̸ [(10)]
 - 2  7 [ 5 ]
 ------------
        [ 5 ]
```

↓

Subtract ones' column: 10 - 5 = 5

↓

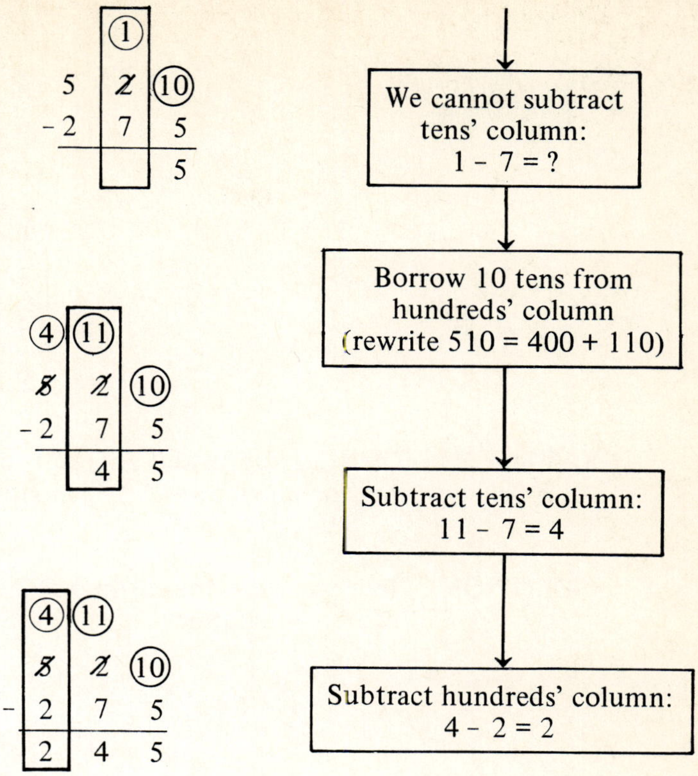

The difference is 245. We check this by adding: 245 + 275 = 520.

EXAMPLE 32 Subtract 705 - 168.

SOLUTION Here, when we go to borrow from the tens' column, there is a zero, so we go on to the hundreds' column.

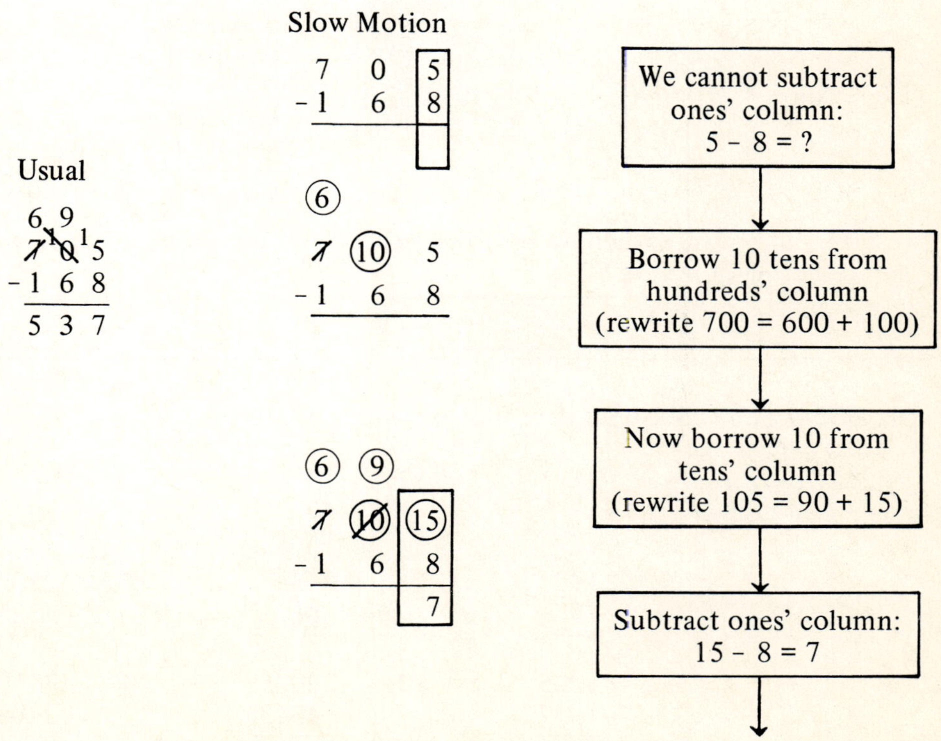

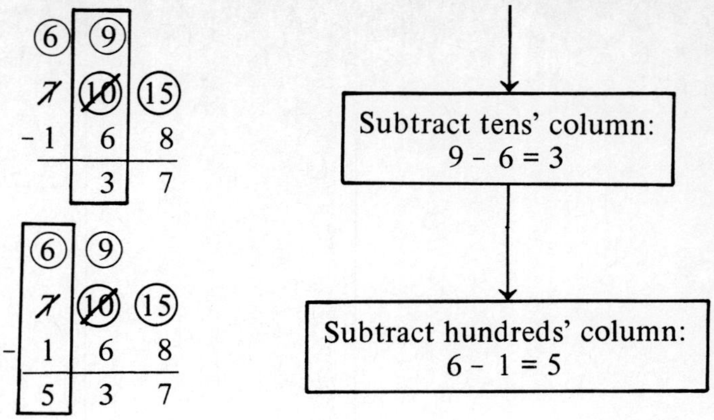

EXAMPLE 33 The Jones borrowed \$35,000 to buy their house. Their mortgage statement says that they now owe \$32,867. How much have they paid off?

SOLUTION Here we want the *difference* between what they borrowed and what they still owe, so we subtract.

$$\begin{array}{r} 4\ \ 9\ \ 9\ \ \ \\ 3\ \not{5}\ {}^1\not{0}\ {}^1\not{0}\ {}^1 0 \\ -3\ \ 2\ \ 8\ \ 6\ \ 7 \\ \hline 2\ \ 1\ \ 3\ \ 3 \end{array}$$

Thus they have paid \$2133. We can check: $2133 + 32{,}867 = 35{,}000$.

> **Why Does It Work?**
>
> This rule (like the one for addition) works because we are subtracting common terms: ones, tens, hundreds, and so on.
>
> $$\begin{array}{rcl} 652 & = & 6\ \textit{hundreds} + 5\ \textit{tens} + 2\ \textit{ones} \\ -\underline{227} & = & -\underline{(2\ \textit{hundreds} + 2\ \textit{tens} + 7\ \textit{ones})} \\ & \Rightarrow & 6\ \textit{hundreds} + 4\ \textit{tens} + 12\ \textit{ones} \\ & & -\underline{(2\ \textit{hundreds} + 2\ \textit{tens} + 7\ \textit{ones})} \\ 425 & = & 4\ \textit{hundreds} + 2\ \textit{tens} + 5\ \textit{ones} \end{array}$$
>
> When we borrow, we are rewriting 5 *tens* + 2 *ones* as 4 *tens* + 12 *ones*. Both are 52.

PROBLEM SET Subtract the following whole numbers.

See Examples 30 to 33

1.	$\begin{array}{r} 18 \\ -13 \\ \hline \end{array}$	**2.**	$\begin{array}{r} 79 \\ -42 \\ \hline \end{array}$	**3.**	$\begin{array}{r} 82 \\ -19 \\ \hline \end{array}$	**4.**	$\begin{array}{r} 90 \\ -47 \\ \hline \end{array}$
5.	$\begin{array}{r} 192 \\ -\ 97 \\ \hline \end{array}$	**6.**	$\begin{array}{r} 243 \\ -\ 86 \\ \hline \end{array}$	**7.**	$\begin{array}{r} 2704 \\ -\ 521 \\ \hline \end{array}$	**8.**	$\begin{array}{r} 4815 \\ -\ 629 \\ \hline \end{array}$
9.	$\begin{array}{r} 4035 \\ -1917 \\ \hline \end{array}$	**10.**	$\begin{array}{r} 22005 \\ -\ 4843 \\ \hline \end{array}$	**11.**	$\begin{array}{r} 74123 \\ -37785 \\ \hline \end{array}$	**12.**	$\begin{array}{r} 1000000 \\ -\ 52368 \\ \hline \end{array}$

13. $\begin{array}{r} 486 \\ -\ 92 \\ \hline \end{array}$ 14. $\begin{array}{r} 451 \\ -\ 69 \\ \hline \end{array}$ 15. $\begin{array}{r} 453 \\ -289 \\ \hline \end{array}$ 16. $\begin{array}{r} 500 \\ -\ 76 \\ \hline \end{array}$

17. $\begin{array}{r} 642 \\ -278 \\ \hline \end{array}$ 18. $\begin{array}{r} 1890 \\ -\ 987 \\ \hline \end{array}$ 19. $\begin{array}{r} 431 \\ -345 \\ \hline \end{array}$ 20. $\begin{array}{r} 6000 \\ -5432 \\ \hline \end{array}$

21. Subtract 106 - 49.
22. Subtract 952 - 666.
23. Subtract 2053 - 738.
24. Subtract 7008 - 5129.
25. Subtract 43,456 - 21,977.
26. Subtract 200,000 - 137,845.
27. Subtract 501 - 334.
28. Subtract 713 - 449.
29. Subtract 5600 - 4832.
30. Subtract 6723 - 4532.
31. Subtract 5682 - 516.
32. Subtract 17,586 - 2359.

ANSWERS **1.** 5 **2.** 37 **3.** 63 **4.** 43 **5.** 95 **6.** 157 **7.** 2183 **8.** 4186 **9.** 2118 **10.** 17,162 **11.** 36,338 **12.** 947,632 **13.** 394 **14.** 382 **15.** 164 **16.** 424 **17.** 364 **18.** 903 **19.** 86 **20.** 568 **21.** 57 **22.** 286 **23.** 1315 **24.** 1879 **25.** 21,479 **26.** 62,155 **27.** 167 **28.** 264 **29.** 768 **30.** 2191 **31.** 5166 **32.** 15,227

Throughout this section we have mentioned that subtraction is the opposite of addition. We now use this fact to find a missing number.

EXAMPLE 34 Find n: $n + 5 = 12$.

SOLUTION What do we have to add to 5 to get 12? We know that the answer is 7, because we use subtraction as the opposite of addition:

$$n + 5 = 12 \qquad \text{means} \qquad n = 12 - 5 = 7$$

EXAMPLE 35 Find n: $n + 49 = 82$.

SOLUTION What number added to 49 gives 82? To find n, we subtract 49 from 82. Thus $n = 82 - 49 = 33$.

EXAMPLE 36 Find x: $x - 4 = 3$.

SOLUTION This problem is slightly different. What number minus 4 gives 3? We can see that the answer is 7. Also, because addition is the opposite of subtraction, notice that $x = 4 + 3 = 7$.

EXAMPLE 37 Find n: $n - 27 = 74$.

SOLUTION This is like Example 36. We reverse the subtraction by adding 27 to 74. Thus $n = 27 + 74 = 101$.

PROBLEM SET Find n or x in the following problems.

See Examples 34 to 37

1. $n + 2 = 7$
2. $x + 6 = 20$
3. $n + 10 = 23$
4. $n + 23 = 31$
5. $n + 35 = 67$
6. $x + 234 = 651$
7. $n - 3 = 5$
8. $n - 6 = 10$
9. $n - 10 = 2$
10. $x - 21 = 11$
11. $n - 34 = 21$
12. $x - 328 = 276$

ANSWERS 1. 5 2. 14 3. 13 4. 8 5. 32 6. 417 7. 8 8. 16 9. 12 10. 32 11. 55 12. 604

Miscellaneous Exercises 1.3

Subtract the following whole numbers.

See Example 27

1. $8 - 3$
2. $12 - 8$
3. $9 - 7$

See Examples 28 and 29

4. $\begin{array}{r} 69 \\ -54 \\ \hline \end{array}$
5. $\begin{array}{r} 583 \\ -361 \\ \hline \end{array}$
6. $\begin{array}{r} 7836 \\ -4520 \\ \hline \end{array}$

See Examples 30 to 33

7. $\begin{array}{r} 53 \\ -28 \\ \hline \end{array}$
8. $\begin{array}{r} 62 \\ -35 \\ \hline \end{array}$
9. $\begin{array}{r} 123 \\ -\ \ 56 \\ \hline \end{array}$
10. $\begin{array}{r} 429 \\ -285 \\ \hline \end{array}$
11. $\begin{array}{r} 512 \\ -289 \\ \hline \end{array}$
12. $\begin{array}{r} 6430 \\ -2368 \\ \hline \end{array}$
13. $75 - 38$
14. $59 - 28$
15. $673 - 575$
16. $361 - 68$
17. $3002 - 2135$
18. $10{,}000 - 3469$

Find n in the following problems.

See Examples 34 and 35

19. $n + 2 = 9$
20. $n + 10 = 19$
21. $n + 23 = 48$
22. $n + 56 = 108$

See Examples 36 and 37

23. $n - 3 = 6$
24. $n - 4 = 9$
25. $n - 37 = 21$
26. $n - 47 = 59$

Business application

27. In business, the profit from the sale of an item is given by

$$\text{profit} = \text{revenue} - \text{cost}$$

Compute the profits in the following table.

Profit	Revenue	Cost
?	$200,000	$ 60,000
?	123,947	98,364
?	586,900	348,902

Health application

28. A patient needs 4000 milliliters (ml) of liquid per day. For breakfast, she had 950 ml of juice and water. For lunch, she had 1225 ml of milk and water. How many more milliliters of liquid does she need to reach her quota of 4000 ml?

Science application

29. The boiling point of water is 212°F, and the freezing point of water is 32°F. What is the difference between these two temperatures?

Consumer applications

30. A certain TV set costs \$451. Ted and Mary can get a set for \$55 off this price. How much will the set cost them?

31. Dave wins a \$10,000 lottery, but must pay \$3671 in taxes. How much does he have left?

32. Bob borrows \$1000. He pays back \$173 of the loan. How much does he still owe?

Miscellaneous applications

33. In 1960, John F. Kennedy defeated Richard M. Nixon in the popular vote for president by 34,227,096 to 34,108,546. What was Kennedy's margin of victory?

34. In a recent year, the world's population was about 3,953,125,000. Of this number, about 944,065,450 were Christian. How many people in the world were not Christian?

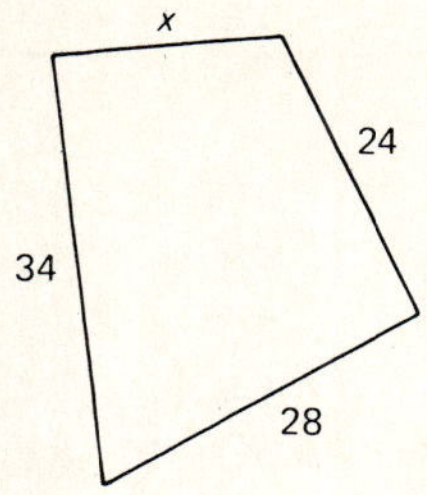

35. The distance (perimeter) around the figure at the left is 105. Find the missing dimension.

Now complete Post-Test 1 for Section 1.3 on page T9.

1.4 MULTIPLICATION OF WHOLE NUMBERS

SECTION OBJECTIVE

After completing this unit, you will be able to multiply any two whole numbers.

TYPICAL PROBLEMS

You will be able to work problems like the ones below.

(a) $\begin{array}{r} 75 \\ \times\ 3 \\ \hline \end{array}$ (b) $\begin{array}{r} 273 \\ \times 139 \\ \hline \end{array}$ (c) $\begin{array}{r} 24000 \\ \times\ \ 80 \\ \hline \end{array}$

(d) Multiply 873 X 548.

(e) Multiply 2000 X 481.

(f) Multiply 72,000 X 890,000.

ANSWERS

(a) 225 (b) 37,947 (c) 1,920,000 (d) 478,404 (e) 962,000
(f) 64,080,000,000

Multiplication of whole numbers can be thought of as "repeated addition." Instead of adding 7 + 7 + 7 + 7 + 7, we write 5 X 7. We sometimes write multiplication with a dot, 5 · 7. This multiplication can be pictured as follows:

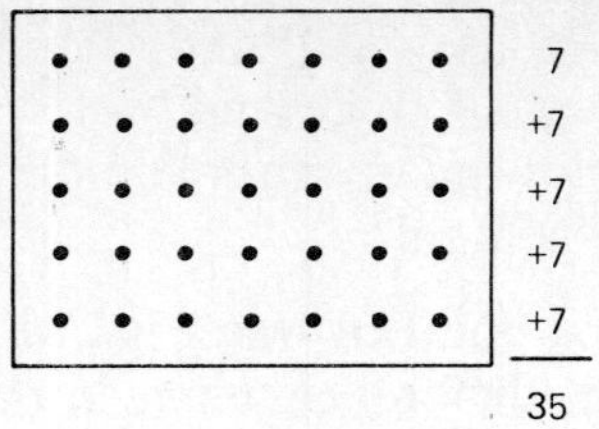

or 5 x 7 = 35

The numbers that we multiply together are called **factors.** The result is called the **product.** As with addition, there is a multiplication table consisting of the basic facts that must be memorized in order to multiply larger numbers. This is given in Table 1.4.

Table 1.4

X	0	1	2	3	4	5	6	7	8	9
0	0	0	0	0	0	0	0	0	0	0
1	0	1	2	3	4	5	6	7	8	9
2	0	2	4	6	8	10	12	14	16	18
3	0	3	6	9	12	15	18	21	24	27
4	0	4	8	12	16	20	24	28	32	36
5	0	5	10	15	20	25	30	35	40	45
6	0	6	12	18	24	30	36	42	48	54
7	0	7	14	21	28	35	42	49	56	63
8	0	8	16	24	32	40	48	56	64	72
9	0	9	18	27	36	45	54	63	72	81

EXAMPLE 38 Let us use Table 1.4 to multiply 6 X 7.

$$\begin{array}{r} 7 \\ \times\ 6 \\ \hline 42 \end{array} \qquad \begin{array}{l} \text{factor} \\ \times\,\text{factor} \\ \hline \text{product} \end{array}$$

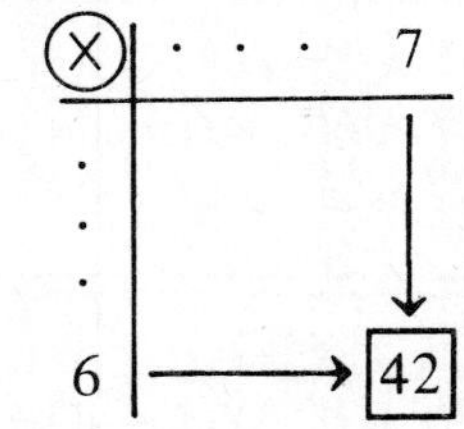

Notice from Table 1.4 that there are two special properties.

1. Zero times any number is zero; in symbols,

$$0 \cdot n = 0$$

2. One times any number is the number; in symbols,

$$1 \cdot n = n$$

PROBLEM SET Multiply the following whole numbers.

See Example 38

1. 4 X 5 **2.** 3 X 7 **3.** 5 X 8 **4.** 6 X 2 **5.** 7 X 9

6. 8 X 8 **7.** 2 X 9 **8.** 4 X 9 **9.** 8 X 6 **10.** 5 X 9

11. 8 X 1 **12.** 3 X 0 **13.** 7 X 7 **14.** 8 X 9 **15.** 3 X 9

16. $\begin{array}{r} 7 \\ \times 4 \\ \hline \end{array}$ 17. $\begin{array}{r} 6 \\ \times 6 \\ \hline \end{array}$ 18. $\begin{array}{r} 7 \\ \times 0 \\ \hline \end{array}$ 19. $\begin{array}{r} 1 \\ \times 6 \\ \hline \end{array}$ 20. $\begin{array}{r} 4 \\ \times 6 \\ \hline \end{array}$

21. $\begin{array}{r} 8 \\ \times 7 \\ \hline \end{array}$ 22. $\begin{array}{r} 5 \\ \times 5 \\ \hline \end{array}$ 23. $\begin{array}{r} 9 \\ \times 9 \\ \hline \end{array}$ 24. $\begin{array}{r} 6 \\ \times 9 \\ \hline \end{array}$ 25. $\begin{array}{r} 5 \\ \times 1 \\ \hline \end{array}$

ANSWERS 1. 20 2. 21 3. 40 4. 12 5. 63 6. 64 7. 18 8. 36 9. 48 10. 45 11. 8 12. 0 13. 49 14. 72 15. 27 16. 28 17. 36 18. 0 19. 6 20. 24 21. 56 22. 25 23. 81 24. 54 25. 5

The column values of our base-10 system are 1, 10, 100, 1000, and so on. These are called the **powers of 10.** Multiplying by these powers of 10 is very easy.

> To multiply a number by a power of 10 (1, 10, 100, and so on), simply write the number of zeros in the power of 10 after the number.

EXAMPLE 39

5 × 10 = 50

42 × 10 = 420

7 × 100 = 700

298 × 100 = 29,800

30 × 1000 = 30,000

903 × 1000 = 903,000

8 × 10,000 = 80,000

36 × 10,000 = 360,000

245 × 100,000 = 24,500,000

40 × 1,000,000 = 40,000,000

PROBLEM SET Multiply the following whole numbers and powers of 10.

See Example 39

1. 4 × 10
2. 56 × 10
3. 51 × 100
4. 8 × 1000
5. 34 × 100
6. 1000 × 67
7. 209 × 100
8. 31 × 10,000
9. 29 × 10,000
10. 100 × 211
11. 1000 × 100
12. 500 × 100,000

13. $\begin{array}{r} 1000 \\ \times \quad 36 \\ \hline \end{array}$ 14. $\begin{array}{r} 5786 \\ \times \quad 100 \\ \hline \end{array}$ 15. $\begin{array}{r} 10000 \\ \times \quad 230 \\ \hline \end{array}$

16. $\begin{array}{r} 2000 \\ \times \quad 10 \\ \hline \end{array}$ 17. $\begin{array}{r} 2300 \\ \times \quad 100 \\ \hline \end{array}$ 18. $\begin{array}{r} 100000 \\ \times \quad 1200 \\ \hline \end{array}$

19. $\begin{array}{r} 1000000 \\ \times \quad 2000 \\ \hline \end{array}$ 20. $\begin{array}{r} 345000 \\ \times \quad 1000 \\ \hline \end{array}$

ANSWERS 1. 40 2. 560 3. 5100 4. 8000 5. 3400 6. 67,000
7. 20,900 8. 310,000 9. 290,000 10. 21,100 11. 100,000
12. 50,000,000 13. 36,000 14. 578,600 15. 2,300,000
16. 20,000 17. 230,000 18. 120,000,000 19. 2,000,000,000
20. 345,000,000

Numbers such as 2000 and 40,000 are **multiples of powers of 10.** Multiplying these numbers is also very easy.

> To multiply multiples by multiples:
> 1. Multiply the first digits of each multiple.
> 2. Write the total number of zeros after the product.

EXAMPLE 40

200 X 4000 = 800,000

30 X 50,000 = 1,500,000

6000 X 5000 = 30,000,000

700 X 800 = 560,000

300 X 400,000 = 120,000,000

200,000 X 8,000,000 = 1,600,000,000,000

PROBLEM SET Multiply the following multiples of powers of 10.

See Example 40

1. 20 X 40
2. 40 X 700
3. 500 X 3000
4. 6000 X 30,000
5. 400 X 800
6. 30 X 70,000
7. 60 X 60,000
8. 40 X 900
9. 7000 X 3000
10. 800 X 9000
11. 40 X 80,000
12. 300,000 X 80,000
13. 900 X 700,000,000
14. 70,000,000 X 800,000
15. 600,000 X 9,000,000,000

ANSWERS 1. 800 2. 28,000 3. 1,500,000 4. 180,000,000
5. 320,000 6. 2,100,000 7. 3,600,000 8. 36,000
9. 21,000,000 10. 7,200,000 11. 3,200,000
12. 24,000,000,000 13. 630,000,000,000
14. 56,000,000,000,000 15. 5,400,000,000,000,000

We now multiply larger numbers. We start by multiplying a larger number by a one-digit number.

> To multiply a larger number by a one-digit number, we multiply each digit of the larger by the one-digit number, carrying when necessary.

EXAMPLE 41 Multiply 4 X 21.

SOLUTION We multiply the 4 by both digits of the 21.

Usual

$$\begin{array}{r} 21 \\ \times\ 4 \\ \hline 84 \end{array}$$

Slow Motion

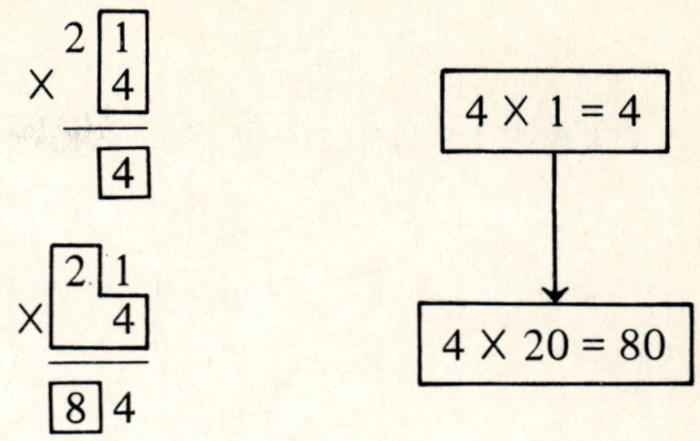

The product is 84.

EXAMPLE 42 Multiply 6 X 348.

SOLUTION In this problem, we have to carry.

Usual

$$\begin{array}{r} {\scriptstyle 24} \\ 348 \\ \times\ \ 6 \\ \hline 2088 \end{array}$$

Slow Motion

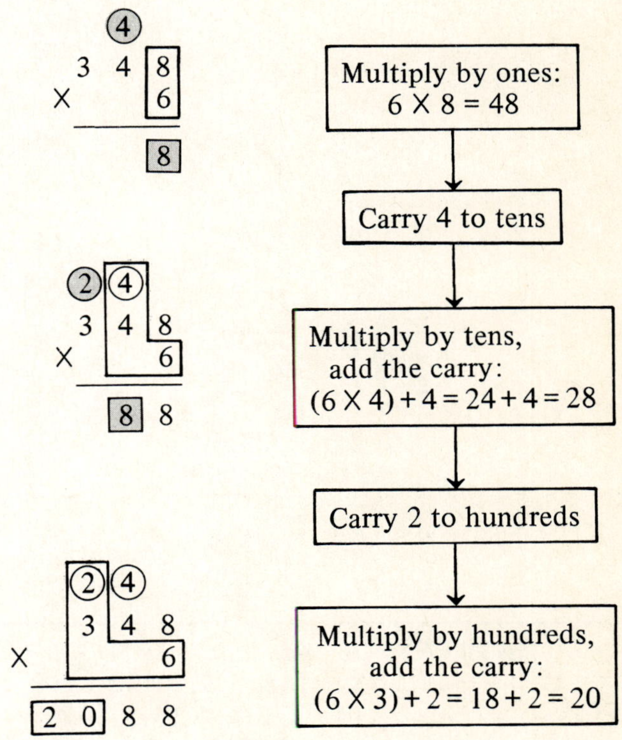

The product is 2088.

EXAMPLE 43 Multiply 700 X 592.

SOLUTION 700 is a multiple of a power of 10. As we did earlier, we multiply 7 X 592 and then write two zeros after the product. In fact, it is easier to write this if we "hang the zeros" off to the right. Then we multiply 7 X 592 as usual.

$$\begin{array}{r} 592 \\ \times 700 \\ \hline \end{array} \Longrightarrow \begin{array}{r} {}^{61} \\ 592 \\ \times \quad 700 \\ \hline 414400 \end{array}$$

EXAMPLE 44 Multiply 8000 X 369.

SOLUTION We again "hang the zeros" off to the right and then multiply 8 X 369.

$$\begin{array}{r} 369 \\ \times 8000 \\ \hline \end{array} \Longrightarrow \begin{array}{r} {}^{57} \\ 369 \\ \times \quad 8000 \\ \hline 2952000 \end{array}$$

PROBLEM SET Multiply the following whole numbers.

See Examples 41 to 44

1. 23 X 2	**2.** 31 X 3	**3.** 14 X 2	**4.** 312 X 3
5. 201 X 4	**6.** 421 X 2	**7.** 36 X 3	**8.** 62 X 4
9. 94 X 5	**10.** 47 X 4	**11.** 734 X 4	**12.** 507 X 6
13. 493 X 7	**14.** 598 X 6	**15.** 867 X 4	**16.** 836 X 9
17. 409 X 8	**18.** 3759 X 6	**19.** 5839 X 8	**20.** 3862 X 700
21. 38594 X 9000	**22.** 4688 X 60	**23.** 5631 X 70	**24.** 7839 X 300
25. 45792 X 2000			

ANSWERS **1.** 46 **2.** 93 **3.** 28 **4.** 936 **5.** 804 **6.** 842 **7.** 108 **8.** 248 **9.** 470 **10.** 188 **11.** 2936 **12.** 3042 **13.** 3451 **14.** 3588 **15.** 3468 **16.** 7524 **17.** 3272 **18.** 22,554 **19.** 46,712 **20.** 2,703,400 **21.** 347,346,000 **22.** 281,280 **23.** 394,170 **24.** 2,351,700 **25.** 91,584,000

Sometimes, if we are in a hurry or we wish to get a quick estimate to check our work, we will want to get an *approximate* answer. To do this, we first round the numbers off to simple multiples of powers of 10. For example, 213,458 ≈ 200,000 and 4892 ≈ 5000. (The symbol ≈ means "approximately equal to.")

To get a quick estimate to a multiplication problem:

1. Round each factor off to a simple multiple of a power of 10. (Round off to the leftmost place.)
2. Multiply these multiples.

EXAMPLE 45 Estimate 5872 X 413.

SOLUTION We round these off and then multiply.

$$\begin{array}{rr} 5872 \approx & 6000 \\ \times \quad 413 \approx \times & 400 \\ \hline \text{product} \approx & 2400000 \end{array}$$

Thus the product is approximately 2,400,000.

EXAMPLE 46 Squirrelville College has an enrollment of 7213 students. Each student pays \$3149 in tuition. Estimate the revenue from all this tuition.

SOLUTION We round these off to simpler numbers and multiply.

$$\begin{array}{rr} 7213 \approx & 7000 \\ \times \quad 3149 \approx \times & 3000 \\ \hline \text{product} \approx & 21000000 \end{array}$$

Thus their revenue is about \$21,000,000, or 21 million dollars.

PROBLEM SET *Estimate* the following products.

See Examples 45 and 46

1. Estimate 482 X 21,798.
2. Estimate 289 X 7892.
3. Estimate 79 X 6898.
4. Estimate 7891 X 6789.
5. Estimate 60,311 X 312.
6. Estimate 9231 X 22,311.
7. Estimate 389 X 479,811.
8. Estimate 98 X 4231.
9. Estimate 70,322 X 41,237.
10. Estimate 93,241 X 6789.

ANSWERS 1. 10,000,000 2. 2,400,000 3. 560,000 4. 56,000,000 5. 18,000,000 6. 180,000,000 7. 200,000,000 8. 400,000 9. 2,800,000,000 10. 630,000,000

The last type of multiplication is that of any two whole numbers, such as 369 X 5913. Here, we multiply each digit of one factor by the other factor. Since these are the hardest of the problems, it is often helpful to estimate the product first so that we have a "ballpark" idea of what answer to expect.

To multiply larger whole numbers:

1. Estimate the product by rounding.
2. Multiply each digit of one factor by the other factor. (Be careful how you place their partial products: When multiplying by the digit in the tens' column, start that partial product in the tens' column; and so on.)
3. Add the partial products.

EXAMPLE 47 Multiply 42 X 79.

SOLUTION We begin by estimating the product as 40 X 80 = 3200.

Usual

```
   79
X  42
  158
 316
 3318
```

Slow Motion

```
  (1)
   79
X  42
  158
```

Multiply by 2: 2 X 79 = 158

```
  (3)
   79
X  42
  158
 316
```

Multiply by 40: 4 X 79 = 316; start partial product in tens' column

```
   79
X  42
  158
 316
 3318
```

Add partial products

We see that the actual product, 3318, is very close to the estimate, 3200. Also, notice that when we multiply 40 X 79, we simply multiply 4 X 79 and start the product in the tens' column. This shorthand notation gives us the same answer.

EXAMPLE 48 Multiply 279 X 814.

SOLUTION We begin by estimating the product as 300 X 800 = 240,000.

Slow Motion

```
  (3)(3)
   279
X  814
  1116
```

Multiply by 4: 4 X 279 = 1116

Usual

```
   279
×  814
------
  1116
  279
2232
------
227106
```

```
   2 7 9
×  8 1 4
--------
 1 1 1 6
 2 7 9
```

Multiply by 10: 1 × 279 = 279; start in tens' column

```
   ⑥ ⑦
   2 7 9
×  8 1 4
--------
 1 1 1 6
 2 7 9
2 2 3 2
```

Multiply by 800: 8 × 279 = 2232; start in hundreds' column

```
      2 7 9
×     8 1 4
-----------
    1 1 1 6
    2 7 9
  2 2 3 2
-----------
2 2 7 1 0 6
```

Add partial products

The product, 227,106, is very close to the estimate, 240,000.

EXAMPLE 49 Multiply 682 × 293.

SOLUTION We first estimate the product as 700 × 300 = 210,000.

```
   682
×  293
------
  2046 ←—— 3 × 682 = 2046
 6138  ←—— 90 × 682 = 61,380
1364   ←—— 200 × 682 = 136,400
------
199826 ←—— sum of partial products
```

The product is 199,826.

EXAMPLE 50 Multiply 341 × 5892.

SOLUTION We first estimate the product as 300 × 6000 = 1,800,000.

```
   5892
×   341
-------
   5892 ←—— 1 × 5892 = 5892
 23568  ←—— 40 × 5892 = 235,680
17676   ←—— 300 × 5892 = 1,767,600
-------
2009172 ←—— sum of partial products
```

The product is 2,009,172. Notice that we do not need to record the final zeros in the partial products if we start the products in the proper columns.

EXAMPLE 51 Jeni buys a car and contracts to pay \$178 a month for 36 months. How much money will this be over the whole 36-month period?

SOLUTION \$178 a month for 36 months is 36 × 178.

$$
\begin{array}{r l}
178 & \\
\times \;\; 36 & \\
\hline
1068 & \leftarrow 6 \times 178 = 1068 \\
534 & \leftarrow 30 \times 178 = 5340 \\
\hline
6408 & \leftarrow \text{sum of partial products}
\end{array}
$$

Therefore, Jeni will pay \$6408 over the 36-month period.

EXAMPLE 52 Find the area of the rectangle shown.

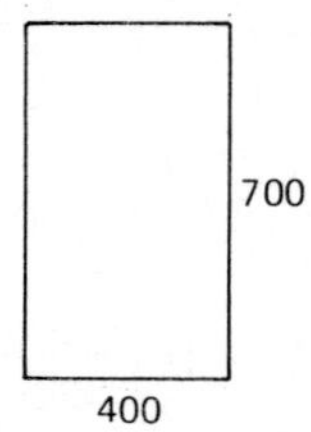

SOLUTION To find the area, we multiply the length times the width, 400 × 700.

$$
\begin{array}{r}
700 \\
\times \;\; 400 \\
\hline
280000
\end{array}
$$

Multiply 4 × 7 = 28 → Count four zeros

The area is 280,000.

Why Does It Work?

The procedure for multiplication works because we are really keeping track of the column values by placing the products in the correct place. Consider 74 × 63.

$$
\begin{array}{r r}
74 = & 7\ \textit{tens} + \ \ 4\ \textit{ones} \\
\times 63 = & 6\ \textit{tens} + \ \ 3\ \textit{ones} \\
\hline
 & 21\ \textit{tens} + 12\ \textit{ones} \\
 & 42\ \textit{hundreds} + 24\ \textit{tens} \phantom{+ 12\ \textit{ones}} \\
\hline
 & 42\ \textit{hundreds} + 45\ \textit{tens} + 12\ \textit{ones}
\end{array}
$$

$$(4\ \textit{thousands} + 2\ \textit{hundreds}) + (4\ \textit{hundreds} + 5\ \textit{tens}) + (1\ \textit{ten} + 2\ \textit{ones})$$

$$4\ \textit{thousands} + (2\ \textit{hundreds} + 4\ \textit{hundreds}) + (5\ \textit{tens} + 1\ \textit{ten}) + 2\ \textit{ones}$$

$$= 4\ \textit{thousands} + 6\ \textit{hundreds} + 6\ \textit{tens} + 2\ \textit{ones}$$

$$= 4662$$

This can also be shown as

$$\begin{array}{r} 74 \\ \times\ \ 63 \\ \hline 12 \\ 210 \\ 240 \\ 4200 \\ \hline 4662 \end{array}$$

Notice that by writing the numbers in expanded form and carefully regrouping products, we can get the answer. This is why it works, but obviously this is not how we would want to work a problem.

YES	NO
$\begin{array}{r} 213 \\ \times\ \ 102 \\ \hline 426 \\ 000 \\ 213 \\ \hline 21726 \end{array}$	$\begin{array}{r} 213 \\ \times\ \ 102 \\ \hline 426 \\ 213 \\ \hline 2556 \end{array}$

We count a place for the zero in the multiplier.

PROBLEM SET Multiply the following whole numbers. (Estimate the product first.)

See Examples 47 to 52

1. $\begin{array}{r} 24 \\ \times\ 6 \\ \hline \end{array}$	**2.** $\begin{array}{r} 73 \\ \times\ 9 \\ \hline \end{array}$	**3.** $\begin{array}{r} 379 \\ \times\ \ 8 \\ \hline \end{array}$	**4.** $\begin{array}{r} 4873 \\ \times\ \ \ 7 \\ \hline \end{array}$
5. $\begin{array}{r} 49 \\ \times 23 \\ \hline \end{array}$	**6.** $\begin{array}{r} 59 \\ \times 62 \\ \hline \end{array}$	**7.** $\begin{array}{r} 94 \\ \times 27 \\ \hline \end{array}$	**8.** $\begin{array}{r} 68 \\ \times 36 \\ \hline \end{array}$
9. $\begin{array}{r} 67 \\ \times 54 \\ \hline \end{array}$	**10.** $\begin{array}{r} 82 \\ \times 17 \\ \hline \end{array}$	**11.** $\begin{array}{r} 348 \\ \times\ \ 46 \\ \hline \end{array}$	**12.** $\begin{array}{r} 219 \\ \times\ \ 32 \\ \hline \end{array}$
13. $\begin{array}{r} 572 \\ \times\ \ 47 \\ \hline \end{array}$	**14.** $\begin{array}{r} 958 \\ \times\ \ 56 \\ \hline \end{array}$	**15.** $\begin{array}{r} 308 \\ \times\ \ 38 \\ \hline \end{array}$	**16.** $\begin{array}{r} 348 \\ \times 571 \\ \hline \end{array}$
17. $\begin{array}{r} 573 \\ \times 429 \\ \hline \end{array}$	**18.** $\begin{array}{r} 947 \\ \times 305 \\ \hline \end{array}$	**19.** $\begin{array}{r} 659 \\ \times 207 \\ \hline \end{array}$	**20.** $\begin{array}{r} 378 \\ \times 496 \\ \hline \end{array}$
21. $\begin{array}{r} 2009 \\ \times 1007 \\ \hline \end{array}$	**22.** $\begin{array}{r} 3108 \\ \times\ \ 208 \\ \hline \end{array}$	**23.** $\begin{array}{r} 6100 \\ \times\ \ 380 \\ \hline \end{array}$	**24.** $\begin{array}{r} 4300 \\ \times\ \ 670 \\ \hline \end{array}$
25. $\begin{array}{r} 39000 \\ \times\ \ 4600 \\ \hline \end{array}$			

26. Multiply 24 X 79.

27. Multiply 39 X 45.

28. Multiply 68 X 57.

29. Multiply 28 X 492.

30. Multiply 47 X 508.

31. Multiply 72 X 439.

32. Multiply 328 X 589.

33. Multiply 274 X 491.

34. Multiply 206 X 509.

35. Multiply 340 X 539.

36. Multiply 2005 X 6701.

ANSWERS 1. 144 2. 657 3. 3032 4. 34,111 5. 1127 6. 3658
7. 2538 8. 2448 9. 3618 10. 1394 11. 16,008
12. 7008 13. 26,884 14. 53,648 15. 11,704 16. 198,708
17. 245,817 18. 288,835 19. 136,413 20. 187,488
21. 2,023,063 22. 646,464 23. 2,318,000 24. 2,881,000
25. 179,400,000 26. 1896 27. 1755 28. 3876 29. 13,776
30. 23,876 31. 31,608 32. 193,192 33. 134,534
34. 104,854 35. 183,260 36. 13,435,505

Miscellaneous Exercises 1.4

Multiply the following whole numbers.

See Example 38 1. 5 X 6 2. 3 X 8 3. 7 X 9

See Example 39 4. 37 X 100 5. 20 X 1000 6. 450 X 10,000

See Example 40 7. 20 X 400 8. 600 X 8000 9. 5000 X 90,000

See Examples 41 to 44 10. 6 X 32 11. 7 X 23 12. 9 X 562

13. 40 X 67 14. 600 X 731 15. 900 X 6531

See Examples 47 to 52 16. 46 X 23 17. 39 X 21 18. 56 X 49

19. 27 X 126 20. 68 X 458 21. 74 X 396

22. 238 X 385 23. 472 X 968 24. 763 X 957

25. 657 X 3857 26. 734 X 2980 27. 837 X 3758

Estimate the following products.

See Examples 45 and 46 28. 4973 X 5784 29. 5873 X 23,587

30. 4936 X 324,574 31. 41,876 X 684,901

Business applications

32. A company has an inventory of 15,200 dingleberry cleaners, each worth $7. What is the total worth of this inventory?

33. H&P Foods has 15 employees, each of whom works 35 hours per week and makes $6 per hour. How much is the weekly payroll?

34. A certain publisher sells 2,000,000 books per year at an average price of $12 per book. How much total revenue is this?

Consumer applications

35. The Stanleys own 435 shares of a stock worth $17 per share. How much is all the stock worth?

36. If Stephanie gives up her daily candy bar (152 calories), how many calories will she save in a 365-day year?

37. A gallon of paint usually covers 450 square feet. How many square feet will 12 gallons cover?

38. Eva buys a color TV set and agrees to pay \$33 per month for 24 months. What is the total payment?

Health applications

39. A patient is to receive 20,000 units of penicillin six times per day. How many units is this per day?

40. A patient is prescribed to take two 350-milligram tablets four times per day for 14 days. How many total milligrams of medicine is this?

Electrical application

41. The power P used to charge a battery is given by

$$P = V \cdot I$$

where V is voltage and I is current. Find the power P needed to charge a car battery if $V = 12$ volts and $I = 20$ amperes.

Miscellaneous applications

42. The number of seconds in a 365-day year is given by

$$60 \times 60 \times 24 \times 365$$

How many seconds in this?

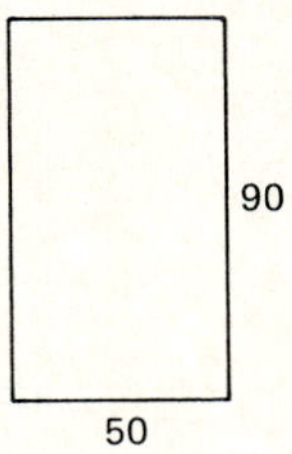

43. Find the area of the rectangle at the left.

Now complete Post-Test 1 on Section 1.4 on page T11.

1.5 DIVISION OF WHOLE NUMBERS

SECTION OBJECTIVE After completing this unit, you will be able to divide whole numbers.

TYPICAL PROBLEMS You will be able to work problems like the ones below.

(a) $72 \div 8$ **(b)** $1357 \div 23$ **(c)** $11{,}201 \div 78$

(d) $76\overline{)4636}$ **(e)** $458\overline{)162132}$ **(f)** Find n: $5 \cdot n = 40$.

ANSWERS **(a)** 9 **(b)** 59 **(c)** 143 R47 **(d)** 61 **(e)** 354 **(f)** 8

Division of numbers means several things. It is the opposite of multiplication. It is repeated subtraction. It is a way to compare two numbers' sizes. It is a way to cut a big set into equal-sized little sets. The expression $a \div b$ is read "a divided by b" and is also written $b\overline{)a}$.

$$a \div b = Q \quad \text{means} \quad b \cdot Q = a$$

For example,

$$36 \div 9 = \boxed{?} \quad \text{means} \quad 9 \times \boxed{?} = 36$$

If we know our multiplication facts well, we know the answer is 4, since $9 \times 4 = 36$. We must know our multiplication facts perfectly to do division.

EXAMPLE 53 $56 \div 8 = \boxed{?}$ means $8 \times \boxed{?} = 56$. So $\boxed{?} = 7$.

$45 \div 5 = \boxed{?}$ means $5 \times \boxed{?} = 45$. So $\boxed{?} = 9$.

Division is also a multiplication problem in which we have the product and one factor, and we want the missing factor. The answer, or missing factor, is called the **quotient**. The other factor is called the **divisor**. The product is called the **dividend**. This is written as

$$42 \div 6 = 7 \quad \text{or} \quad 6\overline{\smash{)}42}^{\;7}$$

$$\text{dividend} \div \text{divisor} = \text{quotient} \qquad \text{divisor}\overline{\smash{)}\text{dividend}}^{\;\text{quotient}}$$

YES	NO
$0 \div a = 0$	~~$a \div 0 = 0$~~

We cannot divide by zero; $a \div 0$ is not defined.

PROBLEM SET Divide the following whole numbers.

See Example 53

1. $9 \div 3$	**2.** $16 \div 8$	**3.** $21 \div 0$	**4.** $25 \div 5$	**5.** $27 \div 9$
6. $36 \div 4$	**7.** $32 \div 8$	**8.** $0 \div 6$	**9.** $48 \div 8$	**10.** $45 \div 9$
11. $54 \div 6$	**12.** $56 \div 7$	**13.** $49 \div 7$	**14.** $63 \div 9$	**15.** $72 \div 8$
16. $81 \div 9$	**17.** $28 \div 7$	**18.** $3\overline{\smash{)}15}$	**19.** $4\overline{\smash{)}24}$	**20.** $5\overline{\smash{)}40}$
21. $6\overline{\smash{)}0}$	**22.** $6\overline{\smash{)}30}$	**23.** $7\overline{\smash{)}63}$	**24.** $9\overline{\smash{)}72}$	**25.** $0\overline{\smash{)}27}$

ANSWERS **1.** 3 **2.** 2 **3.** Not defined **4.** 5 **5.** 3 **6.** 9 **7.** 4 **8.** 0 **9.** 6 **10.** 5 **11.** 9 **12.** 8 **13.** 7 **14.** 7 **15.** 9 **16.** 9 **17.** 4 **18.** 5 **19.** 6 **20.** 8 **21.** 0 **22.** 5 **23.** 9 **24.** 8 **25.** Not defined

To divide larger numbers, we have to use some trial and error.

EXAMPLE 54 Divide $3496 \div 8$.

SOLUTION We start by dividing the 8 into 3496 on the left.

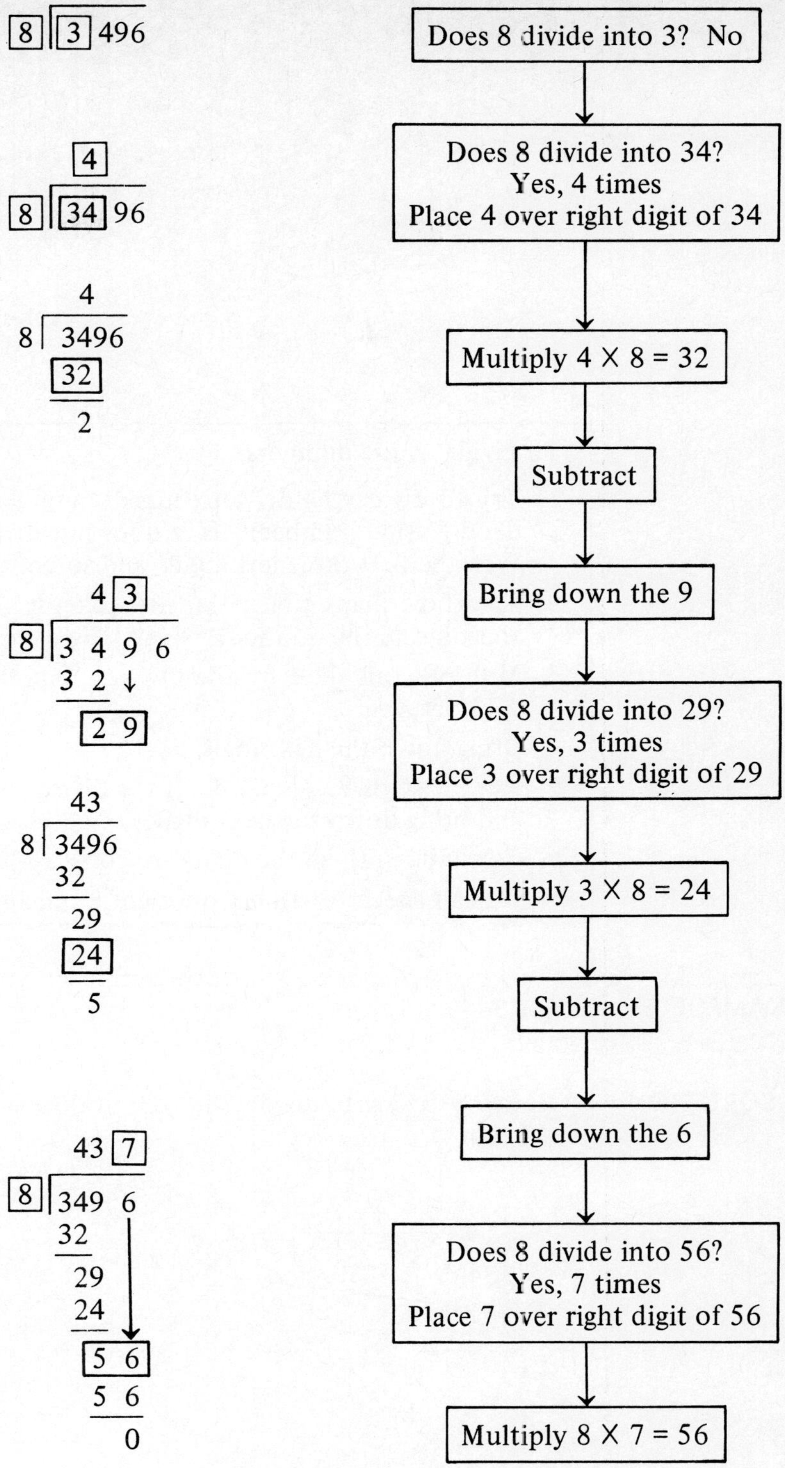

The division comes out evenly, with 437 as the quotient. We can check this by multiplying: $8 \times 437 = 3496$.

Not every division comes out evenly. The following illustration shows how $46 \div 7$ looks. We try to arrange the 46 items into columns of 7, but we get some leftovers. This illustration shows how $46 \div 7$ might look. We start with a set of 46 items. If we straighten them into columns of 7, we get 6 columns, and there will be 4 left over. We say that 6 is the quotient and 4 is the remainder.

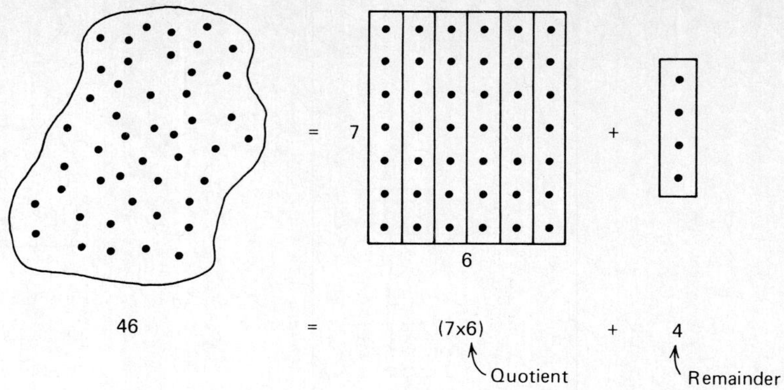

To divide whole numbers:

1. Try to divide the divisor (outside number) into the left digit of the dividend (inside number). If it does not divide, try the first two left digits; then the first three left digits; and so on.
2. Find how many times it divides (largest number without going over). Put this digit in the quotient.
3. Multiply this digit by the divisor. Put the product below dividend, and subtract.
4. Bring down the next digit.
5. Repeat steps 2, 3, and 4. If the divisor is too big, put a 0 in the quotient and bring down the next digit.
6. Continue until all the digits are gone.
7. We can check: divisor × quotient + remainder = dividend.

EXAMPLE 55 Divide 3449 ÷ 7.

SOLUTION We start by trying to divide 7 into 3. It does not divide, so we try 7 into 34. This goes 4 times.

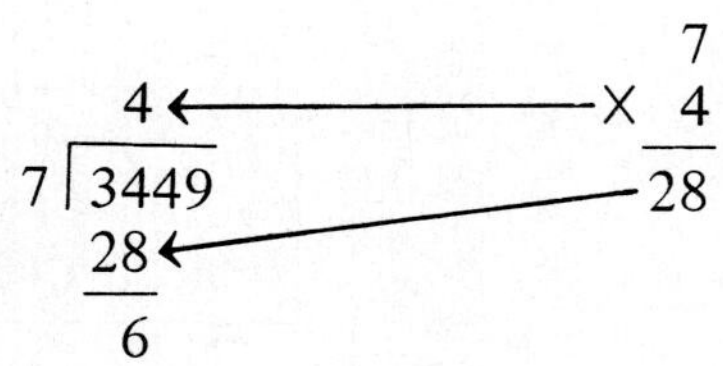

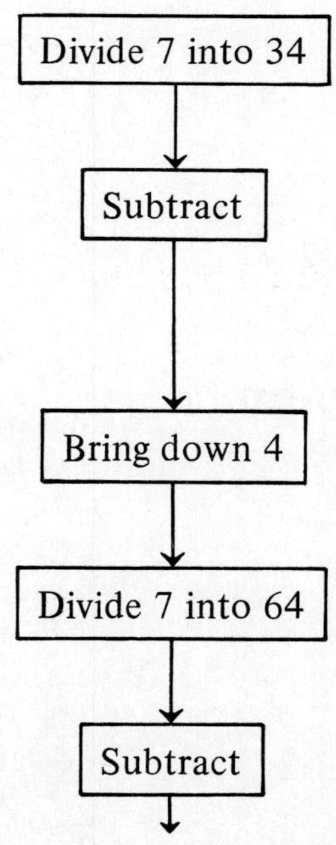

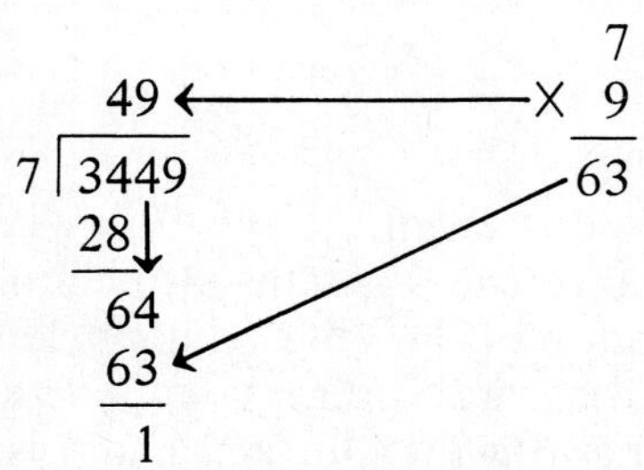

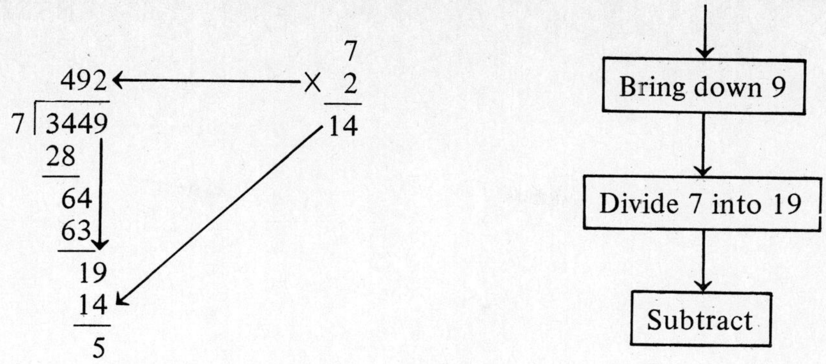

The quotient is 492, and the remainder is 5. We can check this:

$$(7 \times 492) + 5 = 3444 + 5 = 3449$$

PROBLEM SET Divide the following whole numbers. Be sure to indicate the quotient and the remainder (if any).

See Examples 54 and 55

1. $2\overline{)38}$	**2.** $3\overline{)87}$	**3.** $4\overline{)376}$	**4.** $5\overline{)745}$
5. $6\overline{)456}$	**6.** $2\overline{)485}$	**7.** $3\overline{)674}$	**8.** $7\overline{)486}$
9. $9\overline{)678}$	**10.** $6\overline{)385}$	**11.** $3\overline{)489}$	**12.** $8\overline{)496}$
13. $5\overline{)398}$	**14.** $7\overline{)5682}$	**15.** $4\overline{)3768}$	**16.** $9\overline{)2340}$
17. $6\overline{)5831}$	**18.** $8\overline{)48{,}373}$	**19.** $5\overline{)39{,}571}$	**20.** $2\overline{)489{,}325}$

ANSWERS **1.** 19 **2.** 29 **3.** 94 **4.** 149 **5.** 76 **6.** 242 R1 **7.** 224 R2 **8.** 69 R3 **9.** 75 R3 **10.** 64 R1 **11.** 163 **12.** 62 **13.** 79 R3 **14.** 811 R5 **15.** 942 **16.** 260 **17.** 971 R5 **18.** 6046 R5 **19.** 7914 R1 **20.** 244,662 R1

When dividing larger numbers, we follow exactly the same procedure: By trial and error, we divide the divisor into the leftmost digits of the inside number, the dividend.

EXAMPLE 56 Divide 11,414 ÷ 26.

SOLUTION We show this division in both the usual form and in slow motion.

Slow Motion

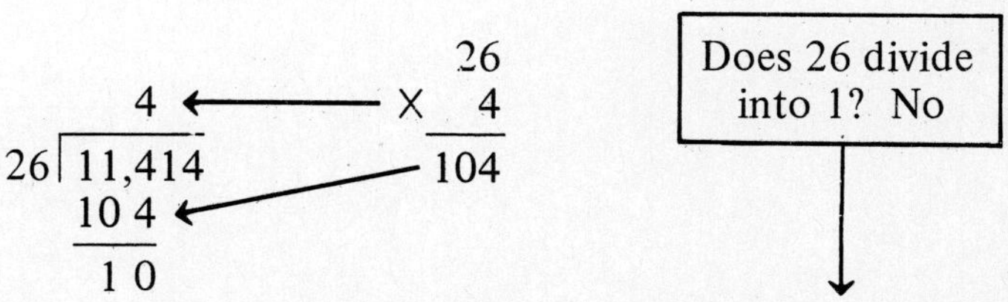

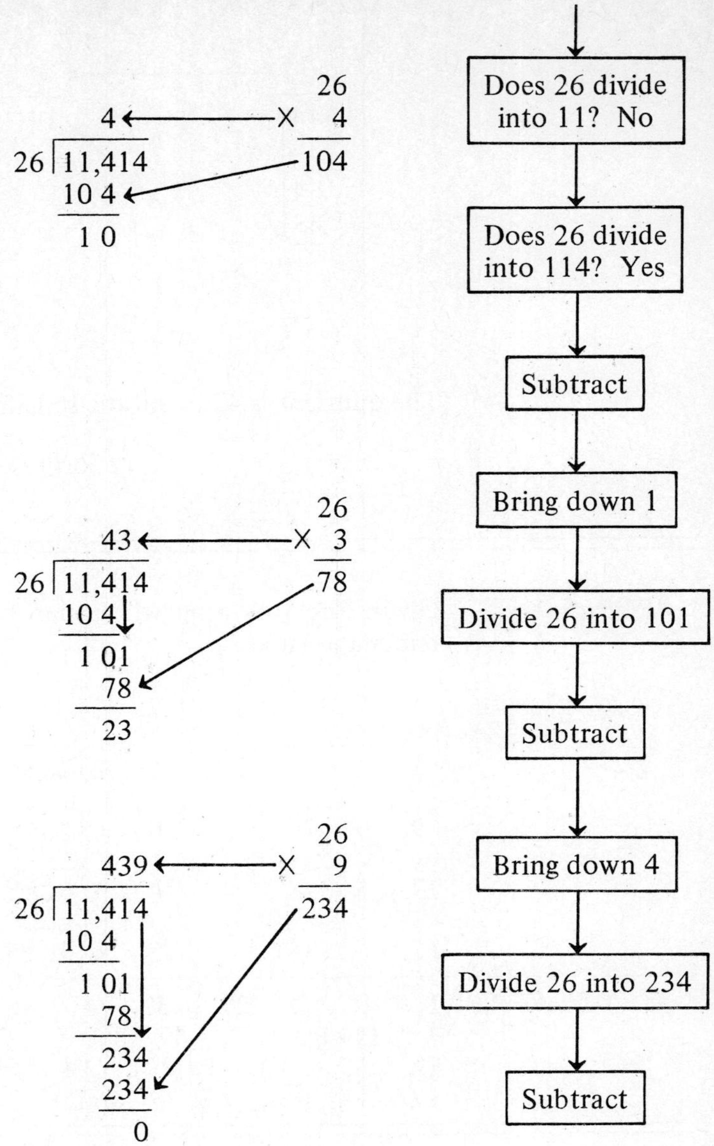

Usual

```
      439
26 ) 11,414
     10 4
      1 01
        78
       234
       234
         0
```

The quotient is 439. The remainder is 0. We can check this:

$$26 \times 439 = 11{,}414$$

EXAMPLE 57 Divide 29,667 ÷ 73.

SOLUTION Again, we do this both in the usual form and in slow motion.

Slow Motion

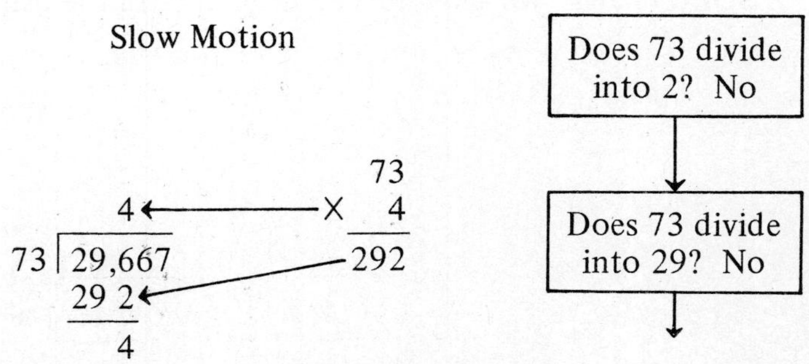

Usual

```
      406
73 | 29,667
     29 2
      467
      438
       29
```

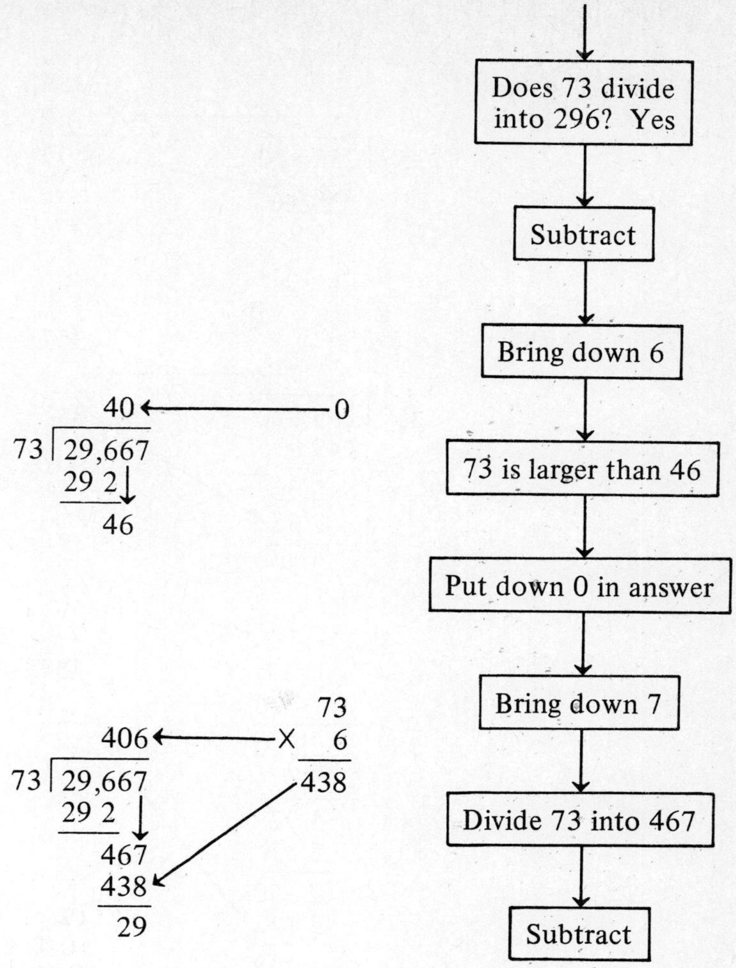

The quotient is 406. The remainder is 29. Notice that since 73 was larger than 46, we put a 0 in the quotient and brought the next digit down. Often, students forget to put in this 0, and they get the wrong answer.

We can check this:

$$(73 \times 406) + 29 = 29{,}638 + 29 = 29{,}667$$

YES	NO
102 213 | 21726 213 426 426	1 2 213 | 21726 213 426 426 (crossed out)

We place a zero in the quotient when the divisor will not divide into the dividend.

EXAMPLE 58 Divide 255,820 ÷ 357.

SOLUTION

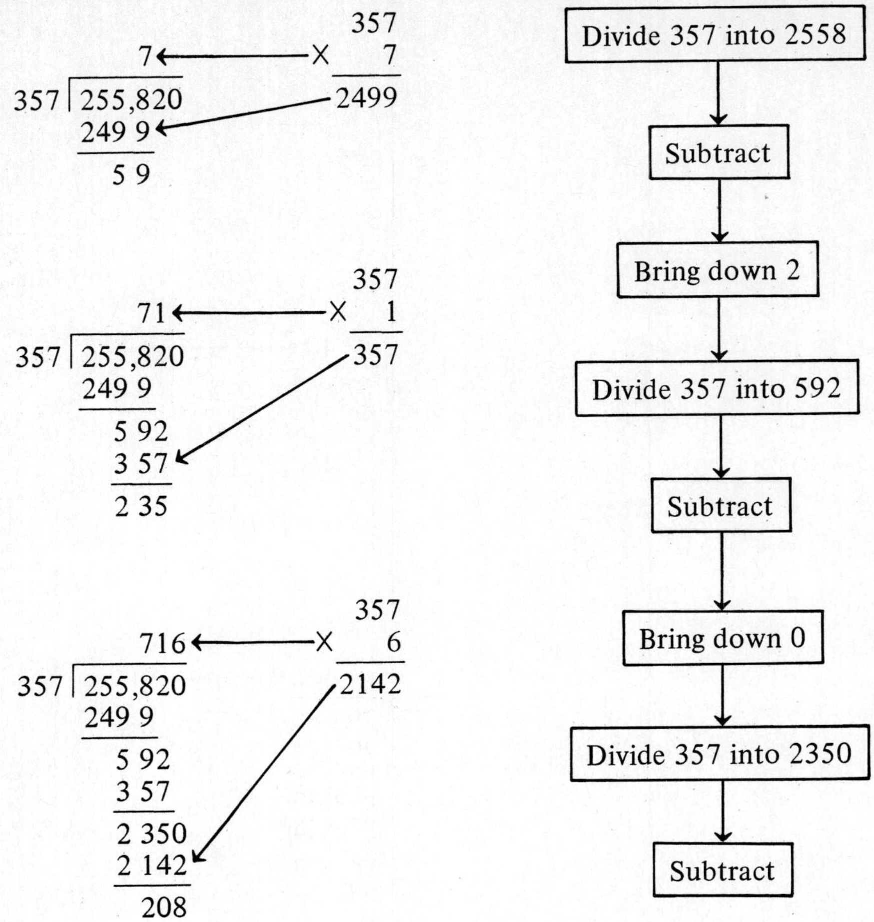

The quotient is 716. The remainder is 208. We can check this:

$$(357 \times 716) + 208 = 255{,}612 + 208 = 255{,}820$$

EXAMPLE 59 Divide 63,156,000 ÷ 760.

SOLUTION

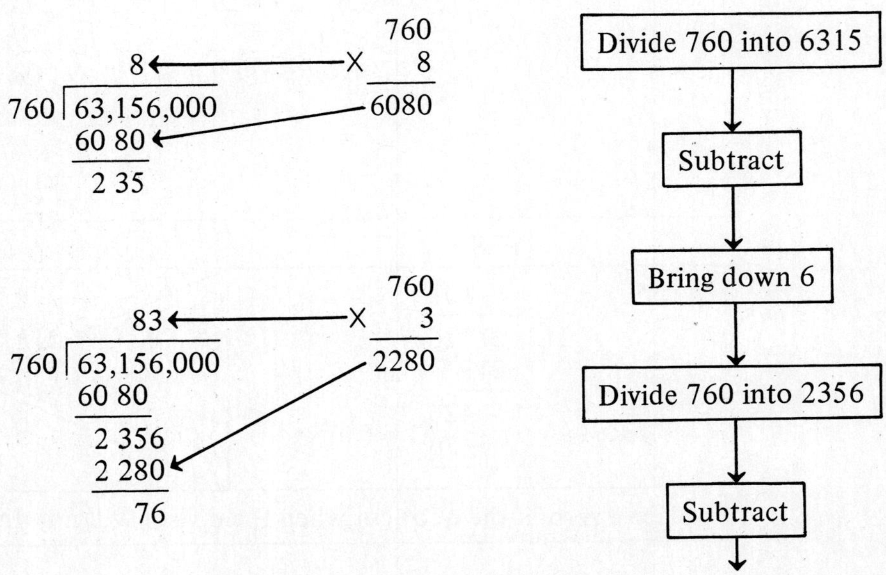

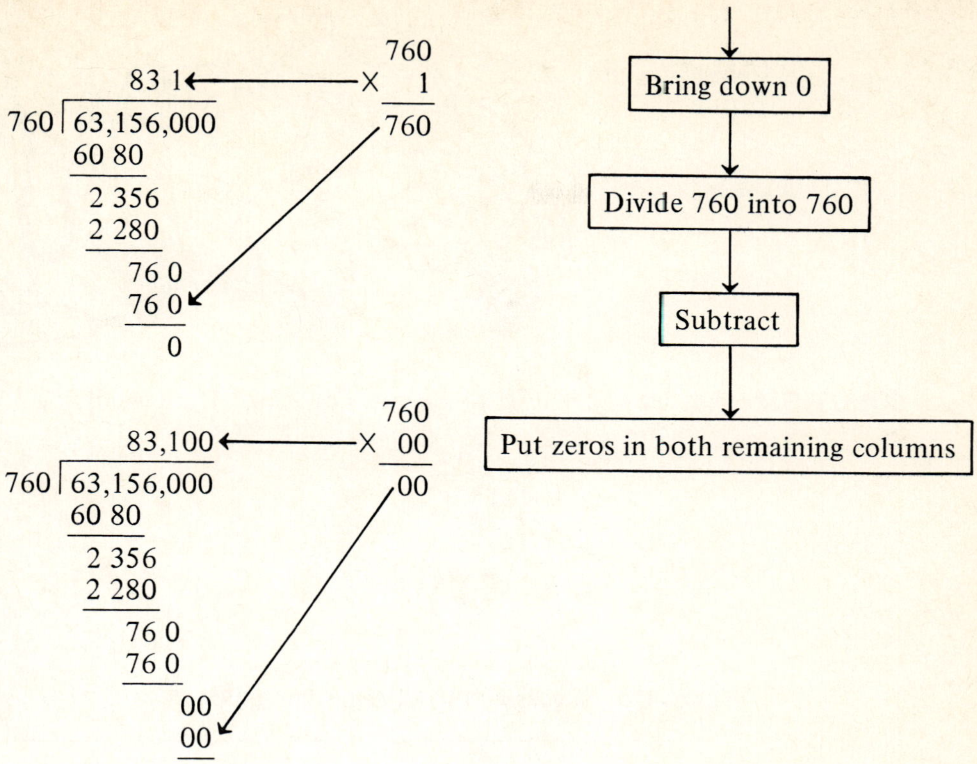

Since the problem came out evenly, we can bring down all the remaining zeros. Be sure to put the zeros in the quotient!

We can check this by multiplying:

$$760 \times 83{,}100 = 63{,}156{,}000$$

EXAMPLE 60 In a recent year, there were about 102,000,000 automobiles in America. In that year, Americans drove a total of 1,036,000,000,000 miles. What was the average yearly mileage per car?

SOLUTION Here is an example where we use division to find an average.

```
                    10 156
102,000,000 | 1,036,000,000,000
              1 020 000 00
                 16 000 000 0
                 10 200 000 0
                   5 800 000 00
                   5 100 000 00
                     700 000 000
                     612 000 000
                      88 000 000
```

Thus the average car is driven about 10,156 miles per year.

EXAMPLE 61 Donna drove 554 miles on 46 gallons of gasoline. How many miles per gallon is this?

SOLUTION The word *per* usually indicates division. Here we divide 554 ÷ 46.

$$\begin{array}{r} 12 \\ 46\,\overline{)\,554} \\ \underline{46} \\ 94 \\ \underline{92} \\ 2 \end{array}$$

Thus Donna gets about 12 miles per gallon.

Why Does It Work?

Division is repeated subtraction. Our division rule allows us to subtract very quickly. Consider 16,827 ÷ 79. We could subtract one 79 at a time:

$$\begin{array}{rlll} & 16{,}827 & & \\ - & \underline{79} & \longleftarrow 1 & 79 \\ & 16{,}748 & & \\ - & \underline{79} & \longleftarrow 1 & 79 \\ & 16{,}669 & & \\ - & \underline{79} & \longleftarrow 1 & 79 \\ & 16{,}590 & & \end{array}$$

and so on

But this is too slow a procedure. Now let us subtract 79 in groups of 100 (7900 at a time), groups of 10 (790 at a time), and then groups of 1 (79 at a time).

$$\begin{array}{rrrl} & 16{,}827 & & \\ - & \underline{7{,}900} & \longleftarrow 100 & 79\text{'s} \\ & 8{,}927 & & \\ - & \underline{7{,}900} & \longleftarrow 100 & 79\text{'s} \\ & 1{,}027 & & \\ - & \underline{790} & \longleftarrow 10 & 79\text{'s} \\ & 237 & & \\ - & \underline{79} & \longleftarrow 1 & 79 \\ & 158 & & \\ - & \underline{79} & \longleftarrow 1 & 79 \\ & 79 & & \\ - & \underline{79} & \longleftarrow \underline{1} & \underline{79} \\ & 0 & 213 & 79\text{'s} \end{array}$$

We see that there are 213 79's in 16,827. We can also see this as follows:

$$
\begin{array}{r l}
 & 213 \\
79 \overline{)} & 16{,}827 \\
 & -15{,}800 \leftarrow 200 \times 79 \\
 & 1{,}027 \\
 & -790 \leftarrow 10 \times 79 \\
 & 237 \\
 & -237 \leftarrow 3 \times 79
\end{array}
$$

PROBLEM SET Divide the following whole numbers, showing both the quotients and remainders, if any.

See Examples 56 to 61

1. $8\overline{)96}$ **2.** $5\overline{)150}$ **3.** $7\overline{)280}$

4. $8\overline{)64{,}000}$ **5.** $23\overline{)1081}$ **6.** $36\overline{)864}$

7. $18\overline{)1872}$ **8.** $29\overline{)783}$ **9.** $48\overline{)892}$

10. $34\overline{)971}$ **11.** $65\overline{)1235}$ **12.** $61\overline{)569}$

13. $81\overline{)27{,}945}$ **14.** $76\overline{)13{,}908}$ **15.** $47\overline{)21{,}308}$

16. $19\overline{)42{,}847}$ **17.** $58\overline{)428{,}475}$ **18.** $38\overline{)32{,}756}$

19. $325\overline{)15{,}600}$ **20.** $547\overline{)37{,}743}$ **21.** $762\overline{)86{,}931}$

22. $489\overline{)847{,}932}$ **23.** $329\overline{)280{,}966}$ **24.** $6892\overline{)3{,}285{,}679}$

25. $150\overline{)7{,}500{,}000}$ **26.** $8000\overline{)320{,}000{,}000}$ **27.** $50\overline{)350{,}000{,}000}$

28. Divide $2668 \div 46$.

29. Divide $2546 \div 67$.

30. Divide $4596 \div 43$.

31. Divide $8961 \div 79$.

32. Divide $45{,}892 \div 58$.

33. Divide $75{,}392 \div 321$.

34. Divide $56{,}947 \div 856$.

35. Divide $48{,}362 \div 457$.

36. Divide $557{,}208 \div 852$.

37. Divide $678{,}493 \div 573$.

38. Divide $400{,}000 \div 2000$.

39. Divide $850{,}000{,}000 \div 5000$.

ANSWERS **1.** 12 **2.** 30 **3.** 40 **4.** 8000 **5.** 47 **6.** 24 **7.** 104 **8.** 27 **9.** 18 R28 **10.** 28 R19 **11.** 19 **12.** 9 R20 **13.** 345 **14.** 183 **15.** 453 R17 **16.** 2,255 R2 **17.** 7,387 R29 **18.** 862 **19.** 48 **20.** 69 **21.** 114 R63 **22.** 1,734 R6 **23.** 854 **24.** 476 R5087 **25.** 50,000 **26.** 40,000 **27.** 7,000,000 **28.** 58 **29.** 38 **30.** 106 R38 **31.** 113 R34 **32.** 791 R14 **33.** 234 R278 **34.** 66 R451 **35.** 105 R377 **36.** 654 **37.** 1,184 R61 **38.** 200 **39.** 170,000

We mentioned before that division is the opposite of multiplication. This means that we can use division to solve multiplication problems such as $3 \cdot n = 12$. (Recall that the dot $\cdot$ can be used to represent multiplication.)

EXAMPLE 62 Find n: $3 \cdot n = 12$.

SOLUTION Three times what number is 12? This is easy and we see that the answer is 4. Also, notice that we could have divided both sides by 3 to find $n = 12 \div 3 = 4$.

EXAMPLE 63 Find n: $5 \cdot n = 100$.

SOLUTION Five times what number is 100? As in Example 62, we divide both sides by 5 to find $n = 100 \div 5 = 20$.

EXAMPLE 64 Find x: $x \div 2 = 7$.

SOLUTION What number divided by 2 is 7? Here we use the fact that multiplication is the opposite of division. Thus, to undo this division, we multiply $x = 2 \cdot 7 = 14$.

EXAMPLE 65 Find n: $n \div 5 = 12$.

SOLUTION What number divided by 5 is 12? We can find n by multiplying both sides by 5:

$$n = 5 \cdot 12 = 60$$

PROBLEM SET Find n or x in the following problems.

See Examples 62 to 65

1. $2 \cdot n = 6$ **2.** $4 \cdot x = 28$ **3.** $10 \cdot n = 90$

4. $12 \cdot n = 108$ **5.** $25 \cdot n = 5000$ **6.** $48 \cdot n = 1200$

7. $x \div 3 = 2$ **8.** $n \div 5 = 3$ **9.** $x \div 10 = 4$

10. $n \div 5 = 14$ **11.** $n \div 10 = 12$ **12.** $x \div 25 = 16$

ANSWERS **1.** 3 **2.** 7 **3.** 9 **4.** 9 **5.** 200 **6.** 25 **7.** 6 **8.** 15 **9.** 40 **10.** 70 **11.** 120 **12.** 400

Miscellaneous Exercises 1.5

Divide the following whole numbers.

See Example 53

1. $16 \div 2$ 2. $48 \div 8$ 3. $63 \div 7$

See Examples 54 and 55

4. $116 \div 4$ 5. $371 \div 7$ 6. $1107 \div 9$
7. $1327 \div 2$ 8. $4738 \div 3$ 9. $8562 \div 5$

See Examples 56 to 61

10. $1392 \div 24$ 11. $4592 \div 56$ 12. $284{,}418 \div 458$
13. $3671 \div 15$ 14. $45{,}673 \div 93$ 15. $321{,}843 \div 219$

Find n in the following problems.

See Examples 62 and 63

16. $3 \cdot n = 21$ 17. $5 \cdot n = 45$ 18. $12 \cdot n = 240$

See Examples 64 and 65

19. $n \div 3 = 4$ 20. $n \div 2 = 8$ 21. $n \div 15 = 12$

Consumer applications

22. Marybeth earns $11,804 per year. How much is this per week?
23. Marty has 304 record albums. He is buying storage cases that hold 24 records each. How many cases does he need?
24. It takes a 3500-calorie reduction to lose 1 pound of body weight. If Roger eats 500 calories fewer than his daily needs, how long will it take him to lose 1 pound?
25. Lucy drives 1218 miles on 58 gallons of gasoline. How many miles per gallon is this?

Health application

26. A patient is to receive 200,000 units of penicillin in eight doses. How large is each dose?

Science application

27. A 130-pound woman is wearing shoes whose heels have an area of 2 square inches. What is the pressure (weight ÷ area) on the floor?

Electrical application

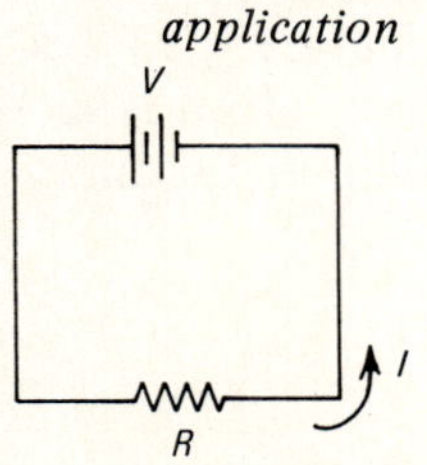

28. Ohm's law states that

$$I = V \div R$$

where I is the current, V is the voltage, and R is the resistance in a circuit. Find I if V is 150 volts and R is 10 ohms.

Business applications

29. Five salespeople had total sales of $175,650. What is the average per person?
30. A company rents 3500 square feet of floor space. The annual rent is $24,500. How much is this rental per square foot?
31. A marketing firm finds that 45 families surveyed watched a total of 270 hours of TV per day. What is the average per family?

Miscellaneous applications

32. Jim Polk leaves $176,205 to his 15 grandchildren. How much does each grandchild receive?

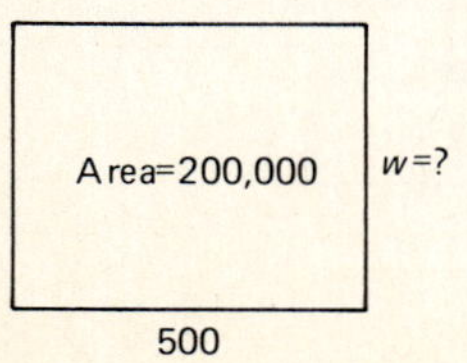

33. The area of the rectangle at the left is 200,000. Find the missing width, w.

Now complete Post-Test 1 for Section 1.5 on page T13.

1.6 PRIME FACTORIZATION AND LEAST COMMON MULTIPLE

SECTION OBJECTIVES

After completing this unit, you will be able to:

1. Write a number as a product of primes.
2. Find the least common multiple of two or more numbers.

TYPICAL PROBLEMS

You will be able to work problems like the ones below.

Factor into primes, or state if the number is prime.

(a) 69 **(b)** 79 **(c)** 42

Find the least common multiple (LCM) of the following numbers.

(d) 10 and 15 **(e)** 8, 12, and 18 **(f)** 3, 4, and 8

ANSWERS **(a)** $3 \cdot 23$ **(b)** Prime **(c)** $2 \cdot 3 \cdot 7$ **(d)** 30 **(e)** 72 **(f)** 24

A **prime number** is a whole number, 2 or more, that can be divided evenly only by 1 and itself. The first primes are:

$$2, 3, 5, 7, 11, 13, 17, 19, 23, 29, 31, \ldots$$

All the other numbers are called **composites.** The first composites are

$$4, 6, 8, 9, 10, 12, 14, 15, 16, 18, 20, 21, 22, 24, 25, 26, 27, 28, \ldots$$

Every composite can be written as a product of primes.

This is a very important statement. We use it when we **factor,** which is the process of writing one number as the product of other numbers. (In this section we write multiplication with a dot $\cdot$.) We use a **factor tree** to help us factor.

EXAMPLE 66 Factor 60 into primes.

At each step, we factor out another prime until all the factors are prime. We could also have done the problem as follows.

Notice that we get the same prime factors: 2, 2, 3, and 5.

> To factor any whole number into primes:
> 1. Factor out all 2's, if any.
> 2. Factor out all 3's, if any.
> 3. Factor out all 5's, if any.
> 4. Continue trying to factor the higher primes (7, 11, 13, 17, and so on) until all factors are prime.
> 5. If the number has only itself as a prime factor, it is prime.

EXAMPLE 67 Factor 84 into primes.

SOLUTION We use the factor tree to help.

Thus $84 = 2 \cdot 2 \cdot 3 \cdot 7$, because all these factors are prime.

EXAMPLE 68 Factor 24 into primes.

SOLUTION

Thus 24 = 2 · 2 · 2 · 3.

EXAMPLE 69 Factor 70 into primes.

SOLUTION

Therefore 70 = 2 · 5 · 7.

EXAMPLE 70 Factor 61 into primes.

SOLUTION We cannot factor 2 out of 61; we cannot factor 3 out; we cannot factor 5 out; and so on. We conclude that 61 is a prime.

EXAMPLE 71 Factor 105 into primes.

SOLUTION

So 105 = 3 · 5 · 7.

EXAMPLE 72 Factor 108 into primes.

SOLUTION

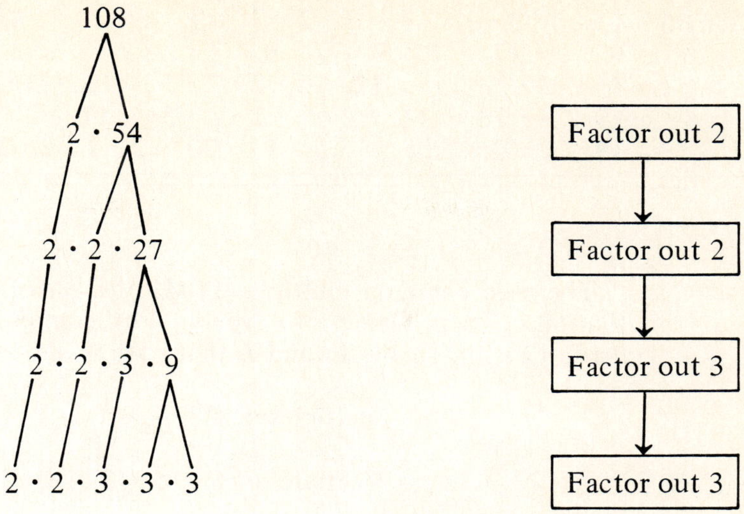

Thus $108 = 2 \cdot 2 \cdot 3 \cdot 3 \cdot 3$.

EXAMPLE 73 Factor 89 into primes.

SOLUTION We try to factor out 2, 3, 5, 7, and so on. But nothing works. So 89 is prime.

Comparison With Chemistry

In chemistry, the basic building blocks are the **elements,** such as hydrogen (H), oxygen (O), carbon (C), nitrogen (N), and so on. We can break chemical compounds into their basic element components. For instance,

water	ammonia	glucose
↓	↓	↓
H_2O	NH_3	$C_6H_{12}O_6$

Similarly, **primes** act like the elements or building blocks for the whole numbers. Just as chemical compounds are composed of elements, the composite numbers can be written as products of primes. As examples,

12	35	198
↓	↓	↓
$2 \cdot 2 \cdot 3$	$5 \cdot 7$	$2 \cdot 3 \cdot 3 \cdot 11$

PROBLEM SET Factor the composites into primes, and state "Prime" for the prime numbers.

See Examples 66 to 73

1. 12	**2.** 28	**3.** 33	**4.** 29	**5.** 54
6. 94	**7.** 49	**8.** 48	**9.** 57	**10.** 67

11. 99	12. 40	13. 75	14. 59	15. 39
16. 120	17. 270	18. 245	19. 5000	20. 100,000

ANSWERS **1.** $2 \cdot 2 \cdot 3$ **2.** $2 \cdot 2 \cdot 7$ **3.** $3 \cdot 11$ **4.** Prime **5.** $2 \cdot 3 \cdot 3 \cdot 3$ **6.** $2 \cdot 47$ **7.** $7 \cdot 7$ **8.** $2 \cdot 2 \cdot 2 \cdot 2 \cdot 3$ **9.** $3 \cdot 19$ **10.** Prime **11.** $3 \cdot 3 \cdot 11$ **12.** $2 \cdot 2 \cdot 2 \cdot 5$ **13.** $3 \cdot 5 \cdot 5$ **14.** Prime **15.** $3 \cdot 13$ **16.** $2 \cdot 2 \cdot 2 \cdot 3 \cdot 5$ **17.** $2 \cdot 3 \cdot 3 \cdot 3 \cdot 5$ **18.** $5 \cdot 7 \cdot 7$ **19.** $2 \cdot 2 \cdot 2 \cdot 5 \cdot 5 \cdot 5 \cdot 5$ **20.** $2 \cdot 2 \cdot 2 \cdot 2 \cdot 2 \cdot 5 \cdot 5 \cdot 5 \cdot 5 \cdot 5$

The **least common multiple (LCM)** of a set of numbers is the smallest number that all of them divide into evenly. For example, the LCM of 4 and 6 is 12, because 12 is the smallest number that both 4 and 6 divide evenly.

Multiples of 4: 4, 8, (12), 16, 20, 24, . . .

Multiples of 6: 6, (12), 18, 24, 30, 36, . . .

EXAMPLE 74 The LCM of 8 and 12 is 24, because 24 is the smallest number that both 8 and 12 divide evenly.

Sometimes, we cannot just look at the numbers and immediately see the LCM. In these cases, we can use the following procedure.

> To find the LCM of a set of numbers:
> 1. Factor each number into primes.
> 2. Line up the common factors, with unlike factors in new columns.
> 3. Multiply one prime from each column to get the LCM.

EXAMPLE 75 Find the LCM of 24 and 18.

SOLUTION We begin by factoring each number into primes.

$$24 = 2 \cdot 2 \cdot 2 \cdot 3$$
$$18 = 2 \qquad\qquad \cdot 3 \cdot 3$$
$$\text{LCM} = 2 \cdot 2 \cdot 2 \cdot 3 \cdot 3 = 72$$

Factor, line up common factors → LCM = product of columns

EXAMPLE 76 Find the LCM of 20 and 28.

SOLUTION We factor each number and line up the primes.

$$20 = 2 \cdot 2 \cdot 5$$
$$28 = 2 \cdot 2 \cdot 7$$
$$\text{LCM} = 2 \cdot 2 \cdot 5 \cdot 7 = 140$$

Factor, line up common factors → LCM = product of columns

EXAMPLE 77 Find the LCM of 10, 12, and 15.

SOLUTION We factor all three numbers into primes.

$$10 = 2 \cdot 5$$
$$12 = 2 \cdot 2 \cdot 3$$
$$15 = 5 \cdot 3$$
$$\text{LCM} = 2 \cdot 5 \cdot 2 \cdot 3 = 60$$

Factor, line up common primes → LCM = product of columns

EXAMPLE 78 Find the LCM of 18, 20, and 16.

SOLUTION As before, we factor all three numbers.

$$18 = 2 \cdot 3 \cdot 3$$
$$20 = 2 \cdot 2 \cdot 5$$
$$16 = 2 \cdot 2 \cdot 2 \cdot 2$$
$$\text{LCM} = 2 \cdot 3 \cdot 3 \cdot 2 \cdot 5 \cdot 2 \cdot 2 = 720$$

Factor, line up common factors → LCM = product of columns

Why Does It Work?

Primes are the building blocks of the whole numbers. By factoring, we are breaking each number into its building blocks. By multiplying all the primes as we do, we are building a new number from the old ones. Because it has all the same primes (and more), it can be divided evenly by all the original numbers.

PROBLEM SET Find the LCM for each of the following sets of numbers.

See Examples 74 to 78

1. 3 and 5
2. 4 and 6
3. 6 and 10
4. 6 and 9
5. 8 and 12
6. 6 and 14

7.	7 and 8	**8.**	9 and 15	**9.**	10 and 12
10.	12 and 15	**11.**	16 and 18	**12.**	9 and 21
13.	15 and 25	**14.**	18 and 24	**15.**	28 and 32
16.	36 and 48	**17.**	4, 6, and 8	**18.**	8, 10, and 12
19.	4, 5, and 6	**20.**	9, 12, and 15	**21.**	7, 8, and 12
22.	4, 8, and 16	**23.**	10, 15, and 20	**24.**	18, 24, and 28
25.	24, 36, and 48				

ANSWERS **1.** 15 **2.** 12 **3.** 30 **4.** 18 **5.** 24 **6.** 42 **7.** 56 **8.** 45 **9.** 60 **10.** 60 **11.** 144 **12.** 63 **13.** 75 **14.** 72 **15.** 224 **16.** 144 **17.** 24 **18.** 120 **19.** 60 **20.** 180 **21.** 168 **22.** 16 **23.** 60 **24.** 504 **25.** 144

Miscellaneous Exercises 1.6

Factor the following numbers in prime factors, or write "Prime" if they are prime.

See Examples 68 to 73

1. 24 **2.** 56 **3.** 75 **4.** 79 **5.** 99 **6.** 129

Find the least common multiple (LCM) for each of the following sets of numbers.

See Examples 74 to 78

7.	12 and 16	**8.**	8 and 20	**9.**	25 and 45
10.	10, 15, and 25	**11.**	12, 16, and 20	**12.**	20, 36, and 48

Now complete Post-Test 1 for Section 1.6 on page T15.

CHAPTER 1 SUMMARY

Important Words and Phrases

addend (1.2)
addition (1.2)
addition table (1.2)
base-10 place-value system (1.1)
borrow (1.3)
carry (1.2)
composite (1.6)
difference (1.3)
dividend (1.5)
division (1.5)
divisor (1.5)
expanded form (1.1)
factoring (1.6)
factors (1.4)
factor tree (1.6)
least common multiple (LCM) (1.6)
minuend (1.3)
multiples of powers of 10 (1.4)
multiplication (1.4)
number line (1.2)
partial product (1.2)
place holders (1.1)
powers of 10 (1.4)
prime number (1.6)
product (1.4)
quotient (1.5)
rounding down (1.1)
rounding up (1.1)
subtraction (1.3)
subtrahend (1.3)
sum (1.2)

Important Ideas

(1.1) Knowing the place value of whole numbers.
Rounding whole numbers.
(1.2) Adding whole numbers.
(1.3) Subtracting whole numbers.
Knowing that addition and subtraction are opposite operations.
(1.4) Multiplying whole numbers.
(1.5) Dividing whole numbers.
Knowing that multiplication and division are opposite operations.
(1.6) Factoring a number into primes (or identifying primes).
Finding the LCM of a set of numbers.

CHAPTER 1 REVIEW TEST

Choose the correct answer.

1. In 475,169, what is the value of the 1?
(a) 10 (b) 100 (c) 1000 (d) 169 (e) 4751

2. In 37,069,428, what digit is in the ten thousands' column?
(a) 7 (b) 6 (c) 0 (d) 9 (e) 4

3. Two million, six hundred thousand, eighty-two is
(a) 2,600,082 (b) 260,082 (c) 2,682
(d) 2,600,820 (e) 200,682

4. Add 495 + 758.
(a) 1144 (b) 1143 (c) 1153 (d) 1243 (e) 1253

5. Add 78 + 319 + 93 + 6207.
(a) 6587 (b) 6687 (c) 6677 (d) 6686 (e) 6697

6. Add
$$\begin{array}{r} 72153 \\ 55314 \\ +87867 \\ \hline \end{array}$$
(a) 214,354 (b) 215,324 (c) 214,334
(d) 215,334 (e) 214,244

7. Subtract
$$\begin{array}{r} 752 \\ -586 \\ \hline \end{array}$$
(a) 256 (b) 266 (c) 276 (d) 176 (e) 166

8. Subtract 35,023 - 8847.
(a) 26,176 (b) 26,186 (c) 26,276
(d) 25,266 (e) 25,286

9. Subtract 200,000 - 49,672.
(a) 250,328 (b) 161,438 (c) 261,438
(d) 151,338 (e) 150,328

10. Find x: $x + 6 = 20$.
(a) 26 (b) 14 (c) 120 (d) 4 (e) 206

11. Multiply 57 × 63.
(a) 3071 (b) 3491 (c) 3591 (d) 3581 (e) 3481

12. Multiply 800 × 2300.
(a) 174,000 (b) 1,740,000 (c) 184,000
(d) 1,840,000 (e) 18,400,000

13. Multiply 842 × 759.
(a) 639,068 (b) 638,088 (c) 639,078
(d) 638,068 (e) 637,178

14. Divide 12,354 ÷ 29.
(a) 417 (b) 426 (c) 432 R17 (d) 438 R2 (e) 460 R14

15. Divide 21,407 ÷ 38.
(a) 544 R31 (b) 536 R34 (c) 561 R7
(d) 572 R11 (e) 563 R13

16. Divide $173\overline{)108{,}945}$.
(a) 621 R87 (b) 608 R106 (c) 629 R128
(d) 625 R58 (e) 631 R56

17. Find n: $9 \cdot n = 45$.
(a) 54 (b) 36 (c) 945 (d) 5 (e) 405

18. Write 131 as a product of primes (if possible).
(a) $1 \cdot 3 \cdot 1$ (b) $3 \cdot 5 \cdot 7$ (c) $13 \cdot 11$
(d) Prime (e) $11 \cdot 31$

19. Write 171 as a product of primes (if possible).
(a) $1 \cdot 7 \cdot 1$ (b) $3 \cdot 3 \cdot 19$ (c) $17 \cdot 11$
(d) Prime (e) $11 \cdot 71$

20. The least common multiple of 50 and 40 is:
(a) 10 (b) 200 (c) 2000 (d) 1000 (e) 2

Now complete Post-Test 1 for Chapter 1 on page T17.

2

FRACTIONS AND MIXED NUMBERS

Before you begin this chapter, complete the Chapter 2 Pre-Test on page T21.

2.1 BASIC IDEAS OF FRACTIONS

SECTION OBJECTIVES After completing this unit, you will be able to:

1. Understand the meaning of fractions and fractional notation.
2. Tell if two fractions are equal.
3. Reduce a fraction to simplest terms.

TYPICAL PROBLEMS You will be able to work problems like the ones below.

(a) For $\frac{2}{3}$, identify the numerator and denominator, and sketch.

(b) What fraction is illustrated by this figure?

(c) Does $\frac{12}{20} = \frac{15}{25}$?

(d) Does $\frac{11}{35} = \frac{9}{31}$?

(e) Reduce $\frac{35}{150}$ to simplest terms.

(f) Reduce $\frac{24}{72}$ to simplest terms.

ANSWERS **(a)** numerator = 2, denominator = 3,

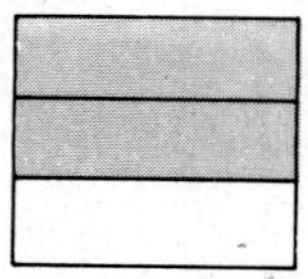

(b) $\frac{3}{4}$ **(c)** Yes **(d)** No

(e) $\frac{7}{30}$ **(f)** $\frac{1}{3}$

Whole numbers are not enough to express every quantity in real life. Very often, we have to deal with parts of whole numbers. We call these parts **fractions** and write them as follows:

$$\frac{a}{b} \begin{array}{l}\leftarrow \text{numerator} \\ \leftarrow \text{denominator}\end{array} \qquad \frac{a}{b} \text{ means } a \div b$$

where a and b are whole numbers (b cannot be zero). The bottom number, b, called the **denominator**, tells how many equal pieces the whole has been cut into. The top number, a, called the **numerator**, tells how many pieces we have. Sometimes, we will call these numbers simply *top* and *bottom*.

EXAMPLE 1 Consider the fraction $\frac{5}{8}$. The denominator is 8; this says that the whole has been cut into eight equal parts. The numerator is 5; this says that we have five of these parts. This can be pictured as follows:

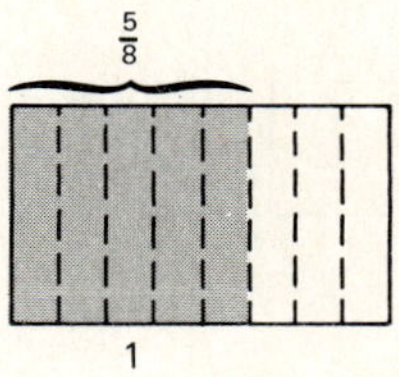

We can also picture the same $\frac{5}{8}$ as 5 units divided by 8, or $5 \div 8$.

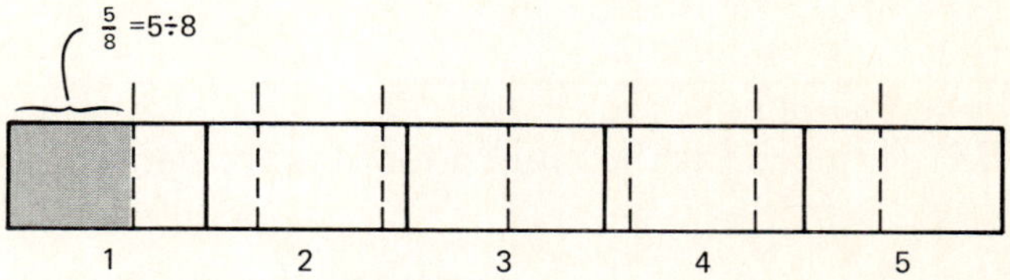

Notice that the same amount is shaded.

EXAMPLE 2 Consider the fraction $\frac{4}{5}$. The numerator is 4, and the denominator is 5. We can picture this fraction as follows:

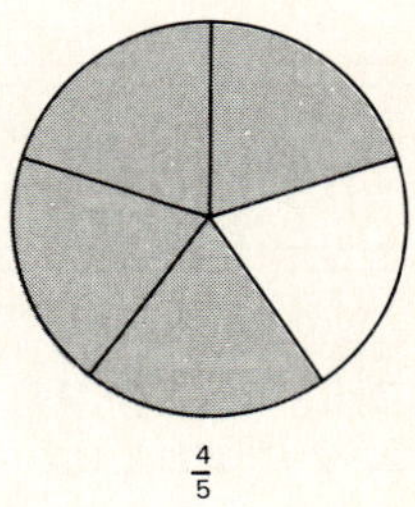

EXAMPLE 3 Consider the fraction $\frac{6}{6}$. The numerator and denominator are both 6. Notice from the picture that $\frac{6}{6}$ is the same as 1. In fact, $\frac{n}{n}$ *is always 1*, if n is not zero.

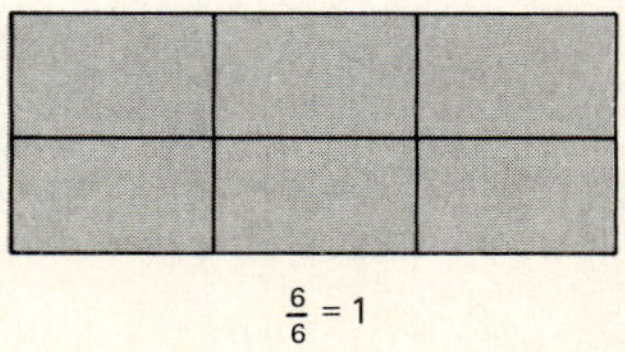

PROBLEM SET For the following illustrations, state the fraction that is given by the shaded region. Identify the numerator and denominator.

See Examples 1 to 3

1. 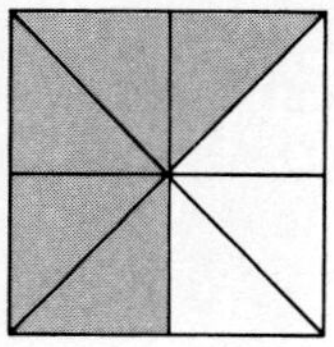

2. 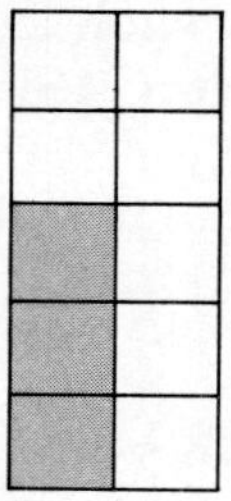

3. 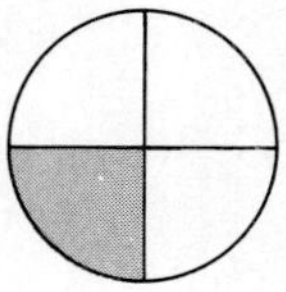

4. 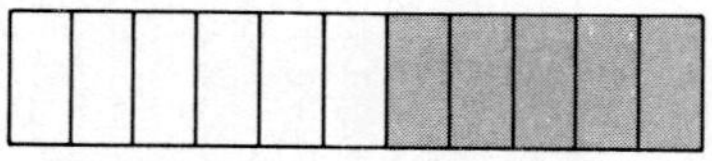

5. 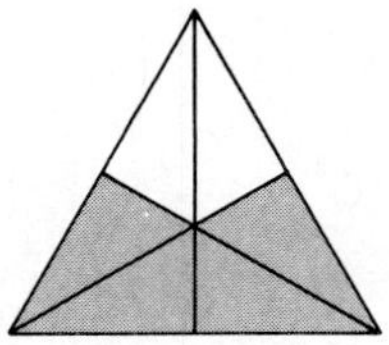

ANSWERS

1. $\frac{5}{8}$; 5 is the numerator, 8 is the denominator.
2. $\frac{3}{10}$; 3 is the numerator, 10 is the denominator.
3. $\frac{1}{4}$; 1 is the numerator, 4 is the denominator.
4. $\frac{5}{11}$; 5 is the numerator, 11 is the denominator.
5. $\frac{4}{6}$; 4 is the numerator, 6 is the denominator.

PROBLEM SET Shade the part of a square, rectangle, or circle that is given by the following fractions. (Use any shape that you wish.)

See Examples 1 to 3

1. $\frac{2}{5}$ 2. $\frac{7}{8}$ 3. $\frac{5}{12}$ 4. $\frac{0}{10}$ 5. $\frac{11}{15}$

ANSWERS

1. 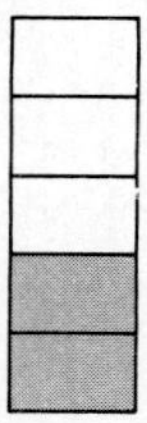

2. 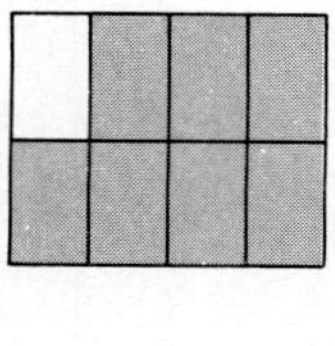

3. 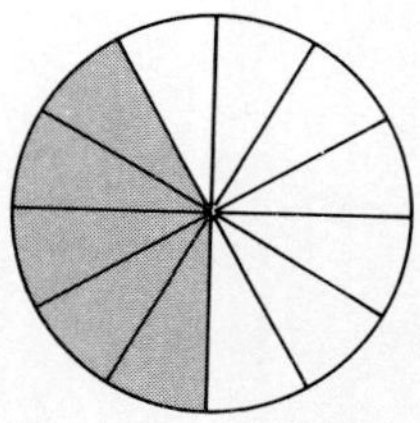

4. 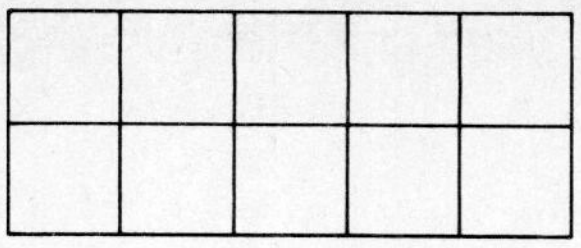

5. 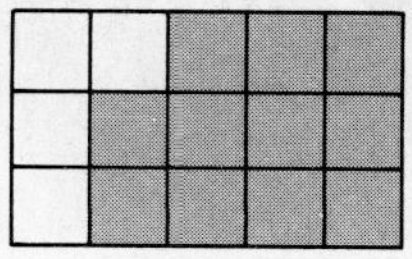

Sometimes, fractions that look different are really **equal**. That is, they represent the same part of a whole. The figures below show how $\frac{3}{4} = \frac{6}{8} = \frac{12}{16}$. They are just different ways to write the same part of the whole.

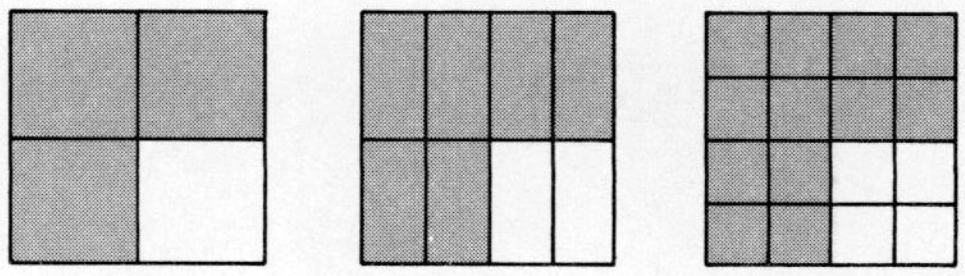

$\frac{3}{4}$ $\frac{6}{8}$ $\frac{12}{16}$

We have an easy way to tell if two fractions are equal: We cross-multiply the fractions.

$$\frac{a}{b} = \frac{c}{d} \quad \text{if} \quad ad = bc$$

To see if two fractions are equal:

1. Cross-multiply the fractions.

2a. If the products are equal, the fractions are equal.

2b. If the products are not equal, the fractions are not equal.

EXAMPLE 4 Does $\frac{15}{25} = \frac{12}{20}$?

SOLUTION

$$\frac{15}{25} \times \frac{12}{20}$$

$$15 \cdot 20 \ ? \ 12 \cdot 25$$

$$300 = 300$$

Thus $\frac{15}{25} = \frac{12}{20}$

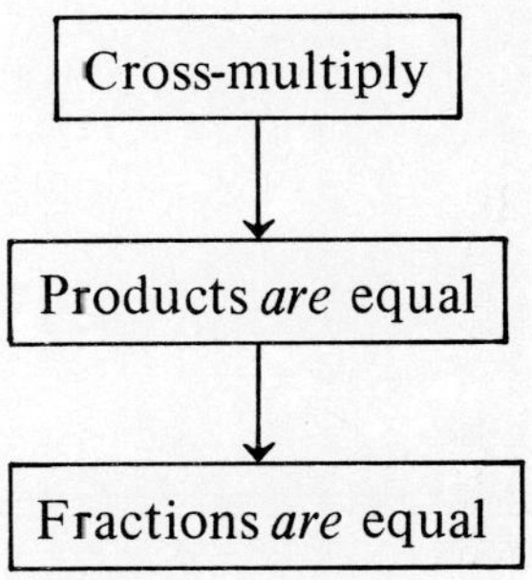

EXAMPLE 5 Does $\frac{8}{14} = \frac{12}{21}$?

SOLUTION $\dfrac{8}{14} \times \dfrac{12}{21}$

$8 \cdot 21 \ ? \ 12 \cdot 14$

$168 = 168$

So $\dfrac{8}{14} = \dfrac{12}{21}$

Cross-multiply → Products *are* equal → Fractions *are* equal

EXAMPLE 6 Does $\dfrac{4}{9} = \dfrac{9}{20}$?

SOLUTION $\dfrac{4}{9} \times \dfrac{9}{20}$

$4 \cdot 20 \ ? \ 9 \cdot 9$

$80 \neq 81$

So $\dfrac{4}{9} \neq \dfrac{9}{20}$

Cross-multiply → Products are *not* equal → Fractions are *not* equal

EXAMPLE 7 Does $\dfrac{2}{3} = \dfrac{14}{21}$?

SOLUTION $\dfrac{2}{3} \times \dfrac{14}{21}$

$2 \cdot 21 \ ? \ 3 \cdot 14$

$42 = 42$

So $\dfrac{2}{3} = \dfrac{14}{42}$

Cross-multiply → Products *are* equal → Fractions *are* equal

---

PROBLEM SET State whether the pairs of fractions below are equal or not equal

See Examples 4 to 7

1. Does $\dfrac{10}{12} = \dfrac{18}{21}$?
2. Does $\dfrac{15}{18} = \dfrac{35}{40}$?
3. Does $\dfrac{16}{25} = \dfrac{64}{100}$?

4. Does $\frac{9}{15} = \frac{21}{35}$? 5. Does $\frac{10}{21} = \frac{19}{39}$? 6. Does $\frac{20}{35} = \frac{44}{77}$?

7. Does $\frac{17}{20} = \frac{30}{35}$? 8. Does $\frac{10}{45} = \frac{26}{117}$? 9. Does $\frac{18}{24} = \frac{21}{28}$?

Draw a sketch to illustrate how the following fractions are equal.

10. $\frac{2}{3} = \frac{6}{9}$ 11. $\frac{1}{6} = \frac{2}{12}$ 12. $\frac{3}{5} = \frac{12}{20}$

ANSWERS 1. No 2. No 3. Yes 4. Yes 5. No 6. Yes 7. No 8. Yes 9. Yes

10. 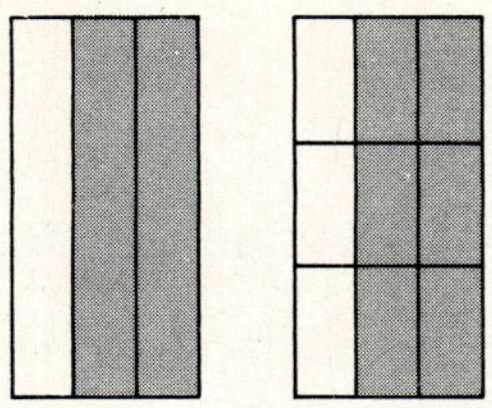

11. 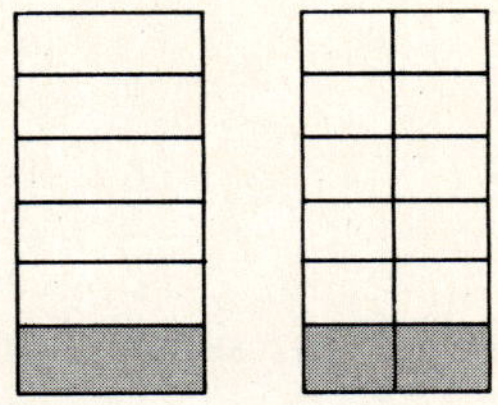

12. 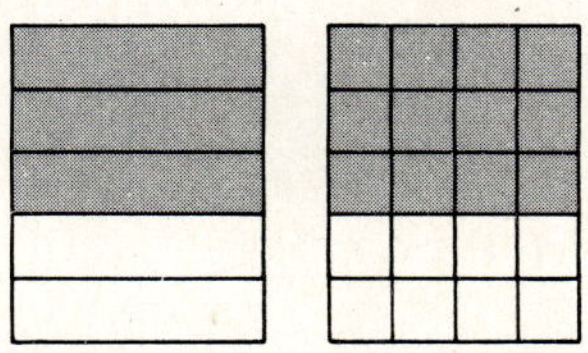

A special equality that we have is one that we have already seen in Example 3: namely,

$$\frac{n}{n} = 1 \qquad \text{for all whole numbers } n \neq 0$$

EXAMPLE 8 The following fractions are all equal to 1.

(a) $\frac{6}{6} = 1$ (b) $\frac{11}{11} = 1$ (c) $\frac{29}{29} = 1$ (d) $\frac{376}{376} = 1$

PROBLEM SET Simplify the following fractions.

See Example 8

1. $\frac{2}{2}$ 2. $\frac{13}{13}$ 3. $\frac{12}{12}$

4. $\frac{54}{54}$ 5. $\frac{102}{102}$ 6. $\frac{3285}{3285}$

ANSWERS 1. 1 2. 1 3. 1 4. 1 5. 1 6. 1

As we will soon see, we multiply fractions by multiplying the numerators together and the denominators together. Recall from Chapter 1 that multiplying by 1 does not change the value of a number; for example, $5 \cdot 1 = 5$ and $23 \cdot 1 = 23$. Let us see what happens when we multiply fractions by 1.

$$\frac{5}{6} = \frac{5}{6} \cdot \boxed{\frac{2}{2}} = \frac{10}{12}$$

$$\text{or} \quad \frac{5}{6} \cdot \boxed{\frac{3}{3}} = \frac{15}{18}$$

$$\text{or} \quad \frac{5}{6} \cdot \boxed{\frac{4}{4}} = \frac{20}{24}$$

and so on

Although $\frac{5}{6} = \frac{10}{12} = \frac{15}{18} = \frac{20}{24}$, the fraction $\frac{5}{6}$ is special. It is **reduced to simplest terms,** since there are no common factors in the numerator or denominator. There are two ways to reduce a fraction. The first method involves factoring the numerator and denominator into primes and canceling common factors.

> To reduce a fraction to simplest terms:
>
> 1. Factor both the numerator and denominator into primes.
> 2. Cancel out any common primes. (If all the factors cancel from the top or bottom, put a 1 there.)
> 3. Multiply the remaining factors.

EXAMPLE 9 Reduce $\frac{24}{42}$ to simplest terms.

SOLUTION

$$\frac{24}{42} = \frac{2 \cdot 2 \cdot 2 \cdot 3}{2 \cdot 3 \cdot 7}$$

Factor top and bottom

$$= \frac{\cancel{2} \cdot 2 \cdot 2 \cdot \cancel{3}}{\cancel{2} \cdot \cancel{3} \cdot 7} = \frac{4}{7}$$

Cancel common factors, multiply remaining factors

EXAMPLE 10 Reduce $\frac{75}{120}$ to simplest terms.

SOLUTION

$$\frac{75}{120} = \frac{3 \cdot 5 \cdot 5}{2 \cdot 2 \cdot 2 \cdot 3 \cdot 5}$$

Factor top and bottom

$$= \frac{\cancel{3} \cdot \cancel{5} \cdot 5}{2 \cdot 2 \cdot 2 \cdot \cancel{3} \cdot \cancel{5}} = \frac{5}{8}$$

Cancel common factors, multiply remaining factors

EXAMPLE 11 Reduce $\dfrac{84}{126}$ to simplest terms.

SOLUTION

$$\frac{84}{126} = \frac{2 \cdot 2 \cdot 3 \cdot 7}{2 \cdot 3 \cdot 3 \cdot 7}$$

Factor numerator and denominator

$$= \frac{\cancel{2} \cdot 2 \cdot \cancel{3} \cdot \cancel{7}}{\cancel{2} \cdot 3 \cdot \cancel{3} \cdot \cancel{7}} = \frac{2}{3}$$

Cancel common factors

EXAMPLE 12 Reduce $\dfrac{30}{120}$ to simplest terms.

SOLUTION

$$\frac{30}{120} = \frac{2 \cdot 3 \cdot 5}{2 \cdot 2 \cdot 2 \cdot 3 \cdot 5}$$

Factor top and bottom

$$= \frac{\cancel{2} \cdot \overset{1}{\cancel{3}} \cdot \cancel{5}}{\cancel{2} \cdot 2 \cdot 2 \cdot \cancel{3} \cdot \cancel{5}} = \frac{1}{4}$$

Cancel common factors (put 1 in numerator), multiply remaining factors

This method suggests another way to reduce fractions that may be more familiar to many students; that is, we can divide the same number from the numerator and denominator.

> Another way to reduce fractions is:
>
> 1. Find a common factor for the numerator and denominator.
> 2. Divide this factor out of both the numerator and denominator.
> 3. Divide out other common factors until there are no more common factors.

EXAMPLE 13 Reduce $\dfrac{12}{30}$ to simplest terms.

SOLUTION We divide out common factors until there are no more common factors.

$$\frac{12}{30} = \frac{12 \div 2}{30 \div 2} = \frac{6}{15}$$

Divide out a 2 from top and bottom

$$= \frac{6 \div 3}{15 \div 3} = \frac{2}{5}$$

Divide out a 3 from top and bottom

We cannot go further, so $\dfrac{2}{5}$ is the answer.

EXAMPLE 14 Reduce $\frac{15}{60}$ to simplest terms.

SOLUTION

$$\frac{15}{60} = \frac{15 \div 5}{60 \div 5} = \frac{3}{12}$$

Divide out a 5 from top and bottom

$$= \frac{3 \div 3}{12 \div 3} = \frac{1}{4}$$

Divide out a 3 from top and bottom

EXAMPLE 15 A marketing survey for a business recently found that $\frac{126}{336}$ of a group of people wear running shoes with their usual wardrobes. Reduce this fraction.

SOLUTION We factor the top and bottom into primes and cancel.

$$\frac{126}{336} = \frac{2 \cdot 3 \cdot 3 \cdot 7}{2 \cdot 2 \cdot 2 \cdot 2 \cdot 3 \cdot 7}$$

$$= \frac{\cancel{2} \cdot 3 \cdot \cancel{3} \cdot \cancel{7}}{\cancel{2} \cdot 2 \cdot 2 \cdot 2 \cdot \cancel{3} \cdot \cancel{7}}$$

$$= \frac{3}{8}$$

Thus $\frac{3}{8}$ of the group wear running shoes with their usual wardrobes.

PROBLEM SET Reduce the following fractions to simplest terms.

See Examples 9 to 15

1. $\frac{40}{120}$ 2. $\frac{18}{28}$ 3. $\frac{9}{42}$ 4. $\frac{100}{250}$ 5. $\frac{24}{32}$

6. $\frac{15}{50}$ 7. $\frac{28}{210}$ 8. $\frac{120}{330}$ 9. $\frac{18}{108}$ 10. $\frac{35}{49}$

11. $\frac{42}{52}$ 12. $\frac{36}{60}$ 13. $\frac{128}{260}$ 14. $\frac{160}{450}$ 15. $\frac{300}{750}$

ANSWERS 1. $\frac{1}{3}$ 2. $\frac{9}{14}$ 3. $\frac{3}{14}$ 4. $\frac{2}{5}$ 5. $\frac{3}{4}$ 6. $\frac{3}{10}$ 7. $\frac{2}{15}$

8. $\frac{4}{11}$ 9. $\frac{1}{6}$ 10. $\frac{5}{7}$ 11. $\frac{21}{26}$ 12. $\frac{3}{5}$ 13. $\frac{32}{65}$ 14. $\frac{16}{45}$

15. $\frac{2}{5}$

Miscellaneous Exercises 2.1

Sketch the following fractions.

See Examples 1 to 3

1. $\frac{4}{7}$ 2. $\frac{7}{10}$ 3. $\frac{11}{12}$

Determine which of the following pairs of fractions are equal.

See Examples 4 to 8

4. Does $\frac{12}{21} = \frac{20}{35}$? 5. Does $\frac{12}{32} = \frac{10}{24}$?

6. Does $\frac{6}{27} = \frac{10}{45}$? 7. Does $\frac{11}{11} = \frac{23}{23}$?

Reduce the following fractions to simplest terms.

See Examples 9 to 15

8. $\frac{15}{25}$ 9. $\frac{12}{30}$ 10. $\frac{24}{40}$

11. $\frac{48}{60}$ 12. $\frac{28}{54}$ 13. $\frac{64}{56}$

Business applications

14. In the field of life insurance, fractions can be used to represent the likelihood of a person's dying within the year. For a 44-year-old man, there is an $\frac{8}{1000}$ chance of dying before reaching 45. Reduce this fraction.

15. In a sample of 120 light bulbs, a factory finds that 15 of them are defective. Reduce the fraction $\frac{15}{120}$.

Health application

16. A patient is to receive 800 milliliters of a medicine. Thus far, she has received 250 milliliters. Write this portion of the medicine as a reduced fraction.

Technical application

17. A mechanic has two bolts: $\frac{23}{32}$-inch and $\frac{7}{10}$-inch. Are these the same size?

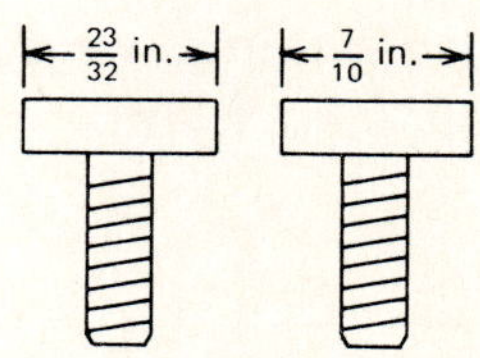

Now complete Post-Test 1 for Section 2.1 on page T23.

2.2 FRACTION/MIXED-NUMBER CONVERSIONS

SECTION OBJECTIVES

After completing this unit, you will be able to:

1. Convert a mixed number to an improper fraction.
2. Convert an improper fraction to a mixed number.

TYPICAL PROBLEMS You will be able to work problems like the ones below.

Write the following mixed numbers as fractions.

(a) $3\frac{1}{9}$ **(b)** $5\frac{7}{10}$ **(c)** $10\frac{1}{4}$

Write the following improper fractions as mixed numbers.

(d) $\frac{21}{4}$ **(e)** $\frac{82}{7}$ **(f)** $\frac{115}{11}$

ANSWERS **(a)** $\frac{28}{9}$ **(b)** $\frac{57}{10}$ **(c)** $\frac{41}{4}$ **(d)** $5\frac{1}{4}$ **(e)** $11\frac{5}{7}$ **(f)** $10\frac{5}{11}$

Often, a fraction has a numerator that is larger than its denominator. We call this an **improper fraction.** Below is pictured the improper fraction $\frac{8}{3}$. (Each square represents 1 unit.)

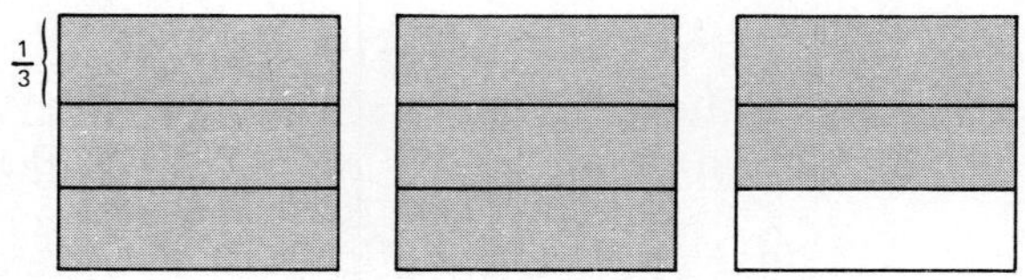

$\frac{8}{3}$ are shaded

In the illustration, each square is cut into thirds. Since 8 of them are shaded, this is $\frac{8}{3}$. But notice that this is also 2 wholes and $\frac{2}{3}$ of another. We can write this as $2\frac{2}{3}$, and we call this a **mixed number.** Notice that the 2 wholes are equal to 6 thirds. This is true since $2 = \frac{2}{1} = \frac{2}{1} \cdot \frac{3}{3} = \frac{6}{3}$. When we add these 6 thirds to the other 2 thirds, we can see how $2\frac{2}{3} = \frac{8}{3}$. (Note that $2\frac{2}{3}$ really means $2 + \frac{2}{3}$.)

> To convert a mixed number to a fraction:
>
> 1. Multiply the whole number by the denominator.
> 2. Add the numerator.
> 3. Put this sum all over the original denominator.
>
> As a formula, this can be written
>
> $$N\frac{a}{b} = \frac{N \cdot b + a}{b}$$

EXAMPLE 16 Write $7\frac{3}{4}$ as a fraction.

SOLUTION $7\frac{3}{4} = 7 + \frac{3}{4} = \frac{28}{4} + \frac{3}{4}$

$= \frac{31}{4}$

$7 = \frac{28}{4}$
↓
Add numerators

EXAMPLE 17 Write $5\frac{1}{9}$ as a fraction.

SOLUTION $5\frac{1}{9} = \frac{45}{9} + \frac{1}{9}$

$= \frac{46}{9}$

$5 = \frac{45}{9}$
↓
Add numerators

EXAMPLE 18 Write $7\frac{3}{10}$ as a fraction.

SOLUTION $7\frac{3}{10} = \frac{70}{10} + \frac{3}{10}$

$= \frac{73}{10}$

$7 = \frac{70}{10}$
↓
Add numerators

EXAMPLE 19 Write $10\frac{1}{14}$ as a fraction.

SOLUTION $10\frac{1}{14} = \frac{140}{14} + \frac{1}{14}$

$= \frac{141}{14}$

$10 = \frac{140}{14}$
↓
Add numerators

Why Does It Work?

This rule works because we are changing the whole-number part to an equivalent fraction when we multiply whole number times denominator. For instance, consider $4\frac{1}{3}$. As we see below, we are changing 4 to $\frac{4}{1} = \frac{4}{1} \cdot \frac{3}{3} = \frac{12}{3}$. Adding the other $\frac{1}{3}$, this is $\frac{13}{3}$.

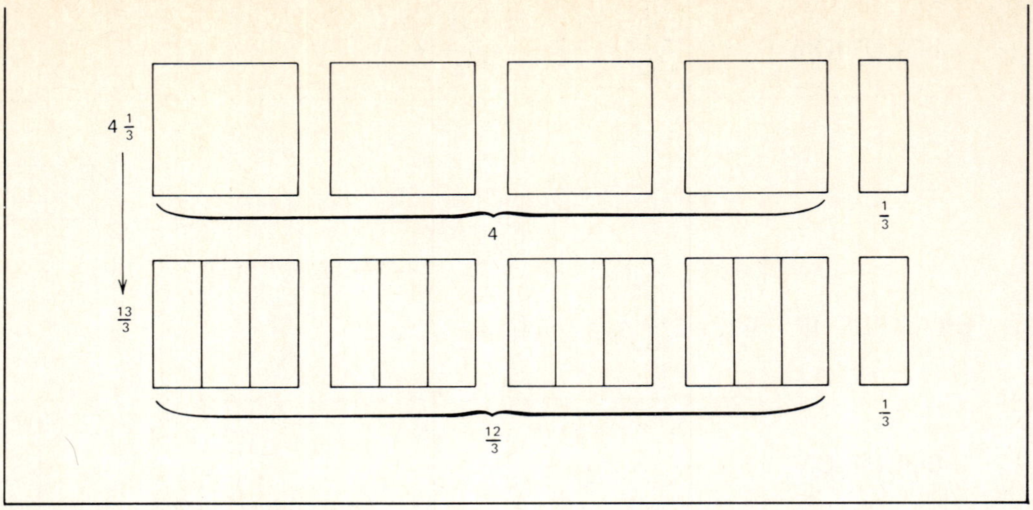

EXAMPLE 20 A doctor feels that it is safe for a certain patient to receive up to $5\frac{1}{2}$ roentgens of X-ray radiation. Each X-ray picture (radiograph) exposes the patient to $\frac{1}{2}$ roentgen. Write $5\frac{1}{2}$ as an improper fraction and determine how many $\frac{1}{2}$-roentgen exposures the patient can have.

SOLUTION We convert $5\frac{1}{2}$ to a fraction in the usual way.

$$5\frac{1}{2} = \frac{10}{2} + \frac{1}{2}$$
$$= \frac{11}{2}$$

Thus, $5\frac{1}{2} = \frac{11}{2}$, and the patient can then receive 11 $\frac{1}{2}$-roentgen exposures.

PROBLEM SET Write the following mixed numbers as fractions.

See Examples 16 to 20

1. $2\frac{1}{2}$ 2. $3\frac{1}{3}$ 3. $1\frac{5}{6}$ 4. $6\frac{2}{9}$ 5. $1\frac{5}{17}$

6. $5\frac{3}{7}$ 7. $8\frac{7}{10}$ 8. $6\frac{5}{12}$ 9. $10\frac{2}{5}$ 10. $12\frac{2}{7}$

11. $15\frac{1}{4}$ 12. $20\frac{5}{8}$ 13. $25\frac{3}{10}$ 14. $37\frac{11}{20}$ 15. $102\frac{4}{25}$

ANSWERS 1. $\frac{5}{2}$ 2. $\frac{10}{3}$ 3. $\frac{11}{6}$ 4. $\frac{56}{9}$ 5. $\frac{22}{17}$ 6. $\frac{38}{7}$ 7. $\frac{87}{10}$

8. $\frac{77}{12}$ 9. $\frac{52}{5}$ 10. $\frac{86}{7}$ 11. $\frac{61}{4}$ 12. $\frac{165}{8}$ 13. $\frac{253}{10}$

14. $\frac{751}{20}$ 15. $\frac{2554}{25}$

Sometimes, we have an improper fraction such as $\frac{11}{3}$ and we wish to change this to a mixed number. We can see that $\frac{11}{3}$ is $\frac{9}{3}$ plus $\frac{2}{3}$. This is the same as 3 plus $\frac{2}{3}$, or $3\frac{2}{3}$.

> To convert an improper fraction to a mixed number:
> 1. Divide the denominator into the numerator.
> 2. The quotient is the whole-number part; the remainder over the denominator is the fraction part.

$$\frac{A}{B} \longrightarrow B\overline{\smash{)}A}\ \text{with quotient } Q,\ \text{subtract } QB,\ \text{remainder } R \longrightarrow \text{quotient} + \frac{\text{remainder}}{\text{denominator}}$$

EXAMPLE 21 Write $\frac{17}{3}$ as a mixed number.

SOLUTION

$$\frac{17}{3} \longrightarrow 3\overline{\smash{)}17}\ \text{quotient } 5;\ 17 - 15 = 2$$

$$= 5\frac{2}{3}$$

Divide → Quotient + $\frac{\text{remainder}}{\text{denominator}}$

EXAMPLE 22 Write $\frac{18}{11}$ as a mixed number.

SOLUTION

$$\frac{18}{11} \longrightarrow 11\overline{\smash{)}18}\ \text{quotient } 1;\ 18 - 11 = 7$$

$$= 1\frac{7}{11}$$

Divide → Quotient + $\frac{\text{remainder}}{\text{denominator}}$

EXAMPLE 23 Write $\frac{19}{7}$ as a mixed number.

SOLUTION $\dfrac{19}{7} \longrightarrow 7\overline{)19}$, quotient 2, -14, remainder 5

$= 2\dfrac{5}{7}$

Divide → Quotient + $\dfrac{\text{remainder}}{\text{denominator}}$

EXAMPLE 24 Write $\dfrac{229}{3}$ as a mixed number.

SOLUTION $\dfrac{229}{3} \longrightarrow 3\overline{)229}$, quotient 76, -21, 19, -18, remainder 1

$= 76\dfrac{1}{3}$

Divide → Quotient + $\dfrac{\text{remainder}}{\text{denominator}}$

Why Does It Work?

This rule works because we are bundling groups together to get wholes. Since we want to know how many groups there are, we divide. For instance, consider $\dfrac{13}{4}$. We want to know how many times we can group 4 fourths together to get a whole. By division, we know that we can get 3 wholes with 1 fourth left over. This is $3\dfrac{1}{4}$.

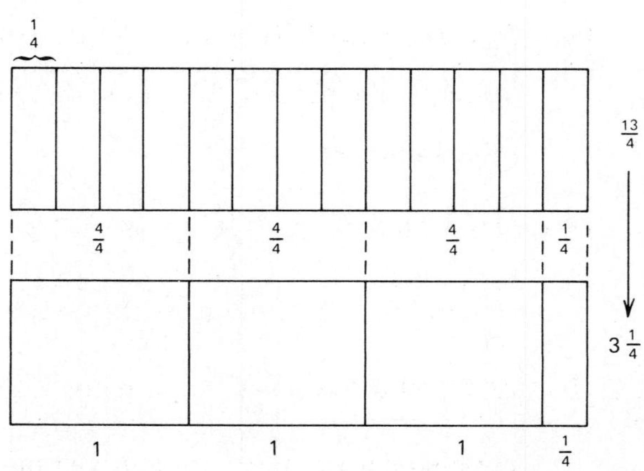

PROBLEM SET Write the following improper fractions as mixed numbers.

See Examples 21 to 24

1. $\frac{9}{5}$ 2. $\frac{11}{2}$ 3. $\frac{17}{6}$ 4. $\frac{21}{8}$ 5. $\frac{31}{9}$

6. $\frac{34}{7}$ 7. $\frac{29}{11}$ 8. $\frac{43}{12}$ 9. $\frac{57}{5}$ 10. $\frac{44}{29}$

11. $\frac{78}{19}$ 12. $\frac{121}{7}$ 13. $\frac{231}{8}$ 14. $\frac{325}{13}$ 15. $\frac{437}{34}$

ANSWERS 1. $1\frac{4}{5}$ 2. $5\frac{1}{2}$ 3. $2\frac{5}{6}$ 4. $2\frac{5}{8}$ 5. $3\frac{4}{9}$ 6. $4\frac{6}{7}$ 7. $2\frac{7}{11}$ 8. $3\frac{7}{12}$ 9. $11\frac{2}{5}$ 10. $1\frac{15}{29}$ 11. $4\frac{2}{19}$ 12. $17\frac{2}{7}$ 13. $28\frac{7}{8}$ 14. 25 15. $12\frac{29}{34}$

Miscellaneous Exercises 2.2

Write the following mixed numbers as improper fractions.

See Examples 16 to 20

1. $3\frac{1}{5}$ 2. $5\frac{2}{7}$ 3. $4\frac{1}{2}$ 4. $10\frac{1}{3}$ 5. $2\frac{1}{15}$ 6. $12\frac{10}{11}$

Write the following improper fractions as mixed numbers.

See Examples 21 to 24

7. $\frac{12}{5}$ 8. $\frac{9}{8}$ 9. $\frac{17}{7}$ 10. $\frac{38}{11}$ 11. $\frac{58}{5}$ 12. $\frac{315}{19}$

Business applications

13. In 1 day, a share of ABC Electric Co. went up $3\frac{1}{4}$ points (dollars). Express this as an improper fraction.

14. A certain marketing study found that 15 families watched a total of 97 hours of TV per day. Write the average number of hours per family, $\frac{97}{15}$, as a mixed number.

Consumer application

15. Mandy has $4\frac{1}{4}$ cups of sugar for a recipe. Write this as an improper fraction to determine how many $\frac{1}{4}$-cups of sugar she has.

Health application

16. A hospital floor has 57 patients in 20 rooms. Write the average number of patients per room, $\frac{57}{20}$, as a mixed number.

Now complete Post-Test 1 for Section 2.2 on page T25.

2.3 MULTIPLICATION OF FRACTIONS AND MIXED NUMBERS

SECTION OBJECTIVES After completing this unit, you will be able to:

1. Multiply fractions.
2. Multiply mixed numbers.

TYPICAL PROBLEMS You will be able to work problems like the ones below.

(a) $3 \times \frac{3}{4}$ (b) $\frac{9}{16} \times \frac{8}{27}$ (c) $\frac{2}{3} \times \frac{9}{8} \times \frac{14}{15}$

(d) $2\frac{1}{2} \times 3\frac{3}{4}$ (e) $7\frac{1}{2} \times 6\frac{2}{5}$ (f) Find n: $n \div \frac{1}{2} = \frac{3}{4}$.

ANSWERS (a) $\frac{9}{4}$ (b) $\frac{1}{6}$ (c) $\frac{7}{10}$ (d) $9\frac{3}{8}$ (e) 48 (f) $\frac{3}{8}$

The first type of **fraction multiplication** that we work is that of a whole number times a fraction, such as $5 \times \frac{2}{7}$. We simply multiply the whole number by the numerator, and put the product over the denominator.

$$N \cdot \frac{a}{b} = \frac{N \cdot a}{b}$$

EXAMPLE 25 Multiply $5 \times \frac{2}{7}$.

SOLUTION

$$5 \times \frac{2}{7} = \frac{5 \times 2}{7} = \frac{10}{7}$$ Whole number times numerator

EXAMPLE 26 Multiply $3 \times \frac{4}{11}$.

SOLUTION

$$3 \times \frac{4}{11} = \frac{3 \times 4}{11} = \frac{12}{11}$$ Whole number times numerator

Why Does It Work?

The numerator tells us how many parts we have in a fraction. When we multiply a fraction by a whole number, we are multiplying the number of parts from the fraction by the whole. Consider $3 \times \frac{2}{7}$.

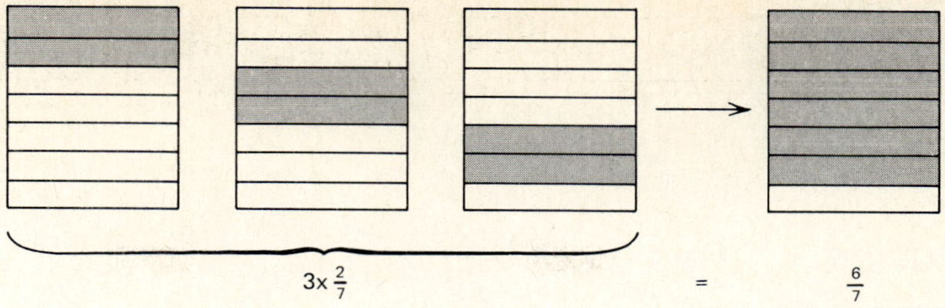

PROBLEM SET Multiply the following.

See Examples 25 and 26

1. $5 \times \frac{1}{8}$ 2. $4 \times \frac{3}{7}$ 3. $7 \times \frac{2}{3}$ 4. $8 \times \frac{2}{11}$ 5. $9 \times \frac{1}{2}$

6. $11 \times \frac{2}{5}$ 7. $15 \times \frac{7}{8}$ 8. $9 \times \frac{4}{17}$ 9. $10 \times \frac{11}{21}$ 10. $7 \times \frac{15}{16}$

ANSWERS 1. $\frac{5}{8}$ 2. $\frac{12}{7}$ 3. $\frac{14}{3}$ 4. $\frac{16}{11}$ 5. $\frac{9}{2}$ 6. $\frac{22}{5}$ 7. $\frac{105}{8}$
8. $\frac{36}{17}$ 9. $\frac{110}{21}$ 10. $\frac{105}{16}$

Multiplying fractions is fairly easy. We simply multiply across: numerator times numerator, denominator times denominator.

$$\frac{a}{b} \cdot \frac{c}{d} = \frac{a \cdot c}{b \cdot d}$$

To multiply fractions:

1. If possible, cancel out any factors common to both the numerator and denominator.
2. Multiply across: numerator times numerator, denominator times denominator.
3. Reduce the fraction, if necessary.

You do not have to cancel first, but it is much easier, since the numbers are still small. Once you multiply, the numbers are bigger, and it is harder to reduce. So we suggest that you cancel first.

EXAMPLE 27 Multiply $\frac{2}{3} \times \frac{5}{7}$.

SOLUTION

$$\frac{2}{3} \times \frac{5}{7} = \frac{10}{21}$$

Multiply numerators; multiply denominators

EXAMPLE 28 Multiply $\frac{12}{5} \times \frac{10}{21}$.

SOLUTION

$$\frac{12}{5} \times \frac{10}{21} = \frac{12}{\cancel{5}_{1}} \times \frac{\cancel{10}^{2}}{21}$$

Cancel

$$= \frac{\cancel{12}^{4}}{1} \times \frac{2}{\cancel{21}_{7}}$$

Cancel

$$= \frac{8}{7}$$

Multiply numerators; multiply denominators

EXAMPLE 29 Multiply $\frac{21}{50} \times \frac{250}{7}$.

SOLUTION

$$\frac{21}{50} \times \frac{250}{7} = \frac{\cancel{21}^{3}}{50} \times \frac{250}{\cancel{7}_{1}}$$

Cancel

$$= \frac{3}{\cancel{50}_{1}} \times \frac{\cancel{250}^{5}}{1}$$

Cancel

$$= \frac{15}{1} = 15$$

Multiply numerators; multiply denominators

EXAMPLE 30 Roger runs 880 yards in 3 minutes. We can write this as miles per hour by multiplying the following fractions:

$$\frac{880 \text{ yd}}{3 \text{ min}} \times \frac{60 \text{ min}}{1 \text{ hr}} \times \frac{1 \text{ mile}}{1760 \text{ yd}} = ? \frac{\text{miles}}{\text{hr}}$$

SOLUTION

$$\frac{880}{3} \times \frac{60}{1} \times \frac{1}{1760} = \frac{880}{\cancel{3}_{1}} \times \frac{\cancel{60}^{20}}{1} \times \frac{1}{1760}$$

Cancel

$$= \frac{\cancel{880}^{1}}{1} \times \frac{20}{1} \times \frac{1}{\cancel{1760}_{2}}$$

Cancel

$$= \frac{20}{2} = 10$$

Multiply numerators; multiply denominators

So Roger can run 10 miles per hour.

Why Does It Work?

When we multiply fractions, we are taking a fraction of a fraction. For example, when we multiply $\frac{4}{5} \times \frac{2}{3}$, we are taking $\frac{4}{5}$ of $\frac{2}{3}$. As we can see here, we are cutting the $\frac{2}{3}$ up as $\frac{10}{15}$. When we take $\frac{4}{5}$, we get $\frac{8}{15}$.

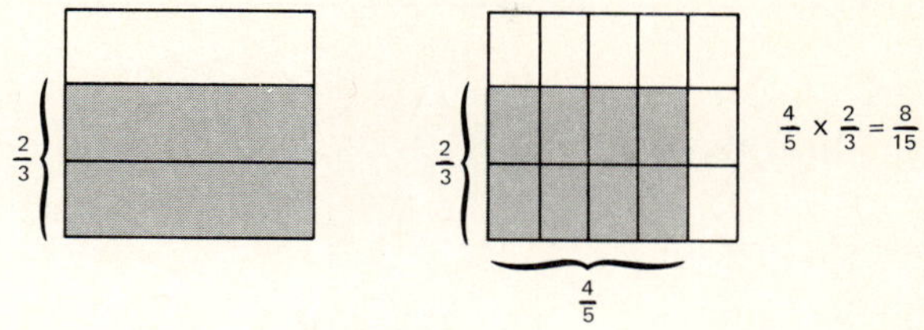

PROBLEM SET Multiply the following fractions.

See Examples 27 to 30

1. $\frac{1}{3} \times \frac{1}{4}$
2. $\frac{2}{5} \times \frac{3}{4}$
3. $\frac{3}{11} \times \frac{4}{5}$
4. $\frac{1}{9} \times \frac{3}{10}$
5. $\frac{2}{7} \times \frac{21}{26}$
6. $\frac{13}{15} \times \frac{35}{39}$
7. $\frac{12}{5} \times \frac{10}{21}$
8. $\frac{3}{10} \times \frac{5}{21}$
9. $\frac{7}{9} \times \frac{27}{28}$
10. $\frac{4}{15} \times \frac{5}{24}$
11. $\frac{2}{7} \times \frac{35}{36}$
12. $\frac{11}{12} \times \frac{45}{44}$
13. $\frac{1}{2} \times \frac{1}{3} \times \frac{1}{4}$
14. $\frac{2}{7} \times \frac{1}{8} \times \frac{14}{15} \times \frac{5}{6}$
15. $\frac{3}{4} \times \frac{6}{7} \times \frac{7}{20} \times \frac{5}{30}$
16. Two-fifths of a group are men. Three-sevenths of these men are overweight. What fraction of the group are overweight men?

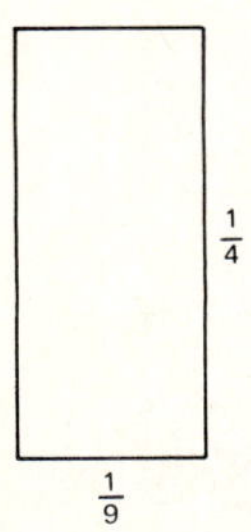

17. At the left is a sketch of Sid's lot: $\frac{1}{4}$-block long by $\frac{1}{9}$-block wide. What is the area in square blocks?
18. In Appleville, $\frac{3}{4}$ of the children graduate from high school; $\frac{1}{6}$ of these graduate from college, $\frac{1}{5}$ of these get master's degrees, and $\frac{1}{100}$ of these get doctorates. What fraction of Appleville's children get a doctorate?

ANSWERS 1. $\frac{1}{12}$ 2. $\frac{3}{10}$ 3. $\frac{12}{55}$ 4. $\frac{1}{30}$ 5. $\frac{3}{13}$ 6. $\frac{7}{9}$ 7. $\frac{8}{7}$ 8. $\frac{1}{14}$ 9. $\frac{3}{4}$ 10. $\frac{1}{18}$ 11. $\frac{5}{18}$ 12. $\frac{15}{16}$ 13. $\frac{1}{24}$ 14. $\frac{1}{36}$ 15. $\frac{3}{80}$ 16. $\frac{6}{35}$ 17. $\frac{1}{36}$ 18. $\frac{1}{4000}$

Multiplying mixed numbers is a bit harder than just multiplying fractions. We must use improper fractions.

> To multiply mixed numbers:
>
> 1. Convert all mixed numbers to improper fractions.
> 2. Multiply as fractions.
> 3. Reduce, if necessary.
> 4. Convert answer back to a mixed number. (This step is optional.)

EXAMPLE 31 Multiply $5\frac{2}{5} \times 6\frac{1}{3}$.

SOLUTION

$$5\frac{2}{5} \times 6\frac{1}{3} = \frac{27}{5} \times \frac{19}{3}$$

$$= \frac{\overset{9}{\cancel{27}}}{5} \times \frac{19}{\underset{1}{\cancel{3}}}$$

$$= \frac{171}{5}$$

$$\rightarrow 5\overline{)171} \quad 34 \text{ R}1$$

$$= 34\frac{1}{5}$$

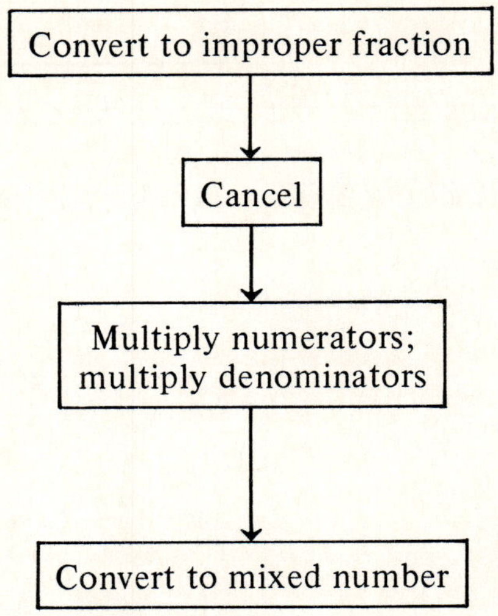

EXAMPLE 32 Multiply $5\frac{1}{4} \times 3\frac{1}{7}$.

SOLUTION

$$5\frac{1}{4} \times 3\frac{1}{7} = \frac{21}{4} \times \frac{22}{7}$$

$$= \frac{\overset{3}{\cancel{21}}}{4} \times \frac{22}{\underset{1}{\cancel{7}}}$$

$$= \frac{3}{\underset{2}{\cancel{4}}} \times \frac{\overset{11}{\cancel{22}}}{1}$$

$$= \frac{33}{2}$$

$$\rightarrow 2\overline{)33} \quad 16 \text{ R}1$$

$$= 16\frac{1}{2}$$

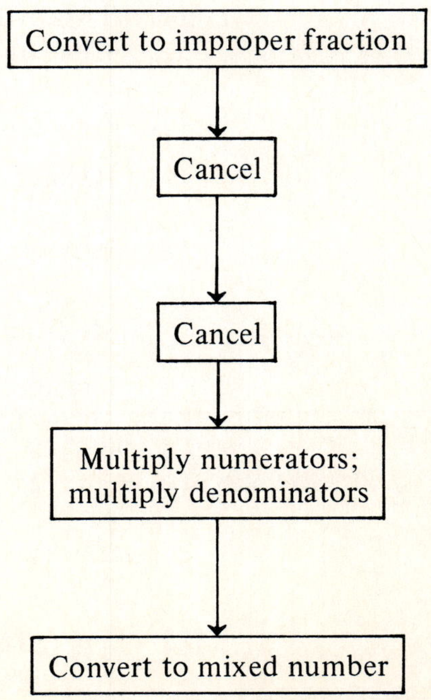

YES	NO
$5\frac{1}{4} \times 3\frac{1}{7} = 16\frac{1}{2}$	~~$5\frac{1}{4} \times 3\frac{1}{7} = 15\frac{1}{28}$~~

We must convert to improper fractions to multiply mixed numbers.

EXAMPLE 33 Multiply $10\frac{2}{5} \times 1\frac{2}{13}$.

SOLUTION

$$10\frac{2}{5} \times 1\frac{2}{13} = \frac{52}{5} \times \frac{15}{13}$$

$$= \frac{52}{\cancel{5}_{1}} \times \frac{\cancel{15}^{3}}{13}$$

$$= \frac{\cancel{52}^{4}}{1} \times \frac{3}{\cancel{13}_{1}}$$

$$= \frac{12}{1} = 12$$

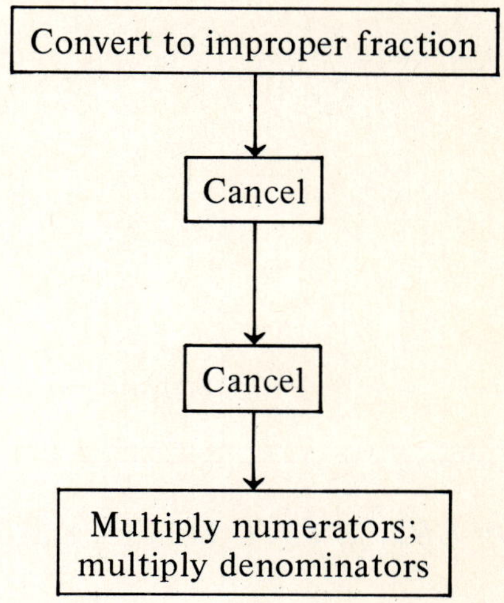

EXAMPLE 34 Multiply $\frac{2}{3} \times 6\frac{3}{4}$.

SOLUTION

$$\frac{2}{3} \times 6\frac{3}{4} = \frac{2}{3} \times \frac{27}{4}$$

$$= \frac{2}{\cancel{3}_{1}} \times \frac{\cancel{27}^{9}}{4}$$

$$= \frac{\cancel{2}^{1}}{1} \times \frac{9}{\cancel{4}_{2}}$$

$$= \frac{9}{2}$$

$$\rightarrow 2\overline{)9} \quad 4 \text{ R}1$$

$$= 4\frac{1}{2}$$

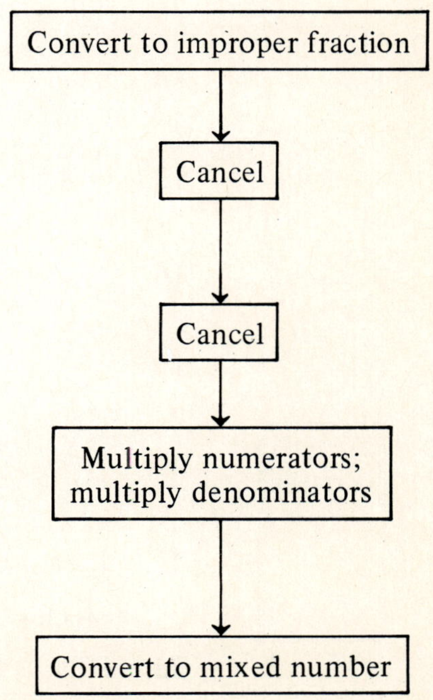

EXAMPLE 35 Andy's car gets $17\frac{1}{3}$ miles to a gallon of gas. How far can he go on $15\frac{1}{2}$ gallons of gas?

SOLUTION We multiply the gallons by the miles per gallon.

$$15\frac{1}{2} \times 17\frac{1}{3} = \frac{31}{2} \times \frac{52}{3}$$

$$= \frac{31}{\cancel{2}_{1}} \times \frac{\cancel{52}^{26}}{3}$$

$$= \frac{806}{3}$$

$$\rightarrow 3\overline{)806} \quad 268 \text{ R2}$$

$$= 268\frac{2}{3}$$

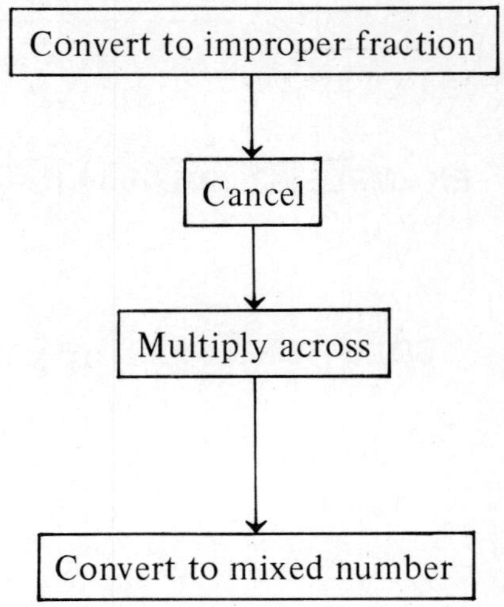

So Andy can drive $268\frac{2}{3}$ miles.

PROBLEM SET Multiply the following fractions and mixed numbers.

See Examples 31 to 35

1. $1\frac{1}{8} \times 2\frac{2}{3}$
2. $3\frac{1}{3} \times 1\frac{4}{5}$
3. $1\frac{1}{14} \times 2\frac{1}{3}$
4. $10\frac{1}{2} \times 1\frac{3}{7}$
5. $7\frac{3}{5} \times 8\frac{1}{3}$
6. $4\frac{2}{7} \times 6\frac{1}{5}$
7. $2\frac{2}{5} \times 3\frac{3}{4}$
8. $2\frac{2}{7} \times 2\frac{5}{8}$
9. $2\frac{6}{11} \times 3\frac{1}{7}$
10. $11\frac{2}{3} \times 1\frac{1}{14}$
11. $8\frac{1}{4} \times 1\frac{7}{9}$
12. $\frac{2}{5} \times 4\frac{1}{6}$
13. $\frac{1}{9} \times 5\frac{2}{5}$
14. $1\frac{1}{2} \times 1\frac{1}{2} \times 1\frac{1}{2}$
15. $2\frac{1}{2} \times 1\frac{1}{15} \times 2\frac{1}{4}$
16. $1\frac{2}{7} \times \frac{2}{3} \times 4\frac{2}{3}$
17. $5\frac{1}{3} \times \frac{5}{6} \times \frac{8}{9} \times 7\frac{1}{2}$
18. $\frac{1}{6} \times 6\frac{1}{4} \times 4\frac{1}{5} \times \frac{2}{7}$
19. $1\frac{2}{5} \times 1\frac{11}{14} \times 2\frac{1}{2} \times 2\frac{1}{25}$
20. $\frac{1}{10} \times 3\frac{1}{3} \times 4\frac{1}{9} \times 6\frac{2}{7}$

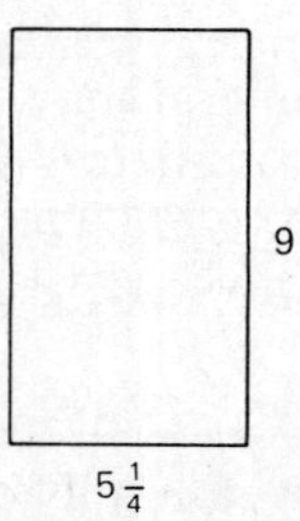

21. Juan uses $1\frac{2}{3}$ teaspoons of powdered cocoa to make a cup of cocoa. How much powdered cocoa does he need to make 25 cups?
22. What is the area of the rectangle at the left?

ANSWERS **1.** 3 **2.** 6 **3.** $2\frac{1}{2}$ **4.** 15 **5.** $63\frac{1}{3}$ **6.** $26\frac{4}{7}$ **7.** 9

8. 6 **9.** 8 **10.** $12\frac{1}{2}$ **11.** $14\frac{2}{3}$ **12.** $1\frac{2}{3}$ **13.** $\frac{3}{5}$ **14.** $3\frac{3}{8}$

15. 6 **16.** 4 **17.** $29\frac{17}{27}$ **18.** $1\frac{1}{4}$ **19.** $12\frac{3}{4}$ **20.** $8\frac{116}{189}$

21. $41\frac{2}{3}$ **22.** 48

Recall from Chapter 1 that *multiplication is the opposite of division.* For example,

$$n \div 2 = 10 \qquad \text{means} \qquad n = 2 \times 10 = 20$$

It is as if we multiply both sides of the equation by 2. We follow the same procedure with fractions: to undo division, we multiply.

EXAMPLE 36 Find n: $n \div 2 = \frac{3}{5}$.

SOLUTION Just as with whole numbers, we multiply by 2 to find n, since multiplication is the opposite of division.

$$n \div 2 = \frac{3}{5} \qquad \text{means} \qquad n = 2 \times \frac{3}{5} = \frac{6}{5}$$

EXAMPLE 37 Find n: $n \div \frac{2}{3} = \frac{5}{9}$.

SOLUTION We undo this division by multiplication.

$$n \div \frac{2}{3} = \frac{5}{9} \qquad \text{means} \qquad n = \frac{2}{3} \times \frac{5}{9} = \frac{10}{27}$$

EXAMPLE 38 Find n: $n \div 1\frac{1}{3} = 3\frac{1}{4}$.

SOLUTION Again we multiply to reverse this division. We are really multiplying both sides by $1\frac{1}{3}$.

$$n \div 1\frac{1}{3} = 3\frac{1}{4} \qquad \text{means} \qquad n = 1\frac{1}{3} \times 3\frac{1}{4}$$

$$n = \frac{\overset{1}{\not 4}}{3} \times \frac{13}{\underset{1}{\not 4}}$$

$$= \frac{13}{4} = 4\frac{1}{3}$$

Thus $n = 4\frac{1}{3}$.

PROBLEM SET Find n or x in the following problems.

See Examples 36 to 38

1. $n \div 3 = \frac{1}{2}$ 2. $x \div 5 = \frac{3}{4}$ 3. $n \div 7 = \frac{3}{7}$ 4. $x \div \frac{2}{3} = \frac{4}{5}$
5. $n \div \frac{1}{5} = \frac{2}{7}$ 6. $x \div \frac{4}{9} = 6$ 7. $n \div 1\frac{1}{2} = 3\frac{2}{5}$ 8. $x \div 2\frac{2}{3} = 4\frac{1}{3}$

ANSWERS 1. $\frac{3}{2}$ 2. $\frac{15}{4}$ 3. 3 4. $\frac{8}{15}$ 5. $\frac{2}{35}$ 6. $\frac{8}{3}$ 7. $5\frac{1}{10}$ 8. $11\frac{5}{9}$

Miscellaneous Exercises 2.3

Multiply the following fractions and mixed numbers.

See Examples 25 and 26

1. $3 \times \frac{4}{5}$ 2. $7 \times \frac{2}{7}$ 3. $5 \times \frac{5}{8}$

See Examples 27 to 30

4. $\frac{3}{4} \times \frac{5}{7}$ 5. $\frac{2}{9} \times \frac{4}{5}$ 6. $\frac{4}{9} \times \frac{5}{6}$
7. $\frac{2}{3} \times \frac{5}{12}$ 8. $\frac{7}{18} \times \frac{5}{2}$ 9. $\frac{1}{15} \times \frac{5}{9}$

See Examples 31 to 35

10. $2\frac{1}{2} \times 4\frac{1}{3}$ 11. $2\frac{1}{7} \times 3\frac{1}{5}$ 12. $6 \times 3\frac{1}{3}$
13. $2\frac{1}{4} \times 3\frac{1}{2} \times \frac{2}{5}$ 14. $5\frac{1}{2} \times 3\frac{1}{3} \times 1\frac{1}{4}$

See Examples 36 to 38

15. Find n: $n \div 6 = \frac{1}{2}$. 16. Find n: $n \div \frac{2}{3} = 1\frac{1}{2}$.

Technical application

17. A sheet of alloy used in manufacturing a certain airplane wing is $3\frac{5}{8}$ pounds. Find the weight of $\frac{3}{4}$ of this sheet.

Business applications

18. A company buys 10,000 shares of another company's stock at $\$21\frac{1}{2}$ per share. How much is the total purchase?

19. An advertising campaign hopes to reach $\frac{3}{4}$ of a 120,000-person market. How many people is this?

20. The annual interest on a loan was \$6000. Since the money was only used for 100 days, the interest was recomputed as $\frac{100}{360} \times 6000$ (a bank year is taken as 360 days). Compute the new interest.

Health applications

21. A patient is given $\frac{1}{2}$ of a $\frac{3}{4}$-grain morphine sulfate tablet. How much of the medicine did the patient receive?

22. The adult dosage of a certain medicine is $7\frac{1}{2}$ grains. The dosage that a 40-pound child would get can be calculated as $\frac{40}{150} \times 7\frac{1}{2}$. Compute this dosage.

Life science application

23. A certain couple has a $\frac{1}{4}$ chance of each child having blue eyes. If they have three children, the chances of exactly two having blue eyes is given by $3 \times \frac{1}{4} \times \frac{1}{4} \times \frac{3}{4}$. Compute this likelihood.

Consumer applications

24. Because of inflation, food costs $\frac{180}{135}$ times what it did two years ago. Two years ago, it cost the Joneses \$81 per week for food. What does it cost them this year to feed the family each week?

25. Judy's car gets $19\frac{1}{3}$ miles per gallon. How far can she go on $14\frac{1}{2}$ gallons of gasoline?

26. Kate buys $6\frac{2}{3}$ yards of material at $1\frac{1}{2}$ dollars per yard. How much does this cost?

Now complete Post-Test 1 for Section 2.3 on page T27.

2.4 DIVISION OF FRACTIONS AND MIXED NUMBERS

SECTION OBJECTIVES

After completing this unit, you will be able to:

1. Divide fractions.
2. Divide mixed numbers.

TYPICAL PROBLEMS You will be able to work problems like the ones below.

(a) $\frac{2}{5} \div \frac{7}{9}$ (b) $\frac{6}{11} \div \frac{3}{13}$ (c) $\frac{7}{10} \div \frac{14}{25}$

(d) $2\frac{1}{7} \div 4\frac{1}{2}$ (e) $10\frac{1}{2} \div 8\frac{2}{5}$ (f) Find n: $7 \cdot n = \frac{2}{3}$

ANSWERS (a) $\frac{18}{35}$ (b) $\frac{26}{11}$ or $2\frac{4}{11}$ (c) $\frac{5}{4}$ or $1\frac{1}{4}$ (d) $\frac{10}{21}$ (e) $1\frac{1}{4}$ (f) $\frac{2}{21}$

The rule for **dividing fractions** is fairly simple: *Invert the divisor* (*the second fraction*) *and multiply.*

$$\frac{a}{b} \div \frac{c}{d} = \frac{a}{b} \cdot \frac{d}{c} = \frac{a \cdot d}{b \cdot c}$$

To divide fractions:

1. Invert the divisor. (Flip the numerator and denominator of the second fraction.)
2. Multiply the resulting fractions.
3. Reduce, if necessary.

EXAMPLE 39 Divide $\frac{7}{3} \div \frac{4}{5}$.

SOLUTION

$$\frac{7}{3} \div \frac{4}{5} = \frac{7}{3} \times \frac{5}{4}$$

Invert divisor and multiply

$$= \frac{35}{12}$$

Multiply

EXAMPLE 40 Divide $\frac{5}{9} \div \frac{2}{3}$.

SOLUTION

$$\frac{5}{9} \div \frac{2}{3} = \frac{5}{9} \times \frac{3}{2}$$

Invert divisor and multiply

$$= \frac{5}{\cancel{9}_{3}} \times \frac{\cancel{3}^{1}}{2}$$

Cancel

$$= \frac{5}{6}$$

Multiply

EXAMPLE 41 Divide $\frac{15}{19} \div \frac{3}{4}$.

SOLUTION

$$\frac{15}{19} \div \frac{3}{4} = \frac{15}{19} \times \frac{4}{3}$$ Invert divisor and multiply

$$= \frac{\overset{5}{\cancel{15}}}{19} \times \frac{4}{\underset{1}{\cancel{3}}}$$ Cancel

$$= \frac{20}{19}$$ Multiply

EXAMPLE 42 Divide $\frac{9}{10} \div \frac{3}{5}$.

SOLUTION

$$\frac{9}{10} \div \frac{3}{5} = \frac{9}{10} \times \frac{5}{3}$$ Invert divisor and multiply

$$= \frac{\overset{3}{\cancel{9}}}{10} \times \frac{5}{\underset{1}{\cancel{3}}}$$ Cancel

$$= \frac{3}{\underset{2}{\cancel{10}}} \times \frac{\overset{1}{\cancel{5}}}{1}$$ Cancel

$$= \frac{3}{2}$$ Multiply

EXAMPLE 43 The area of the rectangle shown is $\frac{3}{11}$ square meter. What is the missing dimension, x?

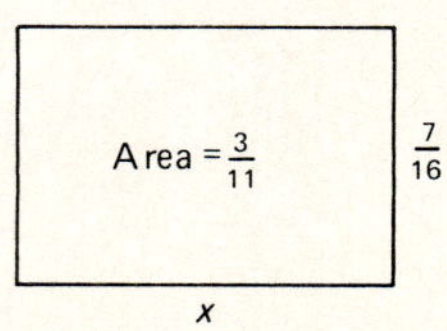

SOLUTION If the area were 20, and the given side were 5, we would know what to do: divide $20 \div 5$. We do the same here: divide area by side, $\frac{3}{11} \div \frac{7}{16}$.

$$\frac{3}{11} \div \frac{7}{16} = \frac{3}{11} \times \frac{16}{7}$$ Invert divisor and multiply

$$= \frac{48}{77}$$ Multiply

The missing side is $\frac{48}{77}$ meter.

Why Does It Work?

Why do we invert and multiply? This is hard to explain well without using algebra. But we can give a sketch that shows this rule. Consider $\frac{2}{5} \div \frac{1}{10}$. We are asking how many $\frac{1}{10}$'s are in $\frac{2}{5}$.

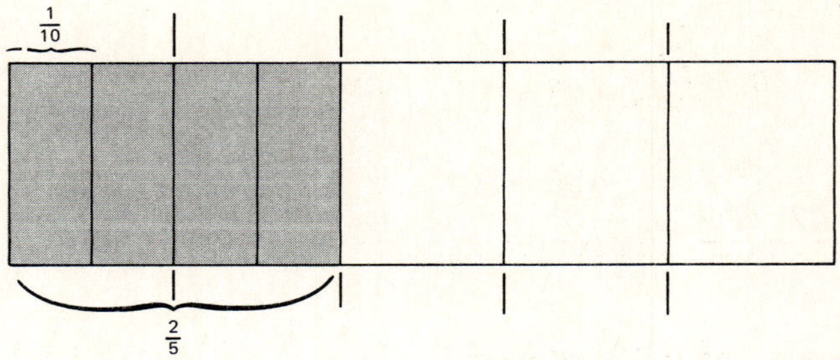

We can see that there are four $\frac{1}{10}$'s in $\frac{2}{5}$. We get the same answer, 4, when we invert and multiply.

$$\frac{2}{5} \div \frac{1}{10} = \frac{2}{5} \times \frac{10}{1} = \frac{20}{5} = 4$$

PROBLEM SET Divide the following fractions.

See Examples 39 to 43

1. $\frac{1}{2} \div \frac{1}{7}$
2. $\frac{2}{5} \div \frac{3}{8}$
3. $\frac{2}{9} \div \frac{1}{12}$
4. $\frac{5}{8} \div \frac{9}{16}$
5. $\frac{8}{15} \div \frac{4}{3}$
6. $\frac{10}{21} \div \frac{5}{14}$
7. $\frac{9}{2} \div \frac{3}{10}$
8. $\frac{16}{17} \div \frac{17}{18}$
9. $\frac{3}{11} \div \frac{1}{9}$
10. $\frac{12}{35} \div \frac{25}{21}$
11. $\frac{2}{45} \div \frac{7}{90}$
12. $\frac{1}{8} \div \frac{5}{24}$
13. $\frac{3}{10} \div \frac{17}{100}$
14. $\frac{4}{39} \div \frac{8}{13}$
15. $\frac{1}{15} \div \frac{1}{20}$
16. $\frac{2}{25} \div \frac{1}{100}$
17. A stretch of land is $\frac{3}{4}$ mile long. A developer wants to subdivide this into lots each $\frac{1}{88}$ mile across. How many lots can she get?
18. Richard has $\frac{5}{8}$ of a tank of gas in his tank. If it takes him $\frac{1}{60}$ of a tank to go to and from work, how many trips can he make with this gas?

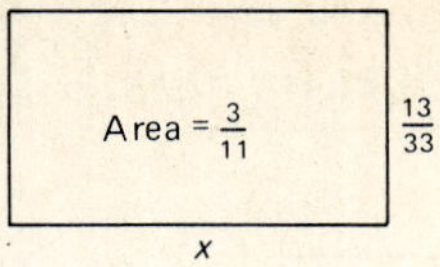

19. The area of the rectangle at the left is $\frac{3}{11}$ square foot. What is the missing width, x?

20. If Helen has $\frac{4}{5}$ pound of coffee, and her family uses $\frac{1}{8}$ pound a day, how many days will this coffee last?

ANSWERS 1. $\frac{7}{2}$ 2. $\frac{16}{15}$ 3. $\frac{8}{3}$ 4. $\frac{10}{9}$ 5. $\frac{2}{5}$ 6. $\frac{4}{3}$ 7. 15 8. $\frac{288}{289}$ 9. $\frac{27}{11}$ 10. $\frac{36}{125}$ 11. $\frac{4}{7}$ 12. $\frac{3}{5}$ 13. $\frac{30}{17}$ 14. $\frac{1}{6}$ 15. $\frac{4}{3}$ 16. 8 17. 66 18. $\frac{75}{2}$ 19. $\frac{9}{13}$ 20. $\frac{32}{5}$

Dividing mixed numbers is similar to multiplying mixed numbers: We convert them to improper fractions.

To divide mixed numbers:

1. Convert the mixed numbers to fractions.
2. Invert the divisor.
3. Multiply the resulting fractions.
4. Reduce, if necessary.
5. Convert the answer to a mixed number. (This step is optional.)

EXAMPLE 44 Divide $9\frac{1}{4} \div 6\frac{2}{5}$.

SOLUTION

$$9\frac{1}{4} \div 6\frac{2}{5} = \frac{37}{4} \div \frac{32}{5}$$

$$= \frac{37}{4} \times \frac{5}{32}$$

$$= \frac{185}{128}$$

$$\rightarrow 128\overline{)185} \quad 1 \text{ R}57$$

$$= 1\frac{57}{128}$$

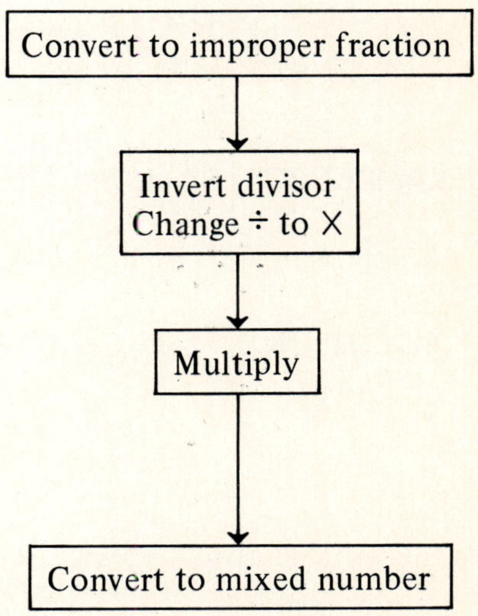

EXAMPLE 45 Divide $1\frac{7}{8} \div 6\frac{1}{5}$.

SOLUTION $1\frac{7}{8} \div 6\frac{1}{5} = \frac{15}{8} \div \frac{31}{5}$

$$= \frac{15}{8} \times \frac{5}{31}$$

$$= \frac{75}{248}$$

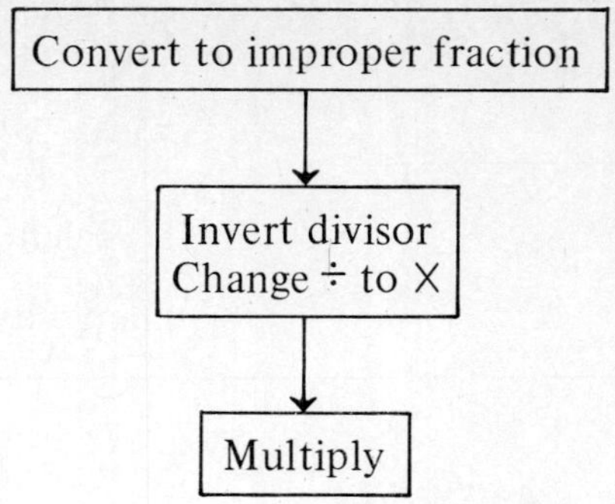

EXAMPLE 46 Divide $10\frac{2}{9} \div 2\frac{1}{3}$.

SOLUTION $10\frac{2}{9} \div 2\frac{1}{3} = \frac{92}{9} \div \frac{7}{3}$

$$= \frac{92}{9} \times \frac{3}{7}$$

$$\frac{92}{\cancel{9}_{3}} \times \frac{\overset{1}{\cancel{3}}}{7}$$

$$= \frac{92}{21}$$

$$\rightarrow 21\overline{)92} \quad 4 \text{ R}8$$

$$= 4\frac{8}{21}$$

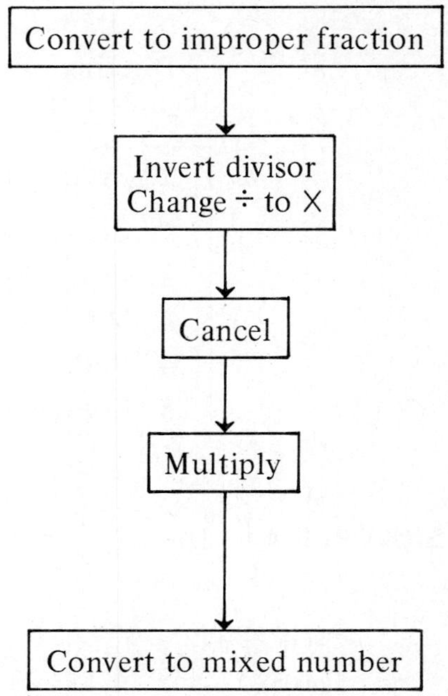

EXAMPLE 47 Divide $6\frac{2}{5} \div 1\frac{3}{10}$.

SOLUTION $6\frac{2}{5} \div 1\frac{3}{10} = \frac{32}{5} \div \frac{13}{10}$

$$= \frac{32}{5} \times \frac{10}{13}$$

$$\frac{32}{\cancel{5}_{1}} \times \frac{\overset{2}{\cancel{10}}}{13}$$

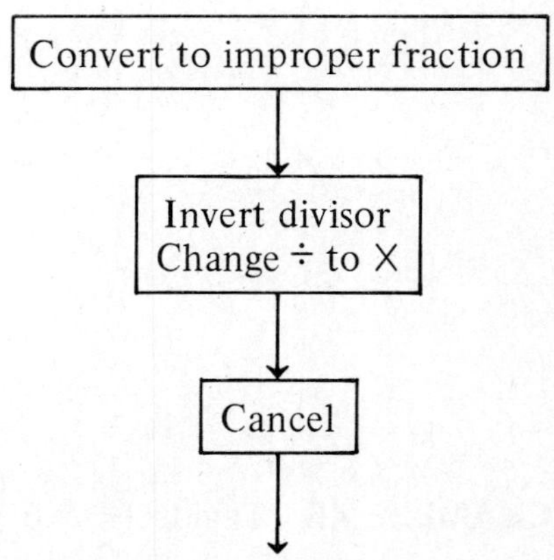

$$= \frac{64}{13}$$

Multiply

$$\rightarrow 13\overline{)64} \quad 4 \text{ R}12$$

$$= 4\frac{12}{13}$$

Convert to mixed number

EXAMPLE 48 Bill was paid \$120 to paint some rooms of a house. He worked $22\frac{1}{2}$ hours. What was his hourly wage?

SOLUTION We divide the pay by the number of hours to find the pay rate.

$$120 \div 22\frac{1}{2} = \frac{120}{1} \div \frac{45}{2}$$

Convert to improper fraction

$$= \frac{120}{1} \times \frac{2}{45}$$

Invert divisor
Change ÷ to X

$$= \frac{\overset{8}{\cancel{120}}}{1} \times \frac{2}{\underset{3}{\cancel{45}}}$$

Cancel

$$= \frac{16}{3}$$

Multiply

$$\rightarrow 3\overline{)16} \quad 5 \text{ R}1$$

$$= 5\frac{1}{3}$$

Convert to mixed number

PROBLEM SET Divide the following fractions and mixed numbers.

See Examples 44 to 48

1. $3\frac{1}{2} \div 2\frac{1}{3}$ **2.** $1\frac{1}{10} \div 2\frac{3}{5}$ **3.** $6\frac{1}{9} \div 4\frac{1}{6}$ **4.** $12\frac{1}{2} \div 1\frac{3}{5}$

5. $1\frac{1}{15} \div 3\frac{1}{3}$ **6.** $20\frac{1}{5} \div 3\frac{3}{10}$ **7.** $4\frac{1}{7} \div 3\frac{3}{5}$ **8.** $18\frac{1}{2} \div 4\frac{1}{4}$

9. $1\frac{7}{9} \div 2\frac{2}{3}$ **10.** $2\frac{6}{7} \div 1\frac{1}{14}$ **11.** $18 \div 2\frac{1}{4}$ **12.** $6\frac{1}{5} \div 4$

13. $25 \div \frac{2}{9}$ **14.** $24 \div 1\frac{5}{7}$ **15.** $2\frac{2}{11} \div 8$ **16.** $1\frac{1}{6} \div 14$

17. $10\frac{1}{4} \div \frac{3}{8}$ 18. $\frac{1}{10} \div 6\frac{1}{5}$ 19. $\frac{7}{9} \div 5\frac{2}{3}$ 20. $7\frac{1}{5} \div \frac{9}{10}$

21. Fran has $1\frac{2}{3}$ pizzas to divide among six children. How much will each child get?

22. Hank has a 60-inch board that he wishes to cut into $1\frac{1}{4}$-inch strips. How many strips can he get?

23. Judy's gas tank holds 22 gallons. How many $\frac{2}{9}$-gallon trips can she make on a tank of gas?

24. Jason runs $2\frac{1}{4}$ miles per day. How many days will it take him to run 500 miles?

25. Penny drove $29\frac{3}{10}$ miles on $2\frac{1}{5}$ gallons of gasoline. How many miles per gallon is this?

ANSWERS 1. $1\frac{1}{2}$ 2. $\frac{11}{26}$ 3. $1\frac{7}{15}$ 4. $7\frac{13}{16}$ 5. $\frac{8}{25}$ 6. $6\frac{4}{33}$ 7. $1\frac{19}{126}$ 8. $4\frac{6}{17}$ 9. $\frac{2}{3}$ 10. $2\frac{2}{3}$ 11. 8 12. $1\frac{11}{20}$ 13. $112\frac{1}{2}$ 14. 14 15. $\frac{3}{11}$ 16. $\frac{1}{12}$ 17. $27\frac{1}{3}$ 18. $\frac{1}{62}$ 19. $\frac{7}{51}$ 20. 8 21. $\frac{5}{18}$ 22. 48 23. 99 24. $222\frac{2}{9}$ 25. $13\frac{7}{22}$

We saw in Chapter 1 that *division is the opposite of multiplication*. For example,

$$7 \cdot n = 35 \qquad \text{means} \qquad n = 35 \div 7 = 5$$

We are really dividing both sides by 7 to isolate n. We follow the same procedure with fractions: We divide to reverse a multiplication problem. (Recall that dot $\cdot$ means multiplication.)

EXAMPLE 49 Find n: $3 \cdot n = 8$.

SOLUTION To reverse this multiplication, we divide by 3.

$$3 \cdot n = 8 \qquad \text{means} \qquad n = 8 \div 3 = \frac{8}{3}$$

Notice that we use the fraction $\frac{8}{3}$ to represent the division $8 \div 3$.

EXAMPLE 50 Find n: $\frac{2}{3} \cdot n = \frac{4}{5}$.

SOLUTION We isolate n by dividing both sides by $\frac{2}{3}$.

$$\frac{2}{3} \cdot n = \frac{4}{5} \quad \text{means} \quad n = \frac{4}{5} \div \frac{2}{3}$$

$$= \frac{4}{5} \cdot \frac{3}{2}$$

$$= \frac{12}{10} = \frac{6}{5}$$

Division is opposite of multiplication → Invert and multiply

Thus, $n = \frac{6}{5}$.

EXAMPLE 51 Three-fifths of what number is $1\frac{1}{5}$?

SOLUTION In word problems, *of* almost always means multiplication, so we translate this as

$$\frac{3}{5} \cdot n = 1\frac{1}{5}$$

We let n be the number that we want. Now we find n by dividing.

$$\frac{3}{5} \cdot n = 1\frac{1}{5} \quad \text{means} \quad n = 1\frac{1}{5} \div \frac{3}{5}$$

$$= 1\frac{1}{5} \cdot \frac{5}{3}$$

$$= \frac{6}{5} \cdot \frac{5}{3}$$

$$= \frac{30}{15} = 2$$

Division is opposite of multiplication → Invert and multiply

We can also work this problem more directly by multiplying both sides by $\frac{5}{3}$ to cancel the $\frac{3}{5}$ on the left.

$$\frac{3}{5} \cdot n = \frac{6}{5}$$

Given

$$\frac{\cancel{5}}{\cancel{3}} \cdot \frac{\cancel{3}}{\cancel{5}} \cdot n = \frac{5}{3} \cdot \frac{6}{5}$$

Multiply both sides by $\frac{5}{3}$

$$n = 2$$

Simplify

PROBLEM SET Find n or x in the following problems.

See Examples 49 to 51

1. $5 \cdot n = 7$ 2. $2 \cdot n = 9$ 3. $\frac{2}{5} \cdot x = 6$
4. $\frac{4}{5} \cdot n = 8$ 5. $7 \cdot x = \frac{3}{5}$ 6. $8 \cdot n = \frac{2}{9}$
7. $\frac{4}{11} \cdot n = \frac{2}{5}$ 8. $\frac{1}{5} \cdot n = \frac{3}{7}$ 9. $1\frac{1}{2} \cdot x = 2\frac{1}{4}$
10. $6\frac{1}{3} \cdot x = 4\frac{1}{3}$ 11. Two-thirds of what number is 12?
12. Two and one-half times what number is $6\frac{1}{2}$?

ANSWERS 1. $\frac{7}{5}$ 2. $\frac{9}{2}$ 3. 15 4. 10 5. $\frac{3}{35}$ 6. $\frac{1}{36}$ 7. $\frac{11}{10}$ 8. $\frac{15}{7}$ 9. $1\frac{1}{2}$ 10. $\frac{13}{19}$ 11. 18 12. $2\frac{3}{5}$

Miscellaneous Exercises 2.4

Divide the following fractions and mixed numbers.

See Examples 39 to 43

1. $\frac{2}{3} \div \frac{5}{7}$ 2. $\frac{3}{8} \div \frac{4}{7}$ 3. $\frac{5}{9} \div \frac{7}{6}$
4. $\frac{1}{12} \div \frac{1}{18}$ 5. $\frac{7}{10} \div \frac{4}{5}$ 6. $\frac{7}{9} \div \frac{5}{18}$

See Examples 44 to 48

7. $5\frac{1}{2} \div 2\frac{1}{4}$ 8. $1\frac{1}{5} \div 2\frac{2}{3}$ 9. $5\frac{1}{3} \div 3\frac{1}{3}$
10. $2\frac{1}{3} \div 3$ 11. $7 \div 4\frac{1}{2}$ 12. $3\frac{1}{4} \div \frac{3}{8}$

Find n in the following problems.

See Examples 49 to 51

13. $6 \cdot n = \frac{2}{3}$ 14. $1\frac{1}{3} \cdot n = 4\frac{1}{2}$
15. Three-fourths of what number is $2\frac{1}{2}$?

Business applications

16. A family has \$1000 to invest in a stock selling at $\$12\frac{1}{2}$ per share. How many shares can they buy?
17. A company has $15\frac{1}{4}$ tons of material to ship out. A truckload can transport $1\frac{3}{4}$ tons. How many truckloads will it take to transport the entire load? (Find both an exact answer and a rounded-up answer.)

Health applications

18. A hospital has 150 liters of an intravenous solution. How many $\frac{1}{3}$-liter containers can a nurse fill with this 150 liters?

19. A patient is to receive a total of $4\frac{1}{2}$ grains of a medicine. How many $\frac{1}{4}$-grain doses is this?

Technical application

20. A plumber has $15\frac{1}{2}$ feet of pipe. How many $1\frac{2}{3}$-feet (20-inch) segments can be cut from the $15\frac{1}{2}$-foot piece?

Consumer applications

21. A certain child's shirt requires $\frac{2}{3}$ yard of material. How many of these shirts can be made with 10 yards of material?

22. Reggie drove 110 miles on $5\frac{1}{2}$ gallons of gasoline. How many miles per gallon is this?

23. A family pays a total of \$4500 for a car. This payment was spread over $2\frac{1}{2}$ years. What was the average per year? What was the average per month?

Science application

24. The density of a substance is given by $d = m \div V$, where m is the mass and V is the volume. Find the density for a substance with $m = \frac{1}{5}$ and $V = \frac{1}{100}$.

Now complete Post-Test 1 for Section 2.4 on page T29.

2.5 ADDITION OF FRACTIONS AND MIXED NUMBERS

SECTION OBJECTIVES

After completing this unit, you will be able to:

1. Add fractions.
2. Add mixed numbers.

TYPICAL PROBLEMS

You will be able to work problems like the ones below.

(a) $\frac{1}{7} + \frac{3}{7}$ **(b)** $\frac{3}{8} + \frac{5}{12}$ **(c)** $\frac{1}{20} + \frac{4}{15} + \frac{11}{12}$

(d) $4\frac{1}{3} + 5\frac{1}{7}$ **(e)** $6\frac{11}{18} + 7\frac{9}{10}$ **(f)** Find n: $n - \frac{1}{2} = 2$.

ANSWERS

(a) $\frac{4}{7}$ **(b)** $\frac{19}{24}$ **(c)** $\frac{37}{30}$ **(d)** $9\frac{10}{21}$ **(e)** $14\frac{23}{45}$ **(f)** $2\frac{1}{2}$

Adding fractions is harder than multiplying them. It is *wrong* to add the numerators and add the denominators. To add fractions, they must have a **common denominator**. When they do, we simply add the numerators over the common denominator.

$$\frac{a}{c} + \frac{b}{c} = \frac{a+b}{c}$$

EXAMPLE 52 Add $\frac{3}{11} + \frac{5}{11}$.

SOLUTION

$$\frac{3}{11} + \frac{5}{11} = \frac{8}{11}$$

Add numerators

EXAMPLE 53 Add $\frac{3}{10} + \frac{1}{10}$.

SOLUTION

$$\frac{3}{10} + \frac{1}{10} = \frac{4}{10}$$

Add numerators

$$= \frac{2}{5}$$

Reduce

Why Does It Work?

Since the denominators are the same, we have the same-sized pieces. When we add the numerators, we are adding the number of pieces together.

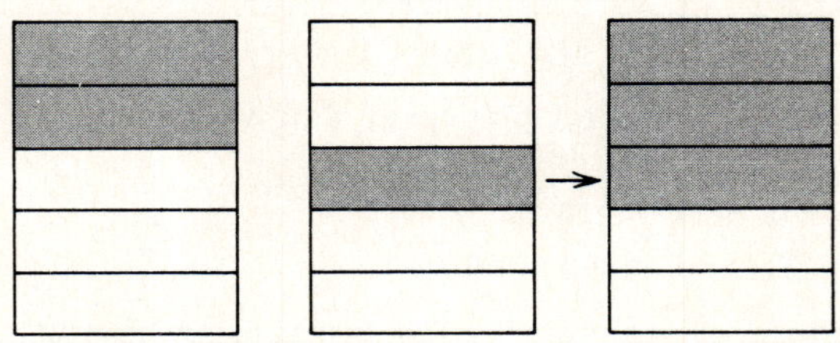

$\frac{2}{5} + \frac{1}{5} = \frac{3}{5}$

PROBLEM SET Add the following fractions and reduce the answers.

See Examples 52 and 53

1. $\frac{1}{3} + \frac{1}{3}$
2. $\frac{1}{7} + \frac{5}{7}$
3. $\frac{4}{5} + \frac{3}{5}$
4. $\frac{2}{13} + \frac{9}{13}$
5. $\frac{1}{6} + \frac{2}{6}$
6. $\frac{3}{8} + \frac{5}{8}$
7. $\frac{3}{10} + \frac{5}{10}$
8. $\frac{7}{20} + \frac{11}{20}$

ANSWERS	1. $\frac{2}{3}$	2. $\frac{6}{7}$	3. $\frac{7}{5}$	4. $\frac{11}{13}$	5. $\frac{1}{2}$	6. 1	7. $\frac{4}{5}$	8. $\frac{9}{10}$

When the denominators are different, it is more difficult to add fractions. We must first rewrite them with the same denominator. To do this, we look for the **least common denominator (LCD)**. This is the least common multiple (LCM) of the denominators (see Section 1.6).

For example, suppose that we wish to add $\frac{1}{3} + \frac{1}{4}$. The LCD is the least common multiple of the denominators, 3 and 4. Thus the LCD is 12. As we saw in Section 2.1, there are many different ways to write the same fraction.

$$\frac{1}{3} = \frac{2}{6} = \frac{3}{9} = \boxed{\frac{4}{12}} = \frac{5}{15} = \cdots$$

$$\frac{1}{4} = \frac{2}{8} = \boxed{\frac{3}{12}} = \frac{4}{16} = \frac{5}{20} = \cdots$$

Notice that each fraction is equal to another fraction with a denominator of 12, the LCD. Now we can add the fractions.

$$\frac{1}{3} + \frac{1}{4}$$

$$= \frac{4}{12} + \frac{3}{12} = \frac{7}{12}$$

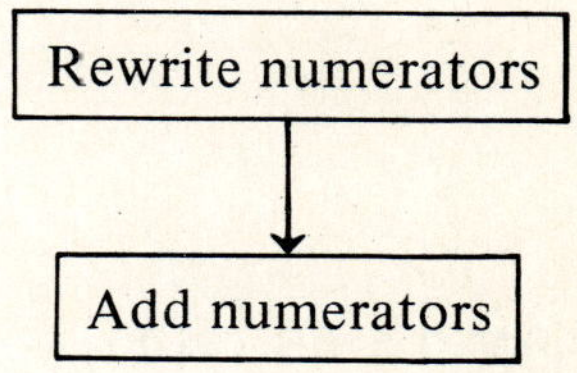

EXAMPLE 54 Add $\frac{1}{6} + \frac{3}{8}$.

SOLUTION We must first find the LCD of 6 and 8. This is 24, since 24 is the smallest number that both 6 and 8 divide into evenly. Now we must rewrite the two fractions so that they both have the same denominator, 24. To do this, we multiply the first fraction by $1 = \frac{4}{4}$ and the second fraction by $1 = \frac{3}{3}$. This will produce denominators of 24 in both fractions without changing their values.

$$\frac{1}{6} + \frac{3}{8}$$

$$\frac{1}{6} \cdot \frac{④}{④} + \frac{3}{8} \cdot \frac{③}{③}$$

$$= \frac{4}{24} + \frac{9}{24}$$

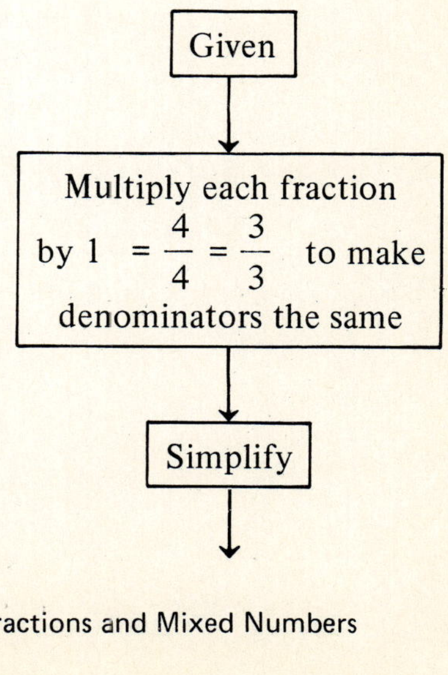

$$= \frac{13}{24}$$

Add numerators

YES	NO
$\frac{1}{6} + \frac{3}{8} = \frac{13}{24}$	~~$\frac{1}{6} + \frac{3}{8} = \frac{4}{14}$~~

We must first find the LCD to add fractions.

YES	NO
$\frac{1}{6} + \frac{3}{8} = \frac{13}{24}$	~~$\frac{1}{6} + \frac{3}{8} = \frac{4}{24}$~~

After we find the LCD, we must also find the new numerators.

What if the LCD is not obvious? In that case, we use the method for finding the LCM that was shown in Section 1.6. Recall that we can find the LCM by factoring the numbers into primes, lining up the primes, and multiplying a prime from each column. This procedure will also show us what to multiply the original fractions by.

EXAMPLE 55 Add $\frac{5}{12} + \frac{8}{21}$.

SOLUTION We start by factoring the denominators 12 and 21 to find the LCD. Notice that we fill in and circle the factors missing from each column. (We use these circled factors to help find the new numerators.)

$12 = 2 \cdot 2 \cdot 3$

$21 = \quad\quad\;\; 3 \cdot 7$

Factor denominators; line up common factors

$12 \rightarrow 2 \cdot 2 \cdot 3 \cdot (7)$

$21 \rightarrow (2) \cdot (2) \cdot 3 \cdot 7$

Fill in missing factors

$\text{LCD} = 2 \cdot 2 \cdot 3 \cdot 7 = 84$

LCD is product of factors

$$\frac{5}{12} + \frac{8}{21}$$

$$\frac{5}{12} \cdot \frac{(7)}{(7)} + \frac{8}{21} \cdot \frac{(4)}{(4)}$$

Multiply each fraction by $1 = \frac{\text{missing factor}}{\text{missing factor}}$ to get fractions with LCD = 84

$$= \frac{35}{84} + \frac{32}{84}$$

Simplify

$$= \frac{67}{84}$$

Add numerators

Notice that we used the missing (circled) factors to help us find the new fractions. We multiplied the first fraction by $1 = \frac{7}{7}$ and the second fraction by

$$1 = \frac{2 \cdot 2}{2 \cdot 2} = \frac{4}{4}$$

Let us state this procedure.

> To find the LCD and add fractions with unlike denominators:
> 1. Factor the denominators and line up the common primes.
> 2. Fill in (and circle) the missing factors.
> 3. The LCD is the product of the factors.
> 4. Multiply each fraction by $1 = \frac{\text{missing factor}}{\text{missing factor}}$ to make the fractions have the same denominators.
> 5. Add the numerators.
> 6. Reduce, if necessary.

EXAMPLE 56 Add $\frac{1}{4} + \frac{5}{6} + \frac{7}{20}$.

SOLUTION We add three (or more) fractions just like two.

$$\begin{aligned} 4 &= 2 \cdot 2 \\ 6 &= \quad\;\; 2 \cdot 3 \\ 20 &= 2 \cdot 2 \cdot \quad\;\; 5 \end{aligned}$$

Factor denominators
Line up common factors

$$\begin{aligned} 4 &\to 2 \cdot 2 \cdot \textcircled{3} \cdot \textcircled{5} \\ 6 &\to \textcircled{2} \cdot 2 \cdot 3 \cdot \textcircled{5} \\ 20 &\to 2 \cdot 2 \cdot \textcircled{3} \cdot 5 \end{aligned}$$

Fill in missing factors

$$\text{LCD} = 2 \cdot 2 \cdot 3 \cdot 5 = 60$$

Multiply all factors for LCD

$$\frac{1}{4} + \frac{5}{6} + \frac{7}{20}$$

$$\frac{1}{4} \cdot \frac{15}{15} + \frac{5}{6} \cdot \frac{10}{10} + \frac{7}{20} \cdot \frac{3}{3}$$

Multiply each fraction by $1 = \frac{\text{missing factor}}{\text{missing factor}}$ to get fractions with LCD = 60

$$= \frac{15}{60} + \frac{50}{60} + \frac{21}{60}$$

Simplify

$$= \frac{86}{60} = \frac{43}{30}$$

Add numerators; reduce

EXAMPLE 57 Add $\frac{5}{8} + \frac{9}{10} + \frac{7}{12}$.

SOLUTION

$$\begin{aligned} 8 &= 2 \cdot 2 \cdot 2 \\ 10 &= \phantom{2 \cdot 2 \cdot{}} 2 \cdot 5 \\ 12 &= 2 \cdot 2 \cdot \phantom{2 \cdot 5 \cdot{}} 3 \end{aligned}$$

Factor all denominators
Line up common factors

$$\begin{aligned} 8 &\to 2 \cdot 2 \cdot 2 \cdot (5) \cdot (3) \\ 10 &\to (2) \cdot (2) \cdot 2 \cdot 5 \cdot (3) \\ 12 &\to 2 \cdot 2 \cdot (2) \cdot (5) \cdot 3 \end{aligned}$$

Fill in missing factors

$$\text{LCD} = 2 \cdot 2 \cdot 2 \cdot 5 \cdot 3 = 120$$

Multiply all factors to get LCD

$$\frac{5}{8} + \frac{9}{10} + \frac{7}{12}$$

$$\frac{5}{8} \cdot \frac{15}{15} + \frac{9}{10} \cdot \frac{12}{12} + \frac{7}{12} \cdot \frac{10}{10}$$

Multiply each fraction by $1 = \frac{\text{missing factor}}{\text{missing factor}}$ to get fractions with LCD = 120

$$= \frac{75}{120} + \frac{108}{120} + \frac{70}{120}$$

Simplify

$$= \frac{253}{120}$$

Add numerators

EXAMPLE 58 In a group of people, $\frac{1}{15}$ of them drive a Volkswagen, and $\frac{1}{40}$ drive a Toyota. What fraction drive either a Volkswagen or a Toyota?

SOLUTION We simply add these fractions.

$$15 = 3 \cdot 5$$
$$40 = 5 \cdot 2 \cdot 2 \cdot 2$$

Factor denominators. Line up common factors

$$15 \rightarrow 3 \cdot 5 \cdot \textcircled{2} \cdot \textcircled{2} \cdot \textcircled{2}$$
$$40 \rightarrow \textcircled{3} \cdot 5 \cdot 2 \cdot 2 \cdot 2$$

Fill in missing factors

$$\text{LCD} = 3 \cdot 5 \cdot 2 \cdot 2 \cdot 2 = 120$$

Multiply all factors to get LCD

$$\frac{1}{15} + \frac{1}{40}$$

$$\frac{1}{15} \cdot \frac{\textcircled{8}}{\textcircled{8}} + \frac{1}{40} \cdot \frac{\textcircled{3}}{\textcircled{3}}$$

Multiply each fraction by $1 = \dfrac{\text{missing factor}}{\text{missing factor}}$ to get fractions with LCD = 120

$$= \frac{8}{120} + \frac{3}{120}$$

Simplify

$$= \frac{11}{120}$$

Add numerators

Thus $\dfrac{11}{120}$ of them drive Volkswagens or Toyotas.

Why Does It Work?

When we find the LCD, we are cutting the two fractions into the same-sized pieces so that we can add the numerators. Lining up the primes helps us get the LCD. The circled primes in the rule tell us what we have to multiply the numerators and denominators by to get an equal fraction with the LCD as a denominator.

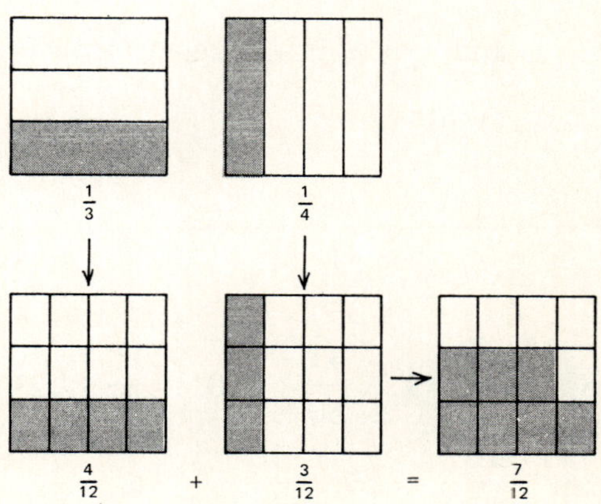

PROBLEM SET Add the following fractions and reduce the answers to simplest terms.

See Examples 54 to 58

1. $\frac{1}{2} + \frac{1}{3}$ 2. $\frac{2}{7} + \frac{3}{5}$ 3. $\frac{1}{6} + \frac{1}{3}$ 4. $\frac{1}{2} + \frac{1}{4}$

5. $\frac{1}{5} + \frac{3}{10}$ 6. $\frac{3}{4} + \frac{1}{6}$ 7. $\frac{1}{9} + \frac{5}{18}$ 8. $\frac{2}{9} + \frac{5}{6}$

9. $\frac{5}{12} + \frac{4}{15}$ 10. $\frac{5}{9} + \frac{7}{6}$ 11. $\frac{7}{9} + \frac{4}{15}$ 12. $\frac{7}{18} + \frac{5}{24}$

13. $\frac{3}{20} + \frac{6}{25}$ 14. $\frac{11}{24} + \frac{13}{30}$

15. $\frac{8}{27} + \frac{11}{18} + \frac{1}{12}$ 16. $\frac{11}{12} + \frac{8}{15} + \frac{3}{20}$

17. $\frac{1}{20} + \frac{1}{30} + \frac{1}{40}$ 18. $\frac{5}{18} + \frac{8}{21} + \frac{11}{24}$

19. $\frac{1}{2} + \frac{1}{3} + \frac{1}{4} + \frac{1}{5} + \frac{1}{6}$ 20. $\frac{1}{2} + \frac{1}{4} + \frac{1}{8} + \frac{1}{16} + \frac{1}{32} + \frac{1}{64}$

21. Albert owns $\frac{4}{15}$ of the family business, Barbara owns $\frac{3}{10}$, and Chuck owns $\frac{7}{20}$. How much of the business do they own together?

22. One week, Becky jogs $\frac{1}{2}$ mile, $\frac{2}{5}$ mile, $\frac{1}{3}$ mile, and $\frac{3}{10}$ mile. How much is this altogether?

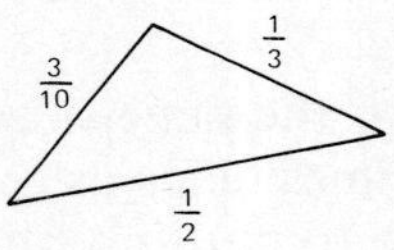

23. What is the perimeter (distance around) the triangle at the left?

24. On her business trip, Jenifer used $\frac{1}{3}$ tank of gas, $\frac{1}{6}$ tank of gas, and $\frac{2}{9}$ tank of gas. How much gas did she use altogether?

25. Charlie is painting his house. He paints $\frac{1}{10}$ of it in June, $\frac{4}{15}$ of it in July, and $\frac{11}{20}$ of it in August. How much of the house did he paint in those three months?

ANSWERS 1. $\frac{5}{6}$ 2. $\frac{31}{35}$ 3. $\frac{1}{2}$ 4. $\frac{3}{4}$ 5. $\frac{1}{2}$ 6. $\frac{11}{12}$ 7. $\frac{7}{18}$ 8. $\frac{19}{18}$ 9. $\frac{41}{60}$ 10. $\frac{31}{18}$ 11. $\frac{47}{45}$ 12. $\frac{43}{72}$ 13. $\frac{39}{100}$ 14. $\frac{107}{120}$ 15. $\frac{107}{108}$ 16. $\frac{8}{5}$ 17. $\frac{13}{120}$ 18. $\frac{563}{504}$ 19. $\frac{29}{20}$ 20. $\frac{63}{64}$ 21. $\frac{11}{12}$ 22. $\frac{23}{15}$ 23. $\frac{17}{15}$ 24. $\frac{13}{18}$ 25. $\frac{11}{12}$

Adding mixed numbers is much like adding the fractions. We add the fraction parts and we add the whole-number parts. If the fractions add to more than 1, we carry.

> To add mixed numbers:
>
> 1. Add the fractions separately, using the procedure given in the first part of this section. Reduce, if necessary.
> 2. If this sum is an improper fraction, convert it to a mixed number, and carry the whole number over to the other whole numbers.
> 3. Add the whole numbers.

EXAMPLE 59 Add $4\frac{1}{7} + 2\frac{3}{7}$.

SOLUTION We add the whole numbers and the fractions separately.

$$4\frac{1}{7} + 2\frac{3}{7}$$

$$= (4 + 2) + \left(\frac{1}{7} + \frac{3}{7}\right)$$

$$= 6 + \frac{4}{7} = 6\frac{4}{7}$$

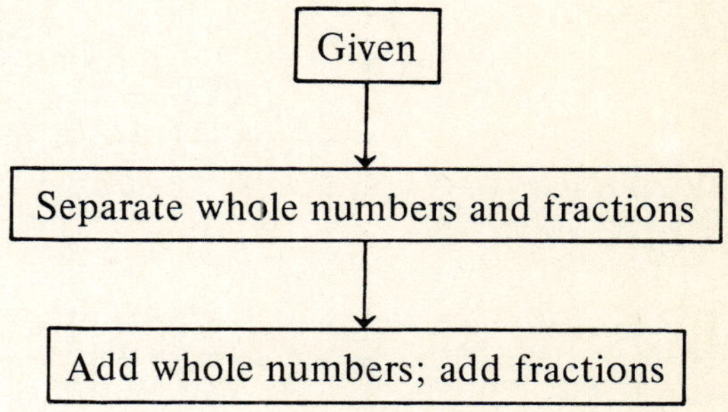

EXAMPLE 60 Add $3\frac{1}{2} + 4\frac{1}{3}$.

SOLUTION The LCD is 6, so we rewrite the fractions with this LCD.

$$3\frac{1}{2} + 4\frac{1}{3}$$

$$= (3 + 4) + \left(\frac{1}{2} + \frac{1}{3}\right)$$

$$= \quad 7 \quad + \left(\frac{3}{6} + \frac{2}{6}\right)$$

$$= 7 + \frac{5}{6} = 7\frac{5}{6}$$

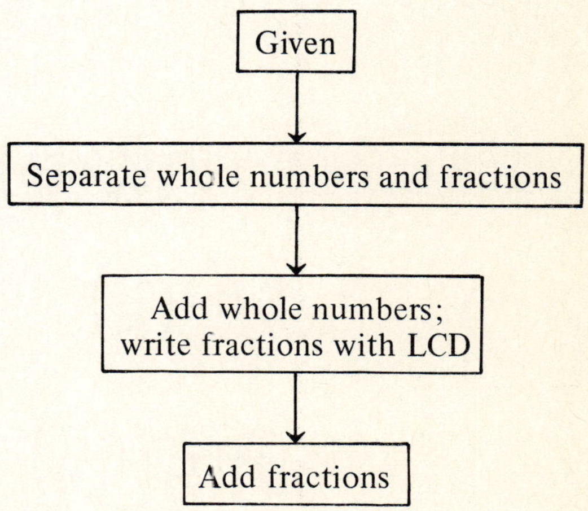

We can also add these vertically:

$$\begin{array}{rcr} 3\frac{1}{2} & = & 3\frac{3}{6} \\ 4\frac{1}{3} & = & 4\frac{2}{6} \\ \hline & & 7\frac{5}{6} \end{array}$$

EXAMPLE 61 Add $7\frac{5}{6} + 4\frac{11}{15}$.

SOLUTION We add the whole numbers and the fractions separately. When adding the fractions, we find the LCD using the procedure given in the first part of this section.

$6 = 2 \cdot 3$ $15 = \quad\; 3 \cdot 5$	Factor into primes; line up common primes
$6 \to 2 \cdot 3 \cdot \textcircled{5}$ $15 \to \textcircled{2} \cdot 3 \cdot 5$	Fill in missing primes
$\text{LCD} = 2 \cdot 3 \cdot 5 = 30$	LCD is product of primes
$7\frac{5}{6} + 4\frac{11}{15}$	Given
$= (7 + 4) + \left(\frac{5}{6} + \frac{11}{15}\right)$	Add whole numbers, add fractions
$= (7 + 4) + \left(\frac{5}{6} \cdot \frac{\textcircled{5}}{\textcircled{5}} + \frac{11}{15} \cdot \frac{\textcircled{2}}{\textcircled{2}}\right)$	Multiply each fraction by $1 = \frac{\text{missing factor}}{\text{missing factor}}$
$= 11 + \left(\frac{25}{30} + \frac{22}{30}\right)$	Simplify
$= 11 + \left(\frac{47}{30}\right)$	Add numerators
$= 11 + \left(1 + \frac{17}{30}\right)$	Write $\frac{47}{30}$ as a mixed number
$= (11 + 1) + \left(\frac{17}{30}\right)$	"Carry" the 1 to the whole numbers
$= 12 + \frac{17}{30} = 12\frac{17}{30}$	Add whole numbers

We can also add these vertically:

$$\begin{array}{r}
7\frac{5}{6} = 7\frac{25}{30} \\
4\frac{11}{15} = 4\frac{22}{30} \\
\hline
11\frac{47}{30} = 11 + 1\frac{17}{30} = 12\frac{17}{30}
\end{array}$$

EXAMPLE 62 Add $3\frac{5}{8} + 5\frac{7}{12} + 9\frac{11}{18}$.

SOLUTION We add the whole numbers and the fractions separately. We first find the LCD for the three denominators.

$$\begin{aligned}
8 &= 2 \cdot 2 \cdot 2 \\
12 &= 2 \cdot 2 \cdot 3 \\
18 &= 2 \cdot 3 \cdot 3
\end{aligned}$$

Factor denominators; line up common primes

$$\begin{aligned}
8 &\to 2 \cdot 2 \cdot 2 \cdot ③ \cdot ③ \\
12 &\to 2 \cdot 2 \cdot ② \cdot 3 \cdot ③ \\
18 &\to 2 \cdot ② \cdot ② \cdot 3 \cdot 3
\end{aligned}$$

Fill in missing factors

$$\text{LCD} = 2 \cdot 2 \cdot 2 \cdot 3 \cdot 3 = 72$$

LCD is product of primes

$$3\frac{5}{8} + 5\frac{7}{12} + 9\frac{11}{18}$$

Given

$$= (3 + 5 + 9) + \left(\frac{5}{8} + \frac{7}{12} + \frac{11}{18}\right)$$

Separate whole numbers and fractions

$$= (3 + 5 + 9) + \left(\frac{5}{8} \cdot \frac{9}{9} + \frac{7}{12} \cdot \frac{6}{6} + \frac{11}{18} \cdot \frac{4}{4}\right)$$

Multiply each fraction by $1 = \frac{\text{missing factor}}{\text{missing factor}}$

$$= 17 + \left(\frac{45}{72} + \frac{42}{72} + \frac{44}{72}\right)$$

Simplify

$$= 17 + \left(\frac{131}{72}\right)$$

Add numerators

$$= 17 + \left(1 + \frac{59}{72}\right)$$

Write $\frac{131}{72}$ as a mixed number

$$= (17 + 1) + \left(\frac{59}{72}\right)$$

"Carry" 1 to the whole numbers

$$= 18 + \frac{59}{72} = 18\frac{59}{72}$$

Add whole numbers

We can also add these vertically:

$$\begin{array}{rcl} 3\frac{5}{8} & = & 3\frac{45}{72} \\ 5\frac{7}{12} & = & 5\frac{42}{72} \\ +9\frac{11}{18} & = & 9\frac{44}{72} \\ \hline & & 17\frac{131}{72} = 17 + 1\frac{59}{72} = 18\frac{59}{72} \end{array}$$

EXAMPLE 63 For a certain marketing survey, the Arbuckles record the following data for four nights of TV watching: $6\frac{1}{4}$, $4\frac{1}{2}$, $5\frac{3}{4}$, and $7\frac{3}{4}$ hours.

(a) Find the total number of hours.
(b) Find the average per night.

SOLUTION **(a)** The word *total* usually means add. Here we add the four different times. The LCD is 4.

$$6\frac{1}{4} + 4\frac{1}{2} + 5\frac{3}{4} + 7\frac{3}{4}$$

Given

$$= (6 + 4 + 5 + 7) + \left(\frac{1}{4} + \frac{1}{2} + \frac{3}{4} + \frac{3}{4}\right)$$

Separate whole numbers and fractions

$$= 22 + \left(\frac{1}{4} + \frac{1}{2}\cdot\frac{②}{②} + \frac{3}{4} + \frac{3}{4}\right)$$

Add whole numbers; give the fractions same denominator

$$= 22 + \left(\frac{1}{4} + \frac{2}{4} + \frac{3}{4} + \frac{3}{4}\right)$$

Simplify

$$= 22 + \left(\frac{9}{4}\right)$$

Add numerators

$$= 22 + \left(2 + \frac{1}{4}\right)$$

Write $\frac{9}{4}$ as a mixed number

$= (22 + 2) + \frac{1}{4}$ — "Carry" 2 to the whole numbers

$= 24 + \frac{1}{4} = 24\frac{1}{4}$ — Add whole numbers

(b) The word *average* usually means find the total and then divide. We already have the total; now we divide by 4.

Average $= 24\frac{1}{4} \div 4$ — Average means divide

$= 24\frac{1}{4} \times \frac{1}{4}$ — Invert and multiply

$= \frac{97}{4} \times \frac{1}{4}$ — Write as improper fraction

$= \frac{97}{16} = 6\frac{1}{16}$ — Multiply and write as a mixed number

---

PROBLEM SET Add the following mixed numbers and fractions. Reduce all answers.

See Examples 59 to 63

1. $4\frac{1}{5} + 7\frac{3}{5}$
2. $6\frac{1}{9} + 8\frac{4}{9}$
3. $3\frac{1}{5} + 9\frac{1}{4}$
4. $2\frac{1}{6} + 7\frac{5}{9}$
5. $12\frac{2}{9} + 19\frac{5}{12}$
6. $14\frac{1}{10} + 46\frac{4}{15}$
7. $7\frac{5}{8} + 9\frac{3}{4}$
8. $8\frac{1}{2} + 7\frac{7}{8}$
9. $24\frac{4}{9} + 17\frac{11}{12}$
10. $6\frac{3}{20} + 21\frac{9}{10}$
11. $67\frac{1}{3} + 112\frac{1}{15}$
12. $45\frac{5}{12} + 19\frac{8}{15}$
13. $15\frac{11}{14} + 34\frac{8}{21}$
14. $101\frac{7}{20} + 74\frac{29}{30}$
15. $4\frac{1}{4} + 8\frac{1}{5} + 6\frac{7}{10}$
16. $10\frac{1}{8} + 4\frac{7}{12} + 2\frac{8}{9}$
17. $\frac{11}{15} + 6\frac{17}{20} + 12\frac{21}{25}$
18. $\frac{13}{60} + 16\frac{7}{30} + 43\frac{14}{15} + 1\frac{7}{20}$

19. $\frac{9}{10} + 6\frac{31}{100} + 8\frac{23}{100} + 13\frac{213}{1000}$ 20. $8\frac{1}{2} + 6\frac{3}{4} + 12\frac{5}{8} + 11\frac{7}{16}$

21. Janet mixed $20\frac{1}{2}$ ounces of orange juice with $17\frac{3}{4}$ ounces of lemonade. How much is this total mixture?

22. Fran goes on a diet and loses $1\frac{1}{2}$, $1\frac{1}{4}$, $2\frac{1}{3}$, and $1\frac{7}{8}$ pounds in 4 weeks. How much did she lose altogether?

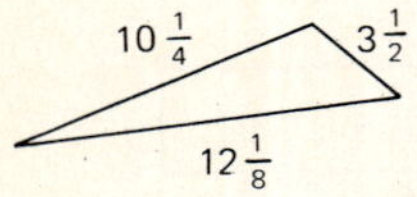

23. On a business trip, Steve travels $37\frac{1}{5}$, $52\frac{1}{2}$, and $101\frac{3}{4}$ miles. How many miles did he travel?

24. Find the perimeter of the triangle at the left.

ANSWERS 1. $11\frac{4}{5}$ 2. $14\frac{5}{9}$ 3. $12\frac{9}{20}$ 4. $9\frac{13}{18}$ 5. $31\frac{23}{36}$ 6. $60\frac{11}{30}$
7. $17\frac{3}{8}$ 8. $16\frac{3}{8}$ 9. $42\frac{13}{36}$ 10. $28\frac{1}{20}$ 11. $179\frac{2}{5}$
12. $64\frac{19}{20}$ 13. $50\frac{1}{6}$ 14. $176\frac{19}{60}$ 15. $19\frac{3}{20}$ 16. $17\frac{43}{72}$
17. $20\frac{127}{300}$ 18. $61\frac{11}{15}$ 19. $28\frac{653}{1000}$ 20. $39\frac{5}{16}$ 21. $38\frac{1}{4}$
22. $6\frac{23}{24}$ 23. $191\frac{9}{20}$ 24. $25\frac{7}{8}$

We saw in Chapter 1 that *addition is the opposite of subtraction.* For example,

$$n - 7 = 2 \qquad \text{means} \qquad n = 2 + 7 = 9$$

Similarly with fractions, to solve subtraction problems of this type, we add to find the missing number.

EXAMPLE 64 Find n: $n - \frac{1}{3} = 4$.

SOLUTION Since addition is the opposite of subtraction, we add $\frac{1}{3}$ to both sides of the equation.

$$n - \frac{1}{3} = 4 \qquad \text{means} \qquad n = 4 + \frac{1}{3} = 4\frac{1}{3}$$

EXAMPLE 65 Find x: $x - \frac{1}{2} = \frac{2}{3}$.

SOLUTION We solve this by adding $\frac{1}{2}$ to both sides of the equation.

$$x - \frac{1}{2} = \frac{2}{3} \quad \text{means} \quad x = \frac{2}{3} + \frac{1}{2}$$
$$= \frac{4}{6} + \frac{3}{6}$$
$$= \frac{7}{6} \text{ or } 1\frac{1}{6}$$

EXAMPLE 66 Find n: $n - 2\frac{1}{4} = 1\frac{5}{6}$.

SOLUTION We solve for n by adding $2\frac{1}{4}$ to both sides.

$$n - 2\frac{1}{4} = 1\frac{5}{6} \quad \text{means} \quad n = 1\frac{5}{6} + 2\frac{1}{4}$$
$$= (1 + 2) + \left(\frac{5}{6} + \frac{1}{4}\right)$$
$$= 3 + \left(\frac{10}{12} + \frac{3}{12}\right)$$
$$= 3 + \left(\frac{13}{12}\right)$$
$$= 3 + \left(1 + \frac{1}{12}\right)$$
$$= (3 + 1) + \frac{1}{12}$$
$$= 4 + \frac{1}{12} = 4\frac{1}{12}$$

Thus $n = 4\frac{1}{12}$.

PROBLEM SET Find n or x in the following problems.

See Examples 64 to 66

1. $n - \frac{1}{4} = 5$
2. $x - \frac{2}{3} = 7$
3. $n - \frac{1}{5} = \frac{2}{5}$
4. $n - \frac{2}{7} = \frac{6}{7}$
5. $x - \frac{2}{5} = \frac{7}{10}$
6. $n - \frac{1}{4} = \frac{3}{16}$
7. $n - 1\frac{1}{2} = 3\frac{1}{4}$
8. $x - \frac{5}{6} = 2\frac{1}{2}$

ANSWERS 1. $5\frac{1}{4}$ 2. $7\frac{2}{3}$ 3. $\frac{3}{5}$ 4. $\frac{8}{7}$ 5. $\frac{11}{10}$ 6. $\frac{7}{16}$ 7. $4\frac{3}{4}$ 8. $3\frac{1}{3}$

Miscellaneous Exercises 2.5

Add the following fractions and mixed numbers.

See Examples 52 and 53

1. $\frac{1}{7}+\frac{4}{7}$ 2. $\frac{2}{9}+\frac{4}{9}$ 3. $\frac{5}{12}+\frac{7}{12}$

See Examples 54 to 58

4. $\frac{1}{2}+\frac{3}{4}$ 5. $\frac{1}{6}+\frac{2}{9}$ 6. $\frac{2}{15}+\frac{7}{10}$

7. $\frac{11}{24}+\frac{13}{18}$ 8. $\frac{1}{8}+\frac{3}{10}+\frac{7}{20}$ 9. $\frac{5}{6}+\frac{7}{10}+\frac{7}{18}$

See Examples 59 to 63

10. $2\frac{1}{5}+4\frac{2}{5}$ 11. $3\frac{1}{3}+6\frac{1}{6}$ 12. $10\frac{1}{5}+3\frac{3}{4}$

13. $4\frac{4}{5}+1\frac{9}{10}$ 14. $1\frac{1}{2}+2\frac{2}{3}+3\frac{3}{4}$ 15. $27\frac{1}{6}+36\frac{1}{8}+9\frac{7}{10}$

See Examples 64 to 66

16. Find n: $n-\frac{1}{2}=\frac{1}{3}$.

17. Find n: $n-2\frac{1}{3}=5\frac{1}{4}$.

Health applications

18. A patient receives three doses of a medicine: $7\frac{1}{2}$, $6\frac{3}{4}$, and $7\frac{3}{4}$ grains.
 (a) Find the total dosage.
 (b) Find the average dosage.

19. One week, a nurse worked $7\frac{1}{2}$ hours on Monday, $8\frac{1}{4}$ hours on Tuesday, $8\frac{1}{2}$ hours on Wednesday, $9\frac{3}{4}$ hours on Thursday, and $7\frac{3}{4}$ hours on Friday.
 (a) Find the total hours worked in the week.
 (b) Find the average number of hours worked per day.

Business applications

20. In the morning one share of Chazery Foods was $56\frac{3}{4}$. It closed $\frac{7}{8}$ point higher. What was the closing price?

21. Burp-O Cola has $\frac{1}{20}$ of the cola market. If they can acquire another $\frac{1}{50}$ of the market, what fraction of the market will they have?

Technical application

22. Four pieces of wire are cut off of a long spool. The pieces cut off were: $21\frac{1}{2}$, $19\frac{3}{4}$, $42\frac{5}{8}$, and $39\frac{7}{16}$ inches. How much total wire was cut from the spool?

Consumer applications

23. Ted is tiling an L-shaped room as shown below. Because of furniture, he can only make the measurements shown. Use these to find L and W.

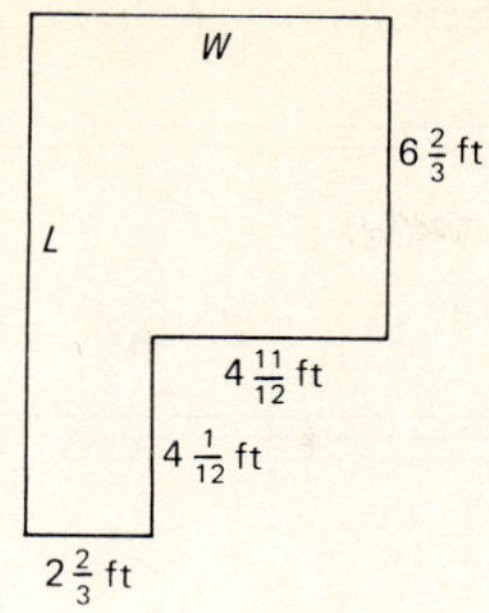

24. One week, Mandy jogs $\frac{3}{4}$, $1\frac{1}{8}$, $1\frac{1}{2}$, and $\frac{5}{8}$ mile.
   - **(a)** What is her total for the week?
   - **(b)** What is her average per run? (Divide by 4.)
   - **(c)** What is her average per day? (Divide by 7.)

25. Larry drove from home to work to a store and back home. The amount of gas used is shown in the sketch. What fraction of a tank did he use?

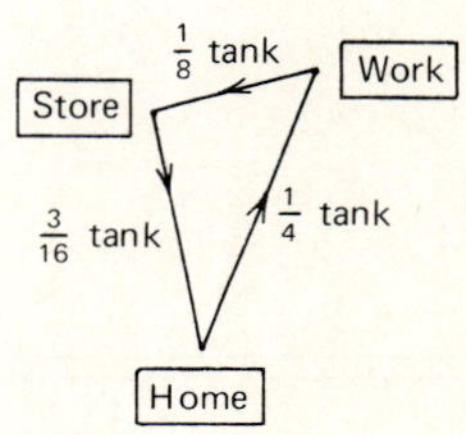

26. Ann got a long piece of material. She used $\frac{3}{4}$ yard on a T-shirt, $1\frac{3}{8}$ yards on a dress, and $1\frac{1}{4}$ yards on a blouse. How much of the material has she used?

Now complete Post-Test 1 for Section 2.5 on page T31.

2.6 SUBTRACTION OF FRACTIONS AND MIXED NUMBERS

SECTION OBJECTIVES

After completing this unit, you will be able to:

1. Subtract fractions.
2. Subtract mixed numbers.
3. Compare the sizes of two or more fractions.

TYPICAL PROBLEMS

You will be able to work problems like the ones below.

(a) $\frac{5}{11} - \frac{2}{11}$ **(b)** $\frac{4}{7} - \frac{1}{2}$ **(c)** $\frac{5}{12} - \frac{1}{18}$

(d) $3\frac{1}{5} - 1\frac{2}{15}$ (e) $6\frac{1}{10} - 2\frac{1}{8}$ (f) Find n: $n + \frac{1}{3} = 2$

(g) Which fraction is larger, $\frac{3}{4}$ or $\frac{11}{16}$?

ANSWERS (a) $\frac{3}{11}$ (b) $\frac{1}{14}$ (c) $\frac{13}{36}$ (d) $2\frac{1}{15}$ (e) $3\frac{39}{40}$ (f) $1\frac{2}{3}$ (g) $\frac{3}{4}$

Subtracting fractions is similar to adding fractions. They must have a common denominator. Then we subtract the numerators.

$$\frac{a}{c} - \frac{b}{c} = \frac{a - b}{c}$$

EXAMPLE 67 (a) $\frac{6}{7} - \frac{4}{7} = \frac{2}{7}$

(b) $\frac{5}{12} - \frac{1}{12} = \frac{4}{12} = \frac{1}{3}$

PROBLEM SET Subtract the following fractions and reduce the answers.

See Example 67

1. $\frac{4}{5} - \frac{1}{5}$
2. $\frac{9}{11} - \frac{5}{11}$
3. $\frac{15}{17} - \frac{13}{17}$
4. $\frac{5}{9} - \frac{1}{9}$
5. $\frac{7}{8} - \frac{3}{8}$
6. $\frac{9}{10} - \frac{7}{10}$
7. $\frac{11}{15} - \frac{8}{15}$
8. $\frac{21}{40} - \frac{11}{40}$

ANSWERS 1. $\frac{3}{5}$ 2. $\frac{4}{11}$ 3. $\frac{2}{17}$ 4. $\frac{4}{9}$ 5. $\frac{1}{2}$ 6. $\frac{1}{5}$ 7. $\frac{1}{5}$ 8. $\frac{1}{4}$

To subtract fractions:

1. Use the procedure given in Section 2.5 to rewrite the fractions with an LCD.
2. Subtract the numerators.
3. Reduce, if necessary.

EXAMPLE 68 Subtract $\frac{7}{9} - \frac{2}{15}$.

SOLUTION We first factor the denominators to find the LCD.

$$9 = 3 \cdot 3$$
$$15 = \phantom{3 \cdot{}} 3 \cdot 5$$

Factor denominators
Line up common factors

$$9 \rightarrow 3 \cdot 3 \cdot \textcircled{5}$$
$$15 \rightarrow \textcircled{3} \cdot 3 \cdot 5$$

Fill in missing factors

$$\text{LCD} = 3 \cdot 3 \cdot 5 = 45$$

Multiply all factors to get LCD

$$\frac{7}{9} - \frac{2}{15}$$

Given

$$= \frac{7}{9} \cdot \frac{\textcircled{5}}{\textcircled{5}} - \frac{2}{15} \cdot \frac{\textcircled{3}}{\textcircled{3}}$$

Multiply each fraction by $1 = \dfrac{\text{missing factor}}{\text{missing factor}}$

$$= \frac{35}{45} - \frac{6}{45}$$

Simplify

$$= \frac{29}{45}$$

Subtract numerators

EXAMPLE 69 Subtract $\dfrac{11}{24} - \dfrac{7}{20}$.

SOLUTION We first find the LCD.

$$24 = 2 \cdot 2 \cdot 2 \cdot 3$$
$$20 = \phantom{2 \cdot{}} 2 \cdot 2 \cdot \phantom{3 \cdot{}} 5$$

Factor denominators
Line up common factors

$$24 \rightarrow 2 \cdot 2 \cdot 2 \cdot 3 \cdot \textcircled{5}$$
$$20 \rightarrow \textcircled{2} \cdot 2 \cdot 2 \cdot \textcircled{3} \cdot 5$$

Fill in missing factors

$$\text{LCD} = 2 \cdot 2 \cdot 2 \cdot 3 \cdot 5 = 120$$

Multiply all factors to get LCD

$$\frac{11}{24} - \frac{7}{20}$$

Given

$$= \frac{11}{24} \cdot \frac{\textcircled{5}}{\textcircled{5}} - \frac{7}{20} \cdot \frac{\textcircled{6}}{\textcircled{6}}$$

Multiply each fraction by $1 = \dfrac{\text{missing factor}}{\text{missing factor}}$

$$= \frac{55}{120} - \frac{42}{120}$$ Simplify

$$= \frac{13}{120}$$ Subtract numerators

YES	NO
$\frac{11}{24} - \frac{7}{20} = \frac{13}{120}$	~~$\frac{11}{24} - \frac{7}{20} = \frac{4}{4}$~~

We must first find the common denominator to subtract fractions.

EXAMPLE 70 Subtract $\frac{7}{18} - \frac{5}{24}$.

SOLUTION

$$18 = 2 \cdot 3 \cdot 3$$
$$24 = 2 \cdot 3 \cdot \qquad 2 \cdot 2$$

Factor denominators
Line up common factors

$$18 \rightarrow 2 \cdot 3 \cdot 3 \cdot (2) \cdot (2)$$
$$24 \rightarrow 2 \cdot 3 \cdot (3) \cdot 2 \cdot 2$$

Fill in missing factors

$$\text{LCD} = 2 \cdot 3 \cdot 3 \cdot 2 \cdot 2 = 72$$

Multiply all factors to get LCD

$$\frac{7}{18} - \frac{5}{24}$$ Given

$$= \frac{7}{18} \cdot \frac{(4)}{(4)} - \frac{5}{24} \cdot \frac{(3)}{(3)}$$

Multiply each fraction by $1 = \frac{\text{missing factor}}{\text{missing factor}}$

$$= \frac{28}{72} - \frac{15}{72}$$ Simplify

$$= \frac{13}{72}$$ Subtract numerators

EXAMPLE 71 Becky owns $\frac{8}{15}$ of the family business. Because she needs cash, she sells $\frac{3}{10}$ to her sister Lisa. How much of the business does Becky now own?

SOLUTION To find the answer, we must subtract $\frac{8}{15} - \frac{3}{10}$.

$15 = 3 \cdot 5$ $10 = \quad 5 \cdot 2$	Factor denominators Line up common factors
$15 \rightarrow 3 \cdot 5 \cdot \textcircled{2}$ $10 \rightarrow \textcircled{3} \cdot 5 \cdot 2$	Fill in missing factors
$\text{LCD} = 3 \cdot 5 \cdot 2 = 30$	Multiply all factors to get LCD
$\frac{8}{15} - \frac{3}{10}$	Given
$= \frac{8}{15} \cdot \frac{\textcircled{2}}{\textcircled{2}} - \frac{3}{10} \cdot \frac{\textcircled{3}}{\textcircled{3}}$	Multiply each fraction by $1 = \frac{\text{missing factor}}{\text{missing factor}}$
$= \frac{16}{30} - \frac{9}{30}$	Simplify
$= \frac{7}{30}$	Subtract numerators

Why Does It Work?

As with addition, we cannot subtract unless the denominators are the same. Consider $\frac{3}{4} - \frac{1}{6}$. We have pieces that are fourths and pieces that are sixths. By further cutting, we can make them both have slices that are twelfths. Then we can subtract. This subtraction, $\frac{3}{4} - \frac{1}{6}$, can be pictured as shown.

$\frac{3}{4} = \frac{9}{12}$

$-\frac{1}{6} = -\frac{2}{12}$

$\frac{7}{12}$

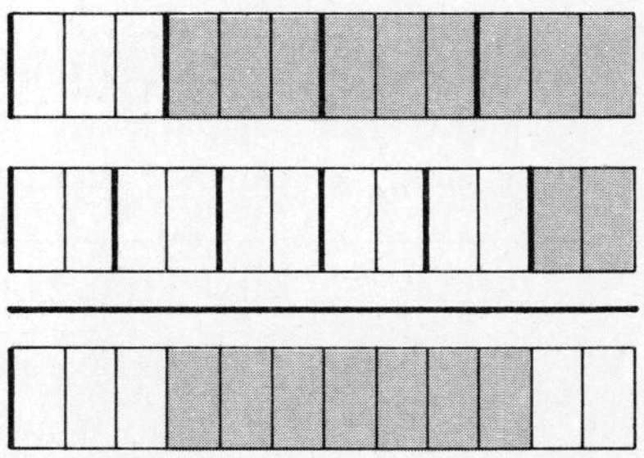

PROBLEM SET Subtract the following fractions and reduce the answers.

See Examples 68 to 71

1. $\frac{1}{3} - \frac{1}{4}$ 2. $\frac{1}{6} - \frac{1}{7}$ 3. $\frac{2}{3} - \frac{3}{5}$ 4. $\frac{3}{5} - \frac{1}{4}$

5. $\frac{1}{5} - \frac{1}{10}$ 6. $\frac{5}{12} - \frac{1}{3}$ 7. $\frac{8}{9} - \frac{5}{6}$ 8. $\frac{7}{20} - \frac{1}{10}$

9. $\frac{5}{18} - \frac{1}{12}$ 10. $\frac{5}{9} - \frac{7}{15}$ 11. $\frac{7}{10} - \frac{4}{15}$ 12. $\frac{17}{21} - \frac{8}{15}$

13. $\frac{9}{20} - \frac{11}{50}$ 14. $\frac{11}{12} - \frac{7}{10}$ 15. $\frac{2}{5} - \frac{7}{40}$ 16. $\frac{4}{15} - \frac{1}{12}$

17. $\frac{7}{20} - \frac{3}{16}$ 18. $\frac{7}{8} - \frac{7}{10}$ 19. $\frac{9}{20} - \frac{3}{20}$ 20. $\frac{23}{100} - \frac{2}{10}$

21. Carla has $\frac{3}{4}$ of a tank of gasoline in her car. It takes her $\frac{3}{16}$ of a tank to drive to work. How much will she have left?

22. Henry owns $\frac{7}{12}$ of the ABC Camper Co. If he sells $\frac{1}{8}$ of the company, what does he have left?

23. It is $\frac{7}{8}$ kilometer from Michele's house to the pizza place. If she already walked $\frac{1}{2}$ kilometer, how much farther does she have to go?

24. Wendy jogged $\frac{11}{4}$ miles, and Fran jogged $\frac{12}{5}$ miles. How much farther than Fran did Wendy jog?

ANSWERS 1. $\frac{1}{12}$ 2. $\frac{1}{42}$ 3. $\frac{1}{15}$ 4. $\frac{7}{20}$ 5. $\frac{1}{10}$ 6. $\frac{1}{12}$ 7 $\frac{1}{18}$ 8. $\frac{1}{4}$ 9. $\frac{7}{36}$ 10. $\frac{4}{45}$ 11. $\frac{13}{30}$ 12. $\frac{29}{105}$ 13. $\frac{23}{100}$ 14. $\frac{13}{60}$ 15. $\frac{9}{40}$ 16. $\frac{11}{60}$ 17. $\frac{13}{80}$ 18. $\frac{7}{40}$ 19. $\frac{3}{10}$ 20. $\frac{3}{100}$ 21. $\frac{9}{16}$ 22. $\frac{11}{24}$ 23. $\frac{3}{8}$ 24. $\frac{7}{20}$

Subtracting mixed numbers is very similar to adding mixed numbers; we deal with the fraction parts and the whole-number parts separately. To make **borrowing** easier, we subtract vertically.

> To subtract mixed numbers:
>
> 1. Subtract the fraction parts (borrow, if necessary).
> 2. Subtract the whole-number parts.

EXAMPLE 72 Subtract $17\frac{4}{5} - 11\frac{1}{5}$.

SOLUTION

$$\begin{array}{r} 17\frac{4}{5} \\ -11\frac{1}{5} \\ \hline 6\frac{3}{5} \end{array}$$

Subtract fraction parts
↓
Subtract whole numbers

EXAMPLE 73 Subtract $12\frac{1}{2} - 8\frac{1}{3}$.

SOLUTION

$$\begin{array}{rcr} 12\frac{1}{2} & = & 12\frac{3}{6} \\ -\ 8\frac{1}{3} & = & -\ 8\frac{2}{6} \\ \hline & & 4\frac{1}{6} \end{array}$$

Write fractions with LCD, 6
↓
Subtract fractions; subtract whole numbers

EXAMPLE 74 Subtract $7\frac{1}{5} - 2\frac{11}{15}$.

SOLUTION After we write the fractions with the LCD, 15, we have to borrow.

$$\begin{array}{rcrcr} 7\frac{1}{5} & = & 7\frac{3}{15} & = & 6\frac{18}{15} \\ -2\frac{11}{15} & = & -2\frac{11}{15} & = & -2\frac{11}{15} \\ \hline & & & & 4\frac{7}{15} \end{array}$$

Write $7\frac{3}{15} = 6 + 1 + \frac{3}{15} = 6 + \frac{18}{15}$
↓
Subtract fractions; subtract whole numbers

Notice that we "borrowed" $1 = \frac{15}{15}$ from the 7. Then we got $\frac{15}{15} + \frac{3}{15} = \frac{18}{15}$.

EXAMPLE 75 Subtract $9\frac{2}{15} - 3\frac{6}{25}$.

SOLUTION We first find the LCD; then we subtract by borrowing.

$$15 = 3 \cdot 5$$
$$25 = \quad\; 5 \cdot 5$$

Factor denominators. Line up common factors

↓

$$15 \to 3 \cdot 5 \cdot \textcircled{5}$$
$$25 \to \textcircled{3} \cdot 5 \cdot 5$$

Fill in missing factors

↓

$$\text{LCD} = 3 \cdot 5 \cdot 5 = 75$$

LCD is product of factors

↓

Write fractions with LCD, 75

$$9\frac{2}{15} = 9\frac{2 \cdot \textcircled{5}}{15 \cdot \textcircled{5}} = 9\frac{10}{75} = 8\frac{85}{75}$$
$$-3\frac{6}{25} = 3\frac{6 \cdot \textcircled{3}}{25 \cdot \textcircled{3}} = 3\frac{18}{75} = 3\frac{18}{75}$$
$$5\frac{67}{75}$$

Write $9\frac{10}{75} = 8 + 1 + \frac{10}{75} = 8 + \frac{85}{75}$

↓

Subtract fractions; subtract whole numbers

Here we "borrowed" $1 = \frac{75}{75}$ from the 9.

EXAMPLE 76 Bruce weighed 250 pounds. Then he joined Weight Losers. Now he weighs $187\frac{3}{4}$ pounds. How much weight did he lose?

SOLUTION The word *lose* means subtract. So we subtract $250 - 187\frac{3}{4}$.

$$250 = 249\frac{4}{4}$$
$$-187\frac{3}{4} = -187\frac{3}{4}$$
$$62\frac{1}{4}$$

Write $250 = 249 + 1 = 249 + \frac{4}{4}$

↓

Subtract fractions; subtract whole numbers

Here we borrowed $1 = \frac{4}{4}$ from the 250. Thus Bruce lost $62\frac{1}{4}$ pounds.

PROBLEM SET Subtract the following mixed numbers and fractions.

See Examples 72 to 76

1. $5\frac{7}{11} - 2\frac{1}{11}$
2. $10\frac{2}{3} - 4\frac{1}{3}$
3. $5\frac{1}{2} - 1\frac{1}{4}$
4. $80\frac{3}{5} - 32\frac{1}{10}$
5. $7\frac{1}{5} - 3\frac{4}{5}$
6. $10\frac{2}{7} - 6\frac{5}{7}$
7. $5\frac{1}{8} - 1\frac{1}{4}$
8. $23\frac{1}{6} - 15\frac{2}{3}$
9. $12\frac{3}{4} - 9\frac{5}{6}$
10. $36\frac{1}{5} - 23\frac{7}{10}$
11. $56\frac{1}{8} - 44\frac{5}{12}$
12. $5\frac{2}{9} - 3\frac{1}{2}$
13. $29\frac{1}{6} - 14\frac{7}{9}$
14. $67\frac{3}{10} - 51\frac{4}{15}$
15. $102\frac{1}{3} - 79\frac{2}{5}$
16. $8 - 6\frac{1}{5}$
17. $100 - 64\frac{4}{7}$
18. $78 - 33\frac{1}{3}$
19. $9\frac{1}{2} - \frac{5}{6}$
20. $23\frac{1}{25} - \frac{3}{20}$
21. Two-sevenths of a group are men. What fraction is women?
22. Becky is $47\frac{1}{2}$ inches tall. Lisa is $39\frac{3}{4}$ inches tall. How much taller is Becky than Lisa?
23. In one day the value of Norton Nickel, Inc., dropped from $117\frac{7}{8}$ to $112\frac{1}{4}$. How much was this drop?
24. On her diet, Sonya goes from $140\frac{1}{4}$ pounds to $132\frac{1}{2}$ pounds. How much weight loss is this?
25. Ron's goal is to run 100 miles over the summer. So far, he has run $61\frac{3}{8}$ miles. How much more does he have to run?
26. Lana has worked $9\frac{2}{5}$ hours this week. If she is allowed 20 hours of work, how many more hours can she work?

ANSWERS 1. $3\frac{6}{11}$ 2. $6\frac{1}{3}$ 3. $4\frac{1}{4}$ 4. $48\frac{1}{2}$ 5. $3\frac{2}{5}$ 6. $3\frac{4}{7}$ 7. $3\frac{7}{8}$ 8. $7\frac{1}{2}$ 9. $2\frac{11}{12}$ 10. $12\frac{1}{2}$ 11. $11\frac{17}{24}$ 12. $1\frac{13}{18}$ 13. $14\frac{7}{18}$ 14. $16\frac{1}{30}$ 15. $22\frac{14}{15}$ 16. $1\frac{4}{5}$ 17. $35\frac{3}{7}$ 18. $44\frac{2}{3}$ 19. $8\frac{2}{3}$

20. $22\frac{89}{100}$ 21. $\frac{5}{7}$ 22. $7\frac{3}{4}$ 23. $5\frac{5}{8}$ 24. $7\frac{3}{4}$ 25. $38\frac{5}{8}$

26. $10\frac{3}{5}$

As we saw in Chapter 1, *subtraction is the opposite of addition*. For example,

$$n + 2 = 6 \quad \text{means} \quad n = 6 - 2 = 4$$

Similarly with fractions, we subtract to solve addition problems of this type.

EXAMPLE 77 Find n: $n + \frac{2}{7} = \frac{6}{7}$.

SOLUTION Since subtraction is the opposite of addition, we subtract $\frac{2}{7}$ from both sides of the equation.

$$n + \frac{2}{7} = \frac{6}{7} \quad \text{means} \quad n = \frac{6}{7} - \frac{2}{7} = \frac{4}{7}$$

EXAMPLE 78 Find x: $x + \frac{1}{2} = \frac{2}{3}$.

SOLUTION We find x by subtracting $\frac{1}{2}$ from both sides of the equation.

$$x + \frac{1}{2} = \frac{2}{3} \quad \text{means} \quad x = \frac{2}{3} - \frac{1}{2}$$
$$= \frac{4}{6} - \frac{3}{6}$$
$$= \frac{1}{6}$$

EXAMPLE 79 What number added to $2\frac{1}{4}$ is 5?

SOLUTION We first translate this into mathematics. Here, we let n be the number that we want. Also, the word *is* always means =.

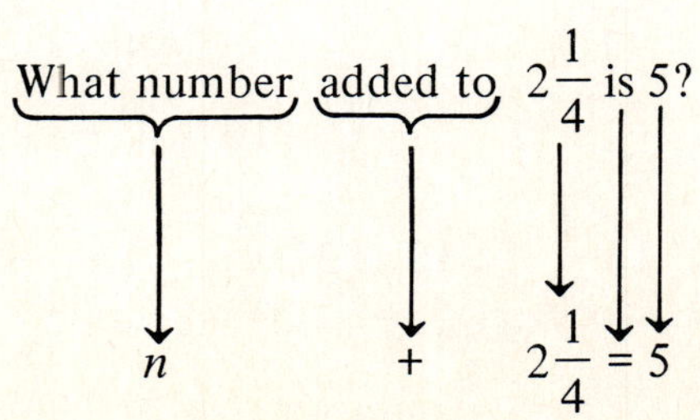

Now we find n by subtracting $2\frac{1}{4}$ from both sides.

$$n + 2\frac{1}{4} = 5 \quad \text{means} \quad n = 5 - 2\frac{1}{4}$$

$$= 4\frac{4}{4} - 2\frac{1}{4}$$

$$= 2\frac{3}{4}$$

PROBLEM SET Find n or x in the following problems.

See Examples 77 to 79

1. $n + \frac{1}{5} = \frac{4}{5}$
2. $x + \frac{3}{11} = \frac{8}{11}$
3. $n + \frac{1}{2} = \frac{3}{4}$
4. $n + \frac{1}{3} = \frac{5}{6}$
5. $x + \frac{2}{5} = \frac{7}{10}$
6. $n + \frac{1}{8} = \frac{3}{4}$
7. $n + 1\frac{1}{2} = 2\frac{2}{3}$
8. $x + 2\frac{1}{5} = 7$

ANSWERS 1. $\frac{3}{5}$ 2. $\frac{5}{11}$ 3. $\frac{1}{4}$ 4. $\frac{1}{2}$ 5. $\frac{3}{10}$ 6. $\frac{5}{8}$ 7. $1\frac{1}{6}$ 8. $4\frac{4}{5}$

We now want to determine which of two (or more) fractions is the larger (largest). For instance, $\frac{5}{7}$ is larger than $\frac{3}{7}$, since $\frac{5}{7}$ has more sevenths. (We simply compare the numerators.)

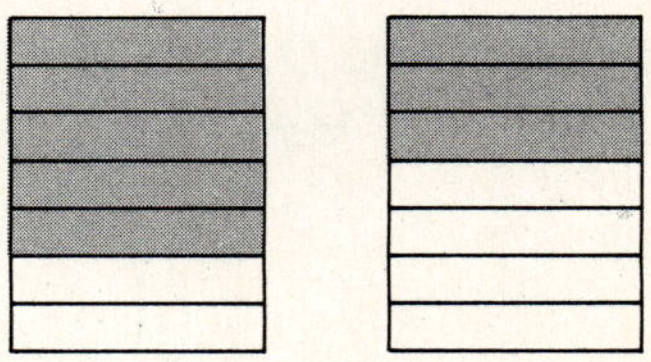

$\frac{5}{7}$ is larger than $\frac{3}{7}$

Now consider $\frac{2}{3}$ and $\frac{3}{4}$. As they are now, they are hard to compare. However, let us rewrite them with their LCD, 12.

$$\frac{2}{3} \rightarrow \frac{8}{12} \qquad \frac{3}{4} \rightarrow \frac{9}{12} \leftarrow \boxed{\text{larger}}$$

Now it is easy to see that $\frac{3}{4} = \frac{9}{12}$ is the larger fraction.

> To compare two (or more) fractions:
> 1. Rewrite all the fractions with their LCD.
> 2. Compare the numerators.

EXAMPLE 80 Which fraction is larger, $\frac{5}{7}$ or $\frac{7}{10}$?

SOLUTION The LCD is 70, so we rewrite both fractions with a common denominator.

$$\boxed{\text{larger}} \rightarrow \frac{50}{70} \leftarrow \frac{5}{7} \qquad \frac{7}{10} \rightarrow \frac{49}{70}$$

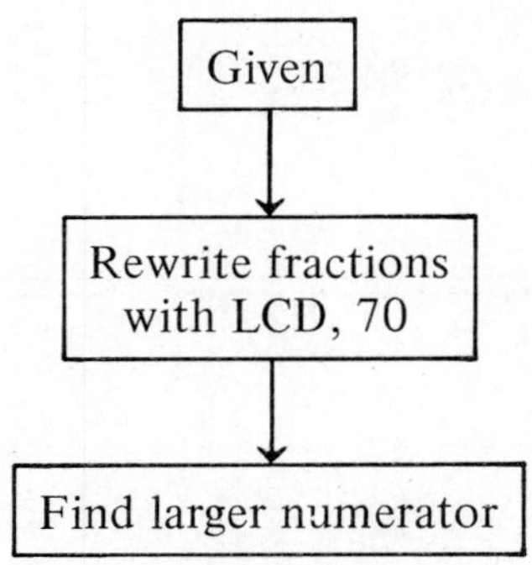

EXAMPLE 81 Which fraction is the largest, $\frac{7}{12}, \frac{9}{16}, \frac{5}{8}$, or $\frac{2}{3}$?

SOLUTION Let us first find the LCD by factoring the denominators.

$$\begin{aligned}
12 &= 2 \cdot 2 \cdot 3 \\
16 &= 2 \cdot 2 \cdot \phantom{3 \cdot{}} 2 \cdot 2 \\
8 &= 2 \cdot 2 \cdot \phantom{3 \cdot{}} 2 \\
3 &= \phantom{2 \cdot 2 \cdot{}} 3 \\
\text{LCD} &= 2 \cdot 2 \cdot 3 \cdot 2 \cdot 2 = 48
\end{aligned}$$

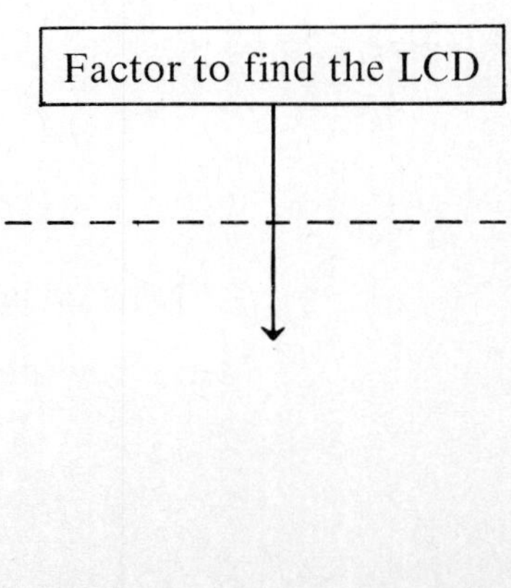

$$\frac{7}{12} \quad \frac{9}{16} \quad \frac{5}{8} \quad \frac{2}{3}$$

$$\downarrow \quad \downarrow \quad \downarrow \quad \downarrow$$

$$\frac{28}{48} \quad \frac{27}{48} \quad \frac{30}{48} \quad \frac{32}{48}$$

Rewrite all fractions with LCD, 48

largest ($\frac{32}{48}$)

Compare numerators

Thus $\frac{2}{3} = \frac{32}{48}$ is the largest. In fact, we can put the fractions in order (largest to smallest) by comparing the numerators of the equivalent fractions:

$$\frac{2}{3}, \frac{5}{8}, \frac{7}{12}, \frac{9}{16}$$

PROBLEM SET For each set of fractions, find the larger.

See Examples 80 and 81

1. $\frac{7}{11}, \frac{3}{11}$ 2. $\frac{4}{9}, \frac{8}{9}$ 3. $\frac{1}{2}, \frac{2}{3}$

4. $\frac{1}{3}, \frac{1}{4}$ 5. $\frac{5}{6}, \frac{9}{10}$ 6. $\frac{5}{12}, \frac{2}{5}$

For each of the following sets of fractions, put them in order (largest to smallest).

7. $\frac{1}{5}, \frac{2}{15}, \frac{2}{9}$ 8. $\frac{5}{18}, \frac{1}{4}, \frac{1}{3}$ 9. $\frac{1}{2}, \frac{2}{5}, \frac{3}{7}$

10. $\frac{1}{3}, \frac{3}{10}, \frac{3}{8}, \frac{5}{12}$

ANSWERS 1. $\frac{7}{11}$ 2. $\frac{8}{9}$ 3. $\frac{2}{3}$ 4. $\frac{1}{3}$ 5. $\frac{9}{10}$ 6. $\frac{5}{12}$ 7. $\frac{2}{9}, \frac{1}{5}, \frac{2}{15}$

8. $\frac{1}{3}, \frac{5}{18}, \frac{1}{4}$ 9. $\frac{1}{2}, \frac{3}{7}, \frac{2}{5}$ 10. $\frac{5}{12}, \frac{3}{8}, \frac{1}{3}, \frac{3}{10}$

Miscellaneous Exercises 2.6

Subtract the following fractions and mixed numbers.

See Example 67

1. $\frac{4}{9} - \frac{1}{9}$ 2. $\frac{6}{5} - \frac{4}{5}$ 3. $\frac{7}{10} - \frac{3}{10}$

See Examples 68 to 71

4. $\frac{1}{2} - \frac{1}{3}$ 5. $\frac{3}{4} - \frac{1}{8}$ 6. $\frac{8}{15} - \frac{3}{10}$

7. $\frac{5}{12} - \frac{1}{18}$ 8. $\frac{7}{8} - \frac{5}{12}$ 9. $\frac{9}{20} - \frac{7}{24}$

See Examples 72 to 76

10. $5\frac{6}{7} - 1\frac{2}{7}$ **11.** $4\frac{2}{3} - 3\frac{1}{6}$ **12.** $16\frac{7}{12} - 9\frac{1}{8}$

13. $6\frac{1}{3} - 2\frac{5}{6}$ **14.** $7\frac{2}{9} - 1\frac{8}{15}$ **15.** $28\frac{1}{10} - 17\frac{11}{15}$

See Examples 77 to 79

16. Find n: $n + \frac{5}{7} = \frac{6}{7}$.

17. Find n: $n + \frac{1}{8} = \frac{3}{4}$.

18. What number added to $1\frac{1}{2}$ is $3\frac{1}{4}$?

For each set of fractions, choose the larger (or largest).

See Examples 80 and 81

19. $\frac{4}{9}, \frac{5}{12}$ **20.** $\frac{1}{2}, \frac{5}{8}, \frac{7}{12}$

Business applications

21. Health-X Drugs opened at the stock exchange at $37\frac{7}{8}$. That night it closed at $40\frac{1}{4}$. How much of an increase was this?

22. An advertising campaign has reached $\frac{4}{15}$ of the market. What fraction of the market has it not reached?

Health applications

23. A patient is prescribed 8 liters of a liquid medicine to be taken intravenously. So far, he has had $5\frac{3}{4}$ liters. How many more liters of medicine will he take?

24. A patient is prescribed $\frac{1}{40}$ grain of atrofine sulfate. She takes one $\frac{1}{200}$-grain tablet. How much more medicine does she need?

Technical application

25. An electrician has an $11\frac{1}{4}$-inch conduit. The size that is needed is $10\frac{7}{8}$. How much must be cut off the end of the longer conduit?

Consumer applications

26. Michele weighs $148\frac{1}{2}$ pounds. After a careful diet, she reduces her weight to $123\frac{7}{8}$. How much weight did she lose?

27. Tony's Volkswagen holds 10 gallons of gasoline. Starting with a full tank, he used $6\frac{3}{4}$ gallons driving to visit a friend. How much gasoline is left in the tank?

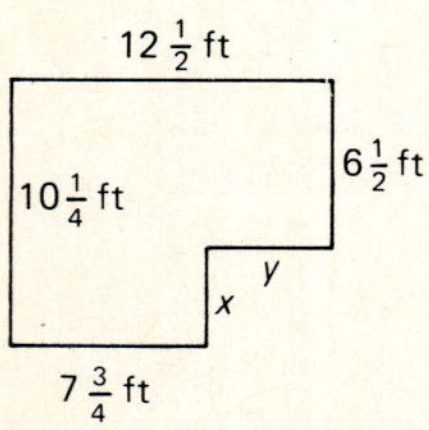

28. Kathy is decorating a room like the one sketched at the left. Use the dimensions given to find the missing dimensions, x and y.

29. Bob buys $1\frac{1}{2}$ yards of material. He uses $1\frac{1}{8}$ yards to make a shirt. How much material does he have left for a bandana?

Now complete Post-Test 1 for Section 2.6 on page T33.

CHAPTER 2 SUMMARY

Important Words and Phrases

adding (2.5)
borrowing (2.6)
canceling (2.3)
carrying (2.5)
common denominator (2.5)
denominator (2.1)
division (2.4)
equal fractions (2.1)
fraction (2.1)
improper fraction (2.2)
invert and multiply (2.4)
least common denominator (LCD) (2.5)
mixed number (2.2)
multiplication (2.3)
numerator (2.1)
reduced to simplest terms (2.1)
subtracting (2.6)

Important Ideas

(2.1) Determining equal fractions.
Reducing fractions to simplest terms.

(2.2) Writing mixed numbers as improper fractions.
Writing improper fractions as mixed numbers.

(2.3) Multiplying fractions and mixed numbers.
Knowing that multiplication is the opposite of division.

(2.4) Dividing fractions and mixed numbers.
Knowing that division is the opposite of multiplication.

(2.5) Adding fractions and mixed numbers.
Knowing that addition is the opposite of subtraction.

(2.6) Subtracting fractions and mixed numbers.
Knowing that subtraction is the opposite of addition.
Comparing the size of two or more fractions.

CHAPTER 2 REVIEW TEST

Choose the correct answer.

1. Which of the following fractions is equal to $\frac{10}{15}$?

   (a) $\frac{9}{14}$ (b) $\frac{20}{25}$ (c) $\frac{15}{22}$ (d) $\frac{21}{30}$ (e) $\frac{12}{18}$

2. Reduce $\frac{85}{95}$ to simplest terms.

   (a) $\frac{8}{9}$ (b) $\frac{15}{17}$ (c) $\frac{18}{19}$ (d) $\frac{13}{14}$ (e) $\frac{17}{19}$

3. Reduce $\frac{120}{144}$ to simplest terms.

(a) $\frac{6}{7}$ (b) $\frac{12}{14}$ (c) $\frac{5}{6}$ (d) $\frac{10}{12}$ (e) $\frac{5}{7}$

4. Write $7\frac{1}{3}$ as an improper fraction.

(a) $\frac{21}{3}$ (b) $\frac{22}{3}$ (c) $\frac{11}{3}$ (d) $\frac{22}{7}$ (e) $\frac{11}{7}$

5. Write $14\frac{3}{10}$ as an improper fraction.

(a) $\frac{140}{10}$ (b) $\frac{27}{10}$ (c) $\frac{143}{10}$ (d) $\frac{143}{14}$ (e) $\frac{27}{14}$

6. Write $\frac{42}{5}$ as a mixed number.

(a) $4\frac{2}{5}$ (b) $8\frac{2}{5}$ (c) $5\frac{2}{5}$ (d) $\frac{5}{42}$ (e) $4\frac{5}{42}$

7. Write $\frac{117}{21}$ as a mixed number.

(a) $1\frac{21}{117}$ (b) $\frac{21}{117}$ (c) $11\frac{7}{21}$ (d) $1\frac{17}{21}$ (e) $5\frac{12}{21}$

8. Multiply $\frac{3}{7} \times \frac{4}{11}$.

(a) $\frac{7}{18}$ (b) $\frac{12}{77}$ (c) $\frac{33}{28}$ (d) $\frac{28}{33}$ (e) $\frac{7}{77}$

9. Multiply $\frac{1}{2} \times \frac{1}{3} \times 2\frac{2}{5}$.

(a) $\frac{14}{10}$ (b) $\frac{14}{30}$ (c) $2\frac{1}{15}$ (d) $1\frac{1}{5}$ (e) $\frac{2}{5}$

10. Multiply $4\frac{1}{2} \times 5\frac{1}{3}$.

(a) $20\frac{2}{5}$ (b) $9\frac{1}{6}$ (c) $20\frac{1}{6}$ (d) 24 (e) 21

11. Divide $\frac{4}{7} \div \frac{9}{11}$.

(a) $\frac{63}{44}$ (b) $\frac{44}{63}$ (c) $\frac{77}{36}$ (d) $\frac{36}{77}$ (e) $\frac{28}{99}$

12. Divide $12\frac{1}{2} \div 3\frac{3}{4}$.

(a) $4\frac{3}{8}$ (b) $3\frac{1}{3}$ (c) $\frac{3}{10}$ (d) $4\frac{2}{3}$ (e) $46\frac{7}{8}$

13. Divide $\frac{2}{5} \div 10$.

(a) $\frac{1}{25}$ (b) $\frac{1}{50}$ (c) 25 (d) 4 (e) $\frac{1}{4}$

14. Add $\frac{1}{12} + \frac{4}{15}$.

(a) $\frac{5}{60}$ (b) $\frac{5}{27}$ (c) $\frac{5}{180}$ (d) $\frac{7}{60}$ (e) $\frac{7}{20}$

15. Add $4\frac{1}{2} + 7\frac{3}{4}$.

(a) $11\frac{7}{8}$ (b) $11\frac{4}{6}$ (c) $11\frac{3}{8}$ (d) $11\frac{1}{4}$ (e) $12\frac{1}{4}$

16. Add $\frac{1}{12} + \frac{1}{3} + 2\frac{5}{6}$.

(a) $2\frac{7}{21}$ (b) $2\frac{11}{12}$ (c) $2\frac{31}{36}$ (d) $3\frac{1}{4}$ (e) $3\frac{1}{12}$

17. Subtract $\frac{7}{25} - \frac{3}{20}$.

(a) $\frac{4}{5}$ (b) $\frac{1}{25}$ (c) $\frac{41}{100}$ (d) $\frac{4}{100}$ (e) $\frac{13}{100}$

18. Subtract $8\frac{1}{12} - 4\frac{5}{8}$.

(a) $3\frac{11}{24}$ (b) $4\frac{11}{24}$ (c) $4\frac{4}{96}$ (d) $4\frac{1}{96}$ (e) $3\frac{23}{24}$

19. Subtract $20 - 5\frac{1}{6}$.

(a) $15\frac{5}{6}$ (b) $15\frac{1}{6}$ (c) $25\frac{1}{6}$ (d) $14\frac{1}{6}$ (e) $14\frac{5}{6}$

20. Find n: $n + \frac{1}{4} = 8$.

(a) $7\frac{3}{4}$ (b) $8\frac{1}{4}$ (c) $8\frac{3}{4}$ (d) 2 (e) 32

Now complete Post-Test 1 for Chapter 2 on page T35.

3

DECIMALS

Before you begin this chapter, complete the Chapter 3 Pre-Test on page T39.

3.1 DECIMAL PLACE VALUE AND ROUNDING OFF

SECTION OBJECTIVES After completing this unit, you will be able to:

1. Determine the place values in a decimal number.
2. Round a decimal number to any accuracy.
3. Compare the sizes of two or more decimals.

TYPICAL PROBLEMS You will be able to work problems like the ones below.

(a) In 237.892, what is the value of the 9?
(b) In 3.14159, what is the value of the 5?
(c) Round 27.173 to the nearest tenth.
(d) Round 108.1753 to three decimal places.
(e) Round 45,185.23 to the nearest hundred.
(f) Determine the larger decimal: 0.02, 0.019.
(g) Put into order (smallest to largest): 0.3, 0.33, 0.03, 0.003.

ANSWERS (a) $\frac{9}{100}$ (b) $\frac{5}{10,000}$ (c) 27.2 (d) 108.175 (e) 45,200 (f) 0.02
(g) 0.003, 0.03, 0.3, 0.33

For thousands of years, fractions were the only way to express a part of a whole. Then, about A.D. 1200 Arab and Jewish traders brought a new numbering system to Europe. This system used **decimals**.

This system is just like the usual base-10 number system except that it includes fractions. Each column has its own value. The **decimal point** is used as a reference point to separate the whole numbers (on the left) from the fractions (on the right).

Figure 3.1 shows the column values. To the left of the decimal point are the usual whole-number place values: 1, 10, 100, 1000, and so on. To the right of the decimal point are the **fractional place values:** $\frac{1}{10}$, $\frac{1}{100}$, $\frac{1}{1000}$, $\frac{1}{10,000}$, and so on.

Ten-thousands	Thousands	Hundreds	Tens	Ones	Decimal Point	Tenths	Hundredths	Thousandths	Ten-thousandths	Hundred-thousandths	Millionths
10000	1000	100	10	1	.	$\frac{1}{10}$	$\frac{1}{100}$	$\frac{1}{1000}$	$\frac{1}{10000}$	$\frac{1}{100{,}000}$	$\frac{1}{1{,}000{,}000}$
	2	7	8	3	.	6	5	1	9		

Figure 3.1

Decimal System

As with whole numbers, each digit has a place value. For example, in the number shown the 6 has a value of $\frac{6}{10}$, the 5 has a value of $\frac{5}{100}$, and so on.

This number, 2783.6519, can be written out in an expanded form.

$$\begin{aligned}
2783.6519 = \quad & 2 \times 1000 &&= 2000 \\
&+ 7 \times 100 &&= 700 \\
&+ 8 \times 10 &&= 80 \\
&+ 3 \times 1 &&= 3 \\
&+ 6 \times \frac{1}{10} &&= \frac{6000}{10{,}000} \\
&+ 5 \times \frac{1}{100} &&= \frac{500}{10{,}000} \\
&+ 1 \times \frac{1}{1000} &&= \frac{10}{10{,}000} \\
&+ 9 \times \frac{1}{10{,}000} &&= \frac{9}{10{,}000} \\
& && \overline{2783\frac{6519}{10{,}000}}
\end{aligned}$$

> To determine the value of a digit in a decimal number:
>
> 1. Determine the column values starting at the decimal.
> 2. Multiply the digit by the column value.

EXAMPLE 1 In 34.8721, what is the value of the 2?

SOLUTION We first determine the column values.

10	1	$\frac{1}{10}$	$\frac{1}{100}$	$\frac{1}{1000}$	$\frac{1}{10{,}000}$
3	4 .	8	7	2 ↑	1

Determine column values ↓

$$\text{value of } 2 = 2 \times \frac{1}{1000} = \frac{2}{1000}$$

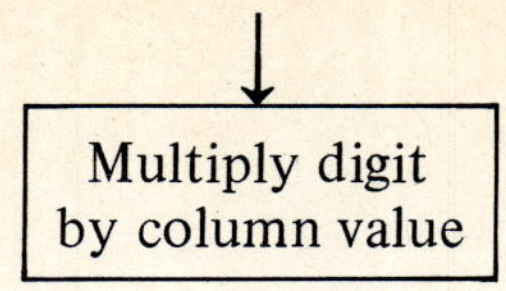

EXAMPLE 2 In 2.9082, what is the value of the 9?

SOLUTION We first determine the column values.

1	$\frac{1}{10}$	$\frac{1}{100}$	$\frac{1}{1000}$	$\frac{1}{10,000}$
2	9	0	8	2

$$\text{value of } 9 = 9 \times \frac{1}{10} = \frac{9}{10}$$

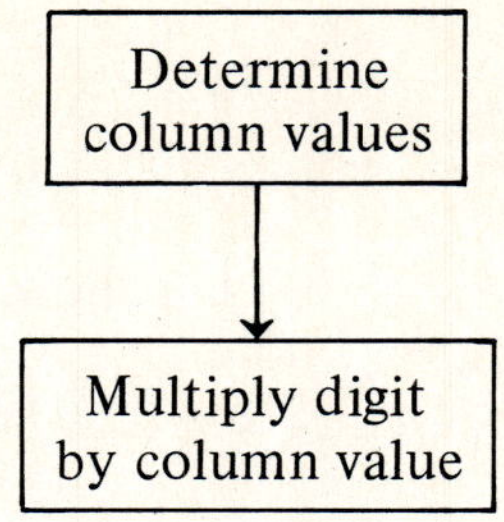

EXAMPLE 3 In 30.852, what is the value of the 5?

SOLUTION We first determine the column values.

10	1	$\frac{1}{10}$	$\frac{1}{100}$	$\frac{1}{1000}$
3	0	8	5	2

$$\text{value of } 5 = 5 \times \frac{1}{100} = \frac{5}{100}$$

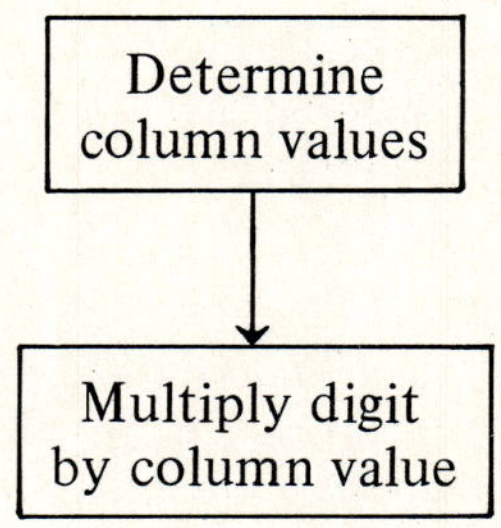

PROBLEM SET

See Examples 1 to 3

1. In 23.759, what is the value of the 7?
2. In 48.9021, what is the value of the 2?
3. In 1.41456, what is the value of the 5?
4. In 467.92, what is the value of the 9?
5. In 9.08321, what is the value of the 2?
6. In 2950.36, what is the value of the 6?
7. In 1.234567, what is the value of the 7?
8. In 360.74832, what is the value of the 8?
9. In 4209.39275, what is the value of the 7?
10. In 10.23965, what is the value of the 2?

ANSWERS 1. $\frac{7}{10}$ 2. $\frac{2}{1000}$ 3. $\frac{5}{10{,}000}$ 4. $\frac{9}{10}$ 5. $\frac{2}{10{,}000}$ 6. $\frac{6}{100}$
7. $\frac{1}{1{,}000{,}000}$ 8. $\frac{8}{1000}$ 9. $\frac{7}{10{,}000}$ 10. $\frac{2}{10}$

Sometimes, we have a decimal that has more digits than we need. When we simplify, we **round off.** For instance, 3.14159 is between two simpler numbers:

3.1
3.14159
3.2

Since our number is closer to 3.10000, so we round 3.14159 to 3.1. (We have already studied the rounding of whole numbers. See page 8.)

> To round off decimal places:
>
> **1.** Determine what place is to be rounded.
>
> **2.** Go one extra place to the right.
>
> **3a.** If this digit is 4 or less (0, 1, 2, 3 or 4), drop this digit and all others to the right. (Round down.)
>
> **3b.** If this digit is 5 or more (5, 6, 7, 8 or 9), make the rounded place one more, and drop all digits to the right. (Round up.)

EXAMPLE 4 Round 85.07276 to two decimal places.

SOLUTION

85.07276 (↑ under the 7 in the second decimal place)	Determine place to be rounded
85.07 [2] 76 (↑)	↓ Go one place right
85.07	↓ Round down since digit is 2

EXAMPLE 5 Round 17.01392 to the nearest thousandth.

SOLUTION The thousandth's place is the third decimal place.

17.01392 (↑ under the 3 in the third decimal place)	Determine place to be rounded
17.013 [9] 2 (↑)	↓ Go one place right
17.014	↓ Round up since digit is 9

EXAMPLE 6 Round 1.95046 to the nearest tenth.

SOLUTION The tenth's place is the first decimal place.

1.95046
↑

1.9 [5] 046
↑

2.0

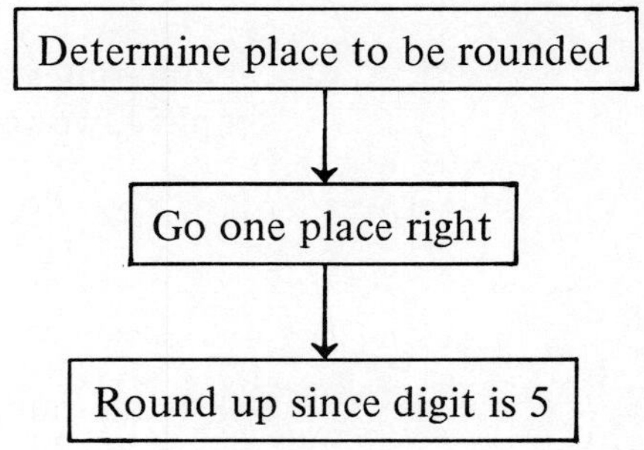

When we round 1.9 up, we get 2.0.

PROBLEM SET

See Examples 4 to 6

1. Round 2.753 to two decimal places.
2. Round 43.7582 to one decimal place.
3. Round 47.90327 to three decimal places.
4. Round 3.27451 to two decimal places.
5. Round 4.18653 to one decimal place.
6. Round 567.89012 to four decimal places.
7. Round 567.83425 to three decimal places.
8. Round 45.938 to the nearest tenth.
9. Round 123.4567 to the nearest thousandth.
10. Round 38.29048 to the nearest hundredth.
11. Round 8438.86734 to the nearest ten.
12. Round 26.36211 to the nearest one.
13. Round 1574.5901 to the nearest hundred.
14. Round 406,947.329747 to the nearest thousand.

ANSWERS **1.** 2.75 **2.** 43.8 **3.** 47.903 **4.** 3.27 **5.** 4.2 **6.** 567.8901 **7.** 567.834 **8.** 45.9 **9.** 123.457 **10.** 38.29 **11.** 8440 **12.** 26 **13.** 1600 **14.** 407,000

When working with decimals, it is often useful to have all the decimals in a particular problem have the same number of decimal places. Fortunately, this is easy to do by simply attaching zeros on the right.

> Any decimal number can have extra zeros inserted on the right.

To write a whole number as a decimal:
1. Put a decimal point to the right of the number.
2. Insert as many zeros to the right as needed.

EXAMPLE 7 (a) $2.6 = 2.60 = 2.600 = 2.6000 = \cdots$
(b) $28.43 = 28.430 = 28.4300 = \cdots$
(c) $36 = 36.0 = 36.00 = 36.000 = \cdots$
(d) $500 = 500.0 = 500.00 = 500.000 = \cdots$

We can use this to help us compare the sizes of decimals. For instance, consider 0.093 and 0.11. Let us rewrite them with the same number of decimal places. This is like finding the LCD.

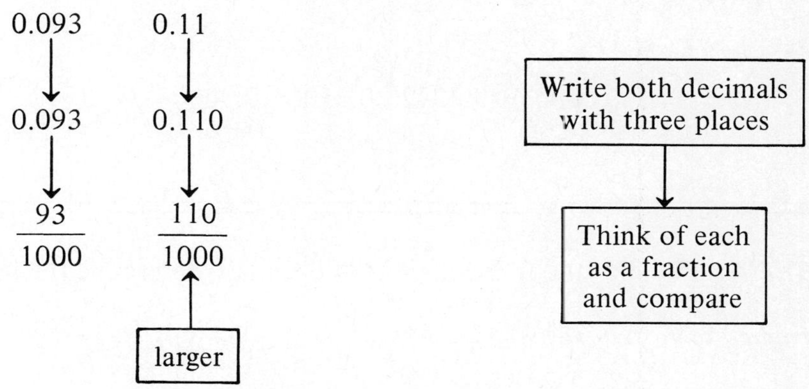

We do not have to write the decimals as fractions. By writing them with the same number of places, we can compare the digits (as if the decimal point were not there).

To compare two (or more) decimals:
1. If necessary, fill in zeros so that all the decimals have the same number of places.
2. Temporarily ignore the decimal point.
3. Compare as you would compare whole numbers.

EXAMPLE 8 Which decimal is larger, 0.5 or 0.444?

SOLUTION We write both of these with three places and then compare.

EXAMPLE 9 Write the following decimals in order (smallest to largest): 0.25, 0.3, 0.088, 0.2242.

SOLUTION We write all these decimals with the same number of places, four.

0.25	0.3	0.088	0.2242
↓	↓	↓	↓
0.2500	0.3000	0.0880	0.2242
↑	↑	↑	↑
(3)	(4)	(1)	(2)

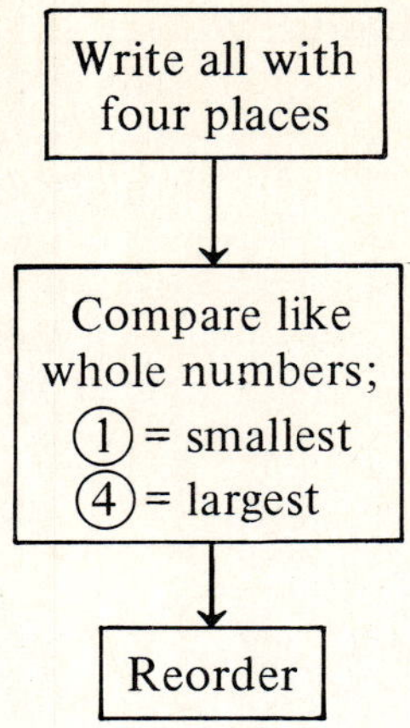

0.0880 (smallest) 0.2242 0.2500 0.3000 (largest)

PROBLEM SET For the following sets of decimals, determine the larger.

See Examples 7 to 9

1. 0.7, 0.66
2. 0.03, 0.004
3. 1.2, 1.09
4. 2.02, 2.007
5. 0.001, 0.0012
6. 0.041, 0.0288

Put the following sets of decimals into order (smallest to largest).

7. 0.5, 0.529, 0.4831
8. 2.1, 2.09, 2.111
9. 0.8, 0.88, 0.888
10. 0.1, 0.01, 0.001, 0.0001

ANSWERS 1. 0.7 2. 0.03 3. 1.2 4. 2.02 5. 0.0012 6. 0.041 7. 0.4831, 0.5, 0.529 8. 2.09, 2.1, 2.111 9. 0.8, 0.88, 0.888 10. 0.0001, 0.001, 0.01, 0.1

Miscellaneous Exercises 3.1

Determine the values of the indicated digits.

See Examples 1 to 3

1. In 34.928, what is the value of the 2?
2. In 7.49201, what is the value of the 2?
3. In 39.1852, what is the value of the 1?
4. In 869.03815, what is the value of the 5?

Round the following numbers to the indicated accuracy.

See Examples 4 to 6

5. Round 4.9372 to a one-place decimal.
6. Round 76.03957 to a two-place decimal.

7. Round 1.38592 to a three-place decimal.

8. Round 47.93752 to the nearest tenth.

9. Round 5.029456 to the nearest thousandth.

10. Round 579.0846523 to the nearest hundredth.

11. Round 4673.8 to the nearest hundred.

12. Round 23,694.73 to the nearest ten thousand.

For the following sets of decimals, determine the largest.

See Examples 8 and 9

13. 0.67, 0.666

14. 3.15, 3.14159

15. 0.02, 0.2, 0.0002

16. 0.9, 0.99, 0.999, 0.9999, 1

Business application

17. A company's gross revenue for a year was $12,732,934.92. Round this number to the nearest hundred thousand dollars.

Technical application

18. A pipe has an inside diameter of 3.067 inches. Round this to the nearest tenth of an inch.

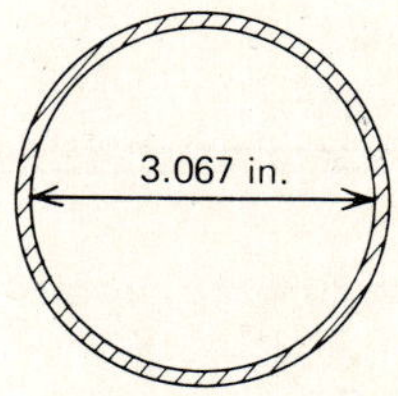

Health application

19. A nurse calculates that a patient needs 0.5555 milliliter of atropine sulfate. Round this to the nearest hundredth of a milliliter.

Consumer application

20. In Section 1.1, we saw how to write check amounts that were whole numbers. Most amounts, however, are two-place decimals such as $438.19. This would be written "four hundred thirty-eight and $\frac{19}{100}$." Fill in the following imaginary checks.

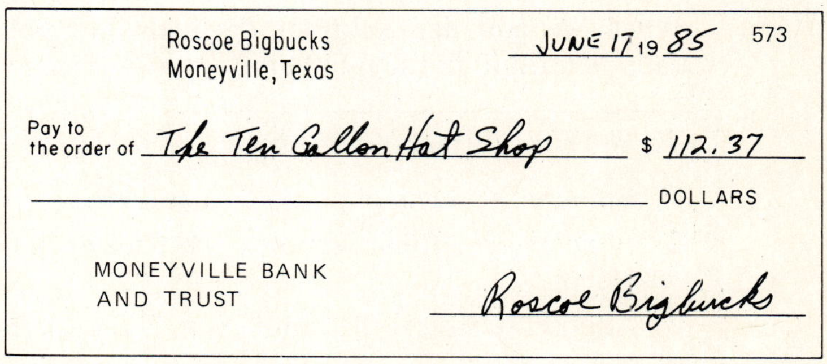

Wendy Wastrel
Spenditall, Ohio

689

Nov 2 19 83

Pay to the order of Nonsense Shop $ 87.42

DOLLARS

SPENDITALL
BANK AND TRUST

Wendy Wastrel

Now complete Post-Test 1 for Section for 3.1 on page T41.

3.2 ADDITION AND SUBTRACTION OF DECIMALS

SECTION OBJECTIVES

After completing this unit, you will be able to:

1. Add decimals.
2. Subtract decimals.

TYPICAL PROBLEMS

You will be able to work problems like the ones below.

(a) Add 4.56 + 39.2.

(b) Add 12.035 + 43 + 0.3781.

(c) Subtract 26.023 − 12.89.

(d) Subtract 500 − 346.73.

(e) Find n: $n - 12.1 = 24.8$.

(f) Find n: $n + 32.5 = 76.89$.

ANSWERS

(a) 43.76 **(b)** 55.4131 **(c)** 13.133 **(d)** 153.27 **(e)** 36.9 **(f)** 44.39

When we **add and subtract decimals**, we line up the decimal points. It is usually helpful if all the numbers contain the same number of decimal places.

> To add (or subtract) decimals:
> 1. Line up the decimal points in a column.
> 2. If necessary, fill in zeros so that all decimals have the same number of places.
> 3. Add (or subtract) the numbers as if they were whole numbers.
> 4. The decimal point for the answer is just below the other decimal points.

EXAMPLE 10 Bob is a big spender and is now up to his ears in bills. Today he is going to write checks for \$125, \$23.69, \$12.38, \$20, and \$89.47. What is the total?

SOLUTION The word *total* almost always means add. Let us add these figures, filling in zeros so that they all have two places.

$$
\begin{array}{r}
\$125.00 \\
23.69 \\
12.38 \\
20.00 \\
+\ \ 89.47 \\
\hline
\$270.54
\end{array}
$$

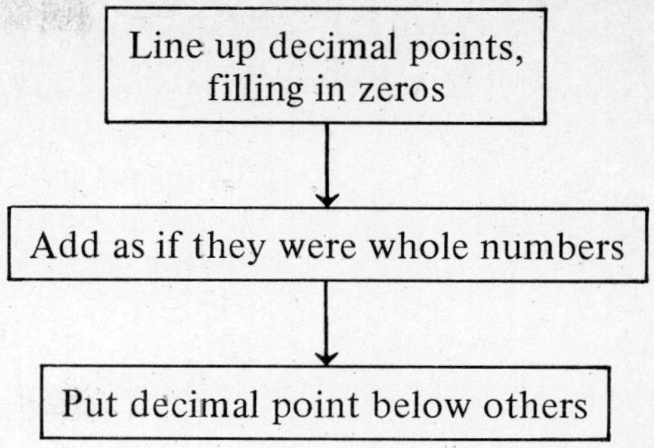

EXAMPLE 11 Add 1.6001 + 21 + 9.04 + 8.731.

SOLUTION We line up the decimal points and fill in the zeros.

$$
\begin{array}{r}
1.6001 \\
21.0000 \\
9.0400 \\
+\ 8.7310 \\
\hline
40.3711
\end{array}
$$

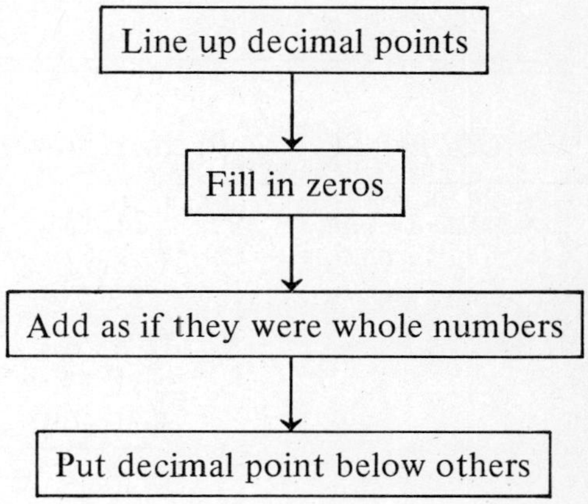

EXAMPLE 12 Find the perimeter of the triangle shown.

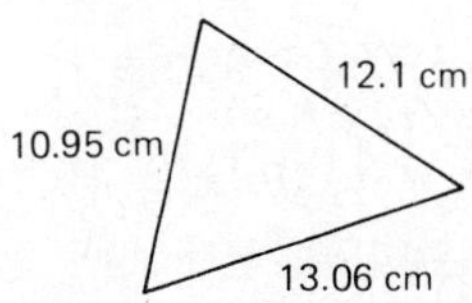

SOLUTION The perimeter of a triangle is the sum of all the lengths of the sides.

$$
\begin{array}{r}
10.95\text{ cm} \\
12.10\text{ cm} \\
+13.06\text{ cm} \\
\hline
36.11\text{ cm}
\end{array}
$$

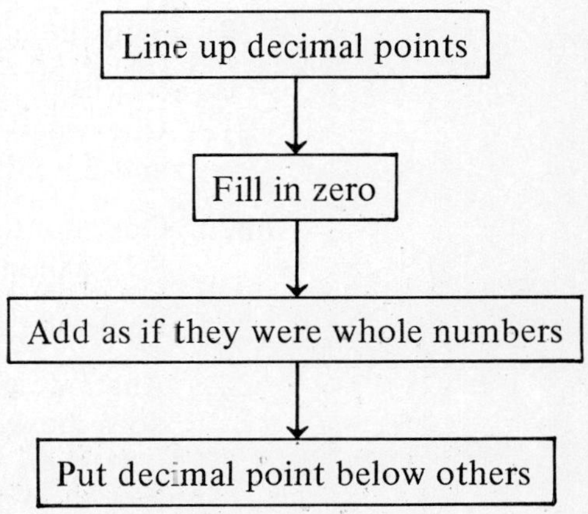

Why Does It Work?

When we add decimals, we are really adding fractions. And when we line up decimal points and fill in zeros, we are really finding the common denominator. Consider 2.3 + 3.17 + 1.214.

$$\begin{array}{rcl} 2.3 & = 2\dfrac{3}{10} & = 2\dfrac{300}{1000} \\ 3.17 & = 3\dfrac{17}{100} & = 3\dfrac{170}{1000} \\ +1.214 & = 1\dfrac{214}{1000} & = 1\dfrac{214}{1000} \\ \hline 6.684 & & 6\dfrac{684}{1000} \end{array}$$

PROBLEM SET Add the following decimals.

See Examples 10 to 12

1. $\begin{array}{r} 28.83 \\ +\ 8.82 \\ \hline \end{array}$

2. $\begin{array}{r} 4.311 \\ +27.41 \\ \hline \end{array}$

3. $\begin{array}{r} 20.2 \\ +\ 6.904 \\ \hline \end{array}$

4. $\begin{array}{r} 72.423 \\ +115.7 \\ \hline \end{array}$

5. $\begin{array}{r} 5.2692 \\ 51.73 \\ +\ 1.169 \\ \hline \end{array}$

6. $\begin{array}{r} 97.766 \\ 630.6235 \\ +\ 78.346 \\ \hline \end{array}$

7. $\begin{array}{r} 431.17 \\ 822.154 \\ +\ 94.74 \\ \hline \end{array}$

8. $\begin{array}{r} 3.9713 \\ 1.46055 \\ +3.335 \\ \hline \end{array}$

9. $\begin{array}{r} 4.83 \\ 4.524 \\ 3.4 \\ +9.328 \\ \hline \end{array}$

10. $\begin{array}{r} 42.56 \\ 37.89 \\ 72.72 \\ +159.61 \\ \hline \end{array}$

11. $\begin{array}{r} 691.96 \\ 49.18 \\ 238.87 \\ +\ 92.91 \\ \hline \end{array}$

12. $\begin{array}{r} 6603 \\ 208.03 \\ 3808.4 \\ +\ 698.87 \\ \hline \end{array}$

13. Add 6.164 + 31.58.

14. Add 577.9 + 15.89 + 7.

15. Add 75.41 + 3.767 + 47.334.

16. Add 8.918 + 41.39 + 0.199.

17. Add 5440 + 5955.67 + 6771.7 + 7026.

18. Add 12.28 + 5.37 + 2.158 + 811.

19. Larry writes checks for \$19.73, \$5.68, \$27.15, \$51.66, and \$108.09. How much is the total?

20. Sally buys the following items at the store: milk, \$1.59; cookies, \$0.83; bread, \$0.67; orange juice, \$0.65; steak, \$2.78. What is the total?

21. One week, Howie runs the following distances: 1.5, 2.35, 0.85, 1.8, 1.15, and 2.9 miles. What is his total for the week?

22. One month, Lynn records her gasoline purchases: 12.1, 13.7, 9, 12.25, and 8.15 gallons. How much gasoline did she buy during the month?

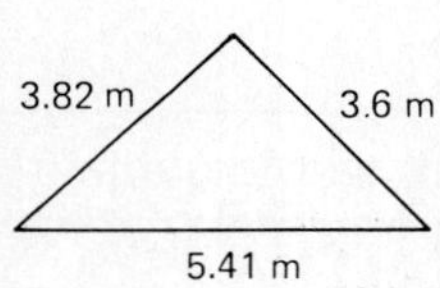

23. Find the perimeter of the triangle at the left.

24. Angie weighs 52.3 kilograms. Her clothes weigh 1.95 kilograms. How much does she weigh fully dressed?

25. Bob needs 0.9 gallon of paint for the kitchen, 1.3 for his bedroom, 1.15 for his child's room, 1.65 for the living room, and 0.85 for the dining room. How much paint does he need?

ANSWERS **1.** 37.65 **2.** 31.721 **3.** 27.104 **4.** 188.123 **5.** 58.1682
6. 806.7355 **7.** 1348.064 **8.** 8.76685 **9.** 22.082 **10.** 312.78
11. 1072.92 **12.** 11,318.3 **13.** 37.744 **14.** 600.79
15. 126.511 **16.** 50.507 **17.** 25,193.37 **18.** 830.808
19. 212.31 **20.** 6.52 **21.** 10.55 **22.** 55.2 **23.** 12.83
24. 54.25 **25.** 5.85

The following **subtraction problems** use the same rule used for addition: *line up the decimal points.* As with addition, we are really finding the common denominator when we subtract like this.

EXAMPLE 13 Subtract 10.02 - 7.593.

SOLUTION

$$\begin{array}{r} 10.020 \\ -\ \ 7.593 \\ \hline 2.427 \end{array}$$

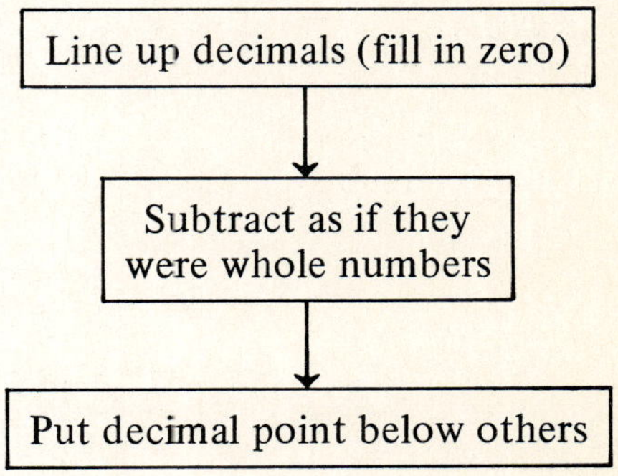

EXAMPLE 14 Subtract 0.00635 - 0.0009.

SOLUTION

$$\begin{array}{r} 0.00635 \\ -0.00090 \\ \hline 0.00545 \end{array}$$

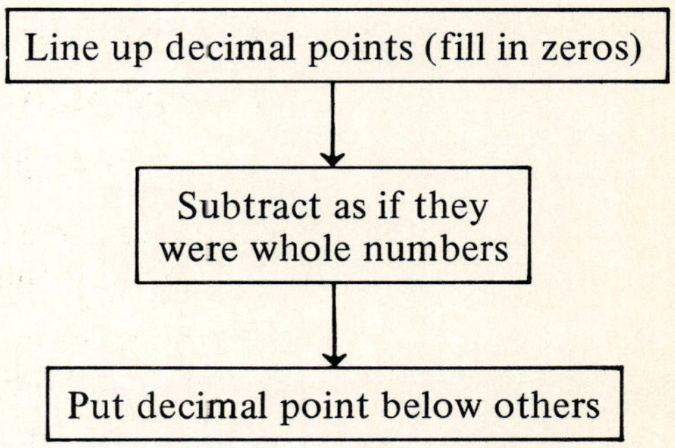

EXAMPLE 15 Subtract 165 - 21.59.

SOLUTION To work the problem, the number 165 is rewritten as 165.00. (All whole numbers have an unwritten decimal point to the right of the ones' column.)

$$\begin{array}{r} 165.00 \\ -\ \ 21.59 \\ \hline 143.41 \end{array}$$

Line up decimal points (fill in zeros)
↓
Subtract as if they were whole numbers
↓
Put decimal point below others

EXAMPLE 16 Tina has \$231.62 in her checking account. She writes a \$51.78 check for a new coat. How much does she have left in her account?

SOLUTION We take the original balance and subtract the amount of the check.

$$\begin{array}{r} \$231.62 \\ -\ \ \ \ 51.78 \\ \hline \$179.84 \end{array}$$

Line up decimal points
↓
Subtract as if they were whole numbers
↓
Put decimal point below others

PROBLEM SET

Subtract the following decimals.

See Examples 13 to 16

1. $\begin{array}{r} 56.64 \\ -\ 1.41 \\ \hline \end{array}$	**2.** $\begin{array}{r} 8.62 \\ -7.013 \\ \hline \end{array}$	**3.** $\begin{array}{r} 62.53 \\ -\ 7.228 \\ \hline \end{array}$	**4.** $\begin{array}{r} 96.612 \\ -71.36 \\ \hline \end{array}$
5. $\begin{array}{r} 181 \\ -\ 93.42 \\ \hline \end{array}$	**6.** $\begin{array}{r} 96.60 \\ -\ 5.234 \\ \hline \end{array}$	**7.** $\begin{array}{r} 5.19 \\ -1.993 \\ \hline \end{array}$	**8.** $\begin{array}{r} 299.51 \\ -178.57 \\ \hline \end{array}$
9. $\begin{array}{r} 73.11 \\ -\ 4.171 \\ \hline \end{array}$	**10.** $\begin{array}{r} 323.61 \\ -\ 96.98 \\ \hline \end{array}$	**11.** $\begin{array}{r} 5.988 \\ -0.3124 \\ \hline \end{array}$	**12.** $\begin{array}{r} 75.81 \\ -75.13 \\ \hline \end{array}$

13. Subtract 5.744 − 3.21.
14. Subtract 31.58 − 23.615.
15. Subtract 577.9 − 158.97.
16. Subtract 9.463 − 0.5352.
17. Subtract 7086 − 7.698.
18. Subtract 9400 − 5373.88.
19. Marion has \$212.18 in her checking account. She writes a check for \$87.53. How much does she have left in her account?
20. Ron has a board 1.73 meters long. He saws 0.8 meters off the end. How much does he have left?
21. The distance between Squirrel Valley and Raccoon Bluff is 25.3 miles. John has cycled 6.9 miles of the trip. How much does he have left?

22. Jeni has \$521.53 in the bank. She withdraws \$273.16 and \$171.42. How much does she have left in the bank?

23. Jane buys a \$39.95 broiler, but gets a \$6.35 discount. What does she pay?

ANSWERS 1. 55.23 2. 1.607 3. 55.302 4. 25.252 5. 87.58
6. 91.366 7. 3.197 8. 120.94 9. 68.939 10. 226.63
11. 5.6756 12. 0.68 13. 2.534 14. 7.965 15. 418.93
16. 8.9278 17. 7078.302 18. 4026.12 19. 124.65 20. 0.93
21. 18.4 22. 76.95 23. 33.60

In Chapters 1 and 2, we saw that *addition and subtraction are the opposites of each other*. As examples,

$$n - 4 = 9 \quad \text{means} \quad n = 9 + 4 = 13$$

$$n + \frac{1}{2} = 6 \quad \text{means} \quad n = 6 - \frac{1}{2} = 5\frac{1}{2}$$

Similarly with decimals, we add to invert subtraction, and vice versa.

EXAMPLE 17 Find n: $n - 12.3 = 41.73$.

SOLUTION Since addition is the opposite of subtraction, we add 12.3 to both sides of the equation to find n.

$$n - 12.3 = 41.73 \quad \text{means} \quad n = 41.73 + 12.3$$
$$= 54.03$$

EXAMPLE 18 Find n: $n - 0.02 = 0.735$.

SOLUTION We invert this subtraction by adding 0.02 to both sides of the equation.

$$n - 0.02 = 0.735 \quad \text{means} \quad n = 0.735 + 0.02$$
$$= 0.755$$

EXAMPLE 19 Find n: $n + 6.29 = 17.1$.

SOLUTION We invert this addition by subtracting 6.29 from both sides of the equation.

$$n + 6.29 = 17.1 \quad \text{means} \quad n = 17.1 - 6.29$$
$$= 10.81$$

EXAMPLE 20 What number added to 123.4 is 259.15?

SOLUTION We let n be the number that we are looking for. The words *added to* mean +, and *is* means =.

What number added to 123.4 is 259.15?

$$n \quad + \quad 123.4 = 259.15$$

Now we solve this by subtracting 123.4 from both sides of the equation:

$$n + 123.4 = 259.15 \quad \text{means} \quad n = 259.15 - 123.4$$
$$= 135.75$$

Thus the number is 135.75.

PROBLEM SET Find n or x in the following problems.

See Examples 17 to 20

1. $n - 2 = 7.8$
2. $x - 4.8 = 11.1$
3. $n - 12.7 = 25.45$
4. $n - 28.03 = 41.75$
5. $x - 42.5 = 82$
6. $x - 0.071 = 1.52$
7. $n + 5 = 8.3$
8. $n + 2.3 = 8.1$
9. $n + 4.8 = 9.72$
10. $x + 6.05 = 10.7$
11. What number added to 307.8 is 1000?
12. What number added to 0.035 is 2.7?

ANSWERS 1. 9.8 2. 15.9 3. 38.15 4. 69.78 5. 124.5 6. 1.591 7. 3.3 8. 5.8 9. 4.92 10. 4.65 11. 692.2 12. 2.665

Miscellaneous Exercises 3.2

Perform the indicated operations.

See Examples 10 to 12

1. $2.3 + 4.69$
2. $5.92 + 47.092$
3. $43 + 59.2 + 58.01$
4. $92.92 + 47.934 + 57$
5. $579 + 46.01 + 2.129$
6. $5.029 + 29.9482 + 3.38 + 0.923$

See Examples 13 to 16

7. $47.93 - 32.958$
8. $34.2 - 19.048$
9. $689 - 483.23$
10. $1000 - 567.473$
11. $57.099 - 12.68$
12. $10.0001 - 9.99998$

Health applications

13. A laboratory technician records the white-blood-cell count (thousands of cells per cubic millimeter of blood): 7.1; 11.3; 8.9; 7.3; 9.9; 5.9; 12.4; 5.1; 9.7; 16.8. Find the total for these counts.
14. A patient is to receive 0.05 grain of atropine sulfate. Thus far, she has received 0.035 grain of atropine sulfate. How much more of the medication does she need?

Electrical application

15. The current in a parallel circuit is the total of the currents through each branch. Find the total current in the circuit sketched at the top of page 149.

Business applications

16. A company's revenue (gross) for the year was \$213,571.48. Its expenses (costs) were \$109,583.69. Find the net profit (revenue − cost).

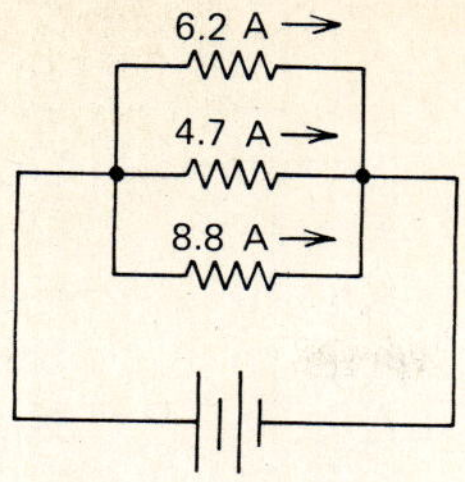

17. The daily sales figures for three salespeople in one week are as follows:

Day of Week	Kotowski	Dohner	Humbert
Monday	$236.93	$312.57	$274.68
Tuesday	210.57	297.47	205.79
Wednesday	296.04	310.39	346.92
Thursday	324.79	362.68	402.58
Friday	329.94	432.58	389.46

(a) Find the sales totals for each person. (Add down.)
(b) Find the sales total for each day. (Add across.)

Technical application

18. The overall length of a bushing is to be 1.25 centimeters with a tolerance of 0.012 centimeter.
(a) Find the greatest allowable length by adding the length and the tolerance.
(b) Find the smallest allowable length by subtracting the length minus the tolerance.

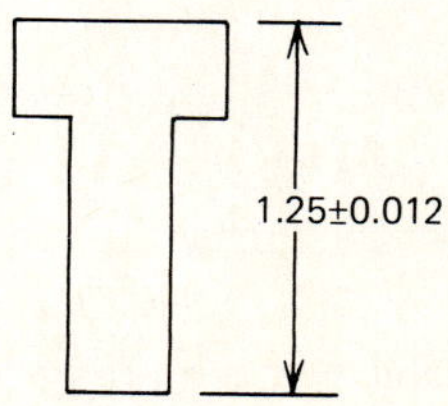

Psychology application

19. While using a chi-square test, a psychologist finds that she must add the following numbers. Find the sum.

f_o^2/f_e
16.333
12.000
10.083
6.750
16.333

Consumer applications

20. The Morisons start the month with a balance in their checking account of $126.94. Use the checks and deposits given below to find their balance at the end of the month.

		Checks	Deposits
Apr 1	Starting balance		$126.94
Apr 2	Deposit		432.59
Apr 4	C&K Foods	$ 64.38	
Apr 6	Phone company	12.68	
Apr 9	Gas company	13.59	

		Checks	Deposits
Apr 11	Goniff Oil Co.	38.93	
Apr 12	MasterDebt Charge	56.82	
Apr 14	C&K Foods	59.37	
Apr 15	Deposit		432.59
Apr 18	Nonsense Shop	32.66	
Apr 21	C&K Foods	72.59	
Apr 24	Auto Finance Co.	145.68	
Apr 26	Safe Form Insurance	23.69	
Apr 29	Bite Mortgage Co.	342.57	

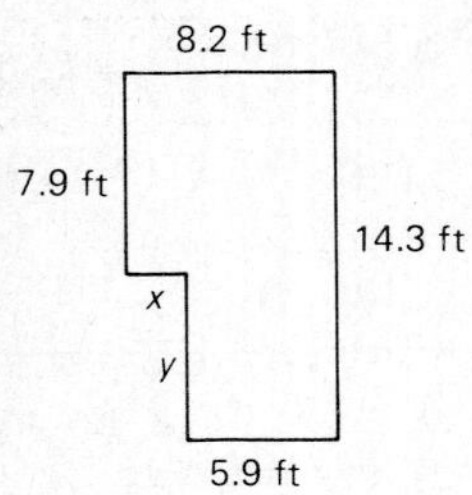

21. In the past month, Janet has put the following amounts of gasoline in her car: 12.3, 15.2, 15, 9.8, and 11.4 gallons. Find the total.

22. Matt buys a \$20 pair of pants, and he gets a \$3.75 discount. How much did he pay?

23. Sandy is putting floor molding around the room shown at the left.
    (a) Find the missing lengths, x and y.
    (b) Find the total distance around the room.

Now complete Post-Test 1 for Section 3.2 on page T43.

3.3 MULTIPLICATION OF DECIMALS

SECTION OBJECTIVE After completing this unit, you will be able to multiply decimals.

TYPICAL PROBLEMS You will be able to work problems like the ones below.

(a) Multiply 0.001 × 45.8. **(b)** Multiply 6.4 × 3.952.

(c) $\begin{array}{r} 0.0713 \\ \times 0.0028 \\ \hline \end{array}$ **(d)** $\begin{array}{r} 2500 \\ \times 0.065 \\ \hline \end{array}$ **(e)** Find n: $n \div 2.5 = 300$.

ANSWERS **(a)** 0.0458 **(b)** 25.2928 **(c)** 0.00019964 **(d)** 162.5 **(e)** 750

Multiplying decimals is different from adding or subtracting them. Instead of lining up decimal points, we count decimal places. We start with the numbers 0.1, 0.01, 0.001, 0.0001, and so on, which we call the **decimal powers of 10.** Recall that as fractions, these are $\frac{1}{10}, \frac{1}{100}, \frac{1}{1000}, \frac{1}{10{,}000}$, and so on.

> To multiply a number by a decimal power of 10:
> 1. Count the number of decimal places in the decimal power of 10.
> 2. Move the decimal point of the other number that many places to the left.

EXAMPLE 21 Multiply 0.01 × 6.23.

SOLUTION The number 0.01 has two places, so move the decimal point two places.

0.01 × 6.23	Count two places
= 006.23	Move two places left
= 0.0623	Fill in zeros

EXAMPLE 22 Multiply 0.001 × 27,000.

SOLUTION Here we move the decimal point three places.

0.001 × 27,000	Count three places
= 27000.	Move three places left
= 27	Simplify

PROBLEM SET Multiply the following decimals.

See Examples 21 and 22

1. 0.1 × 36	2. 0.01 × 572	3. 0.0001 × 290,000
4. 0.001 × 5900	5. 0.00001 × 2,300,000	6. 0.1 × 0.089

ANSWERS 1. 3.6 2. 5.72 3. 29 4. 5.9 5. 23 6. 0.0089

> To multiply any two decimals:
> 1. Total the number of decimal places in both factors.
> 2. Multiply the digits as if they were whole numbers.
> 3. Move the decimal point in the answer to the left the total number of the places of the factors. (Fill in zeros, if necessary.)

EXAMPLE 23 Multiply 8.3 × 6.9.

SOLUTION Each factor has one decimal place, so the answer will have two decimal places.

$$\begin{array}{r} 8.3 \\ \times \quad 6.9 \\ \hline 747 \\ 498 \\ \hline 5727 \end{array}$$

$= 57.27$

8.3 has one decimal place
↓
6.9 has one decimal place
↓
Multiply as if they were whole numbers
↓
Move 1 + 1 = 2 places left

EXAMPLE 24 Multiply 0.7 × 2.403.

SOLUTION The factors have a total of four decimal places.

$$\begin{array}{r} 2.403 \\ \times \quad 0.7 \\ \hline 16821 \end{array}$$

$= 1.6821$

2.403 has three decimal places
↓
0.7 has one decimal place
↓
Multiply as if they were whole numbers
↓
Move 3 + 1 = 4 places to the left

EXAMPLE 25 Multiply 0.0063 × 0.091.

SOLUTION The factors have a total of seven decimal places.

$$\begin{array}{r} 0.0063 \\ \times \quad 0.091 \\ \hline 63 \\ 567 \\ \hline 0005733 \end{array}$$

$= 0.0005733$

0.0063 has four decimal places
↓
0.091 has three decimal places
↓
Multiply as if they were whole numbers
↓
Move 4 + 3 = 7 places left

EXAMPLE 26 Multiply 4000 × 1.62.

SOLUTION The factors have a total of two decimal places.

$$\begin{array}{r} 1.62 \\ \times \quad 4000 \\ \hline 648000 \end{array}$$

$$= 6480.00$$

1.62 has two decimal places

↓

4000 has zero decimal places

↓

Multiply as if they were whole numbers

↓

Move 2 + 0 = 2 places to the left

EXAMPLE 27 Find the area of the rectangle shown.

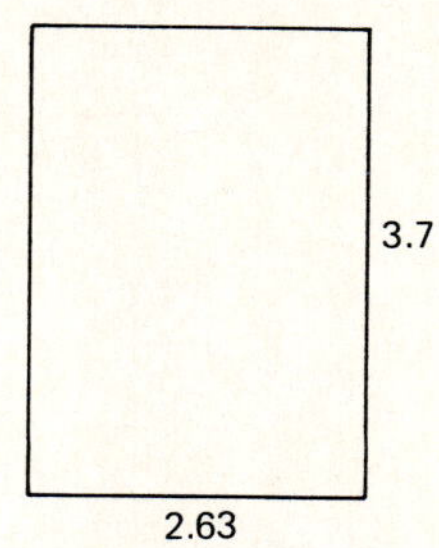

SOLUTION The area of a rectangle is the length times the width. So we multiply 3.7 × 2.63.

$$\begin{array}{r} 2.63 \\ \times \quad 3.7 \\ \hline 1841 \\ 789 \\ \hline 9731 \end{array}$$

$$= 9.731$$

2.63 has two decimal places

↓

3.7 has one decimal place

↓

Multiply as if they were whole numbers

↓

Move 2 + 1 = 3 places left

YES	NO
$\begin{array}{r} 0.06 \\ \times \; 0.7 \\ \hline 0.042 \end{array}$	~~$\begin{array}{r} 0.06 \\ \times 0.7 \\ \hline 0.42 \end{array}$~~

We count the total number of decimal places when we multiply.

EXAMPLE 28 A patient is to receive four injections of a regular Iletin insulin solution. Each injection is 0.65 milliliter. How much total medicine is this?

SOLUTION We find the total by multiplying 4 X 0.65.

$$\begin{array}{r} 0.65 \\ \times \quad 4 \\ \hline 2.60 \end{array}$$

Thus the patient receives a total of 2.6 milliliters of medicine.

Why Does It Work?

When we count the decimal places in a multiplication problem, we are secretly working with fractions. Consider 3.9 X 2.73.

$$3.9 \times 2.73$$

$$= 3\frac{9}{10} \times 2\frac{73}{100}$$

$$= \frac{39}{10} \times \frac{273}{100}$$

$$= \frac{10{,}647}{1000}$$

$$= 10\frac{647}{1000}$$

$$= 10.647$$

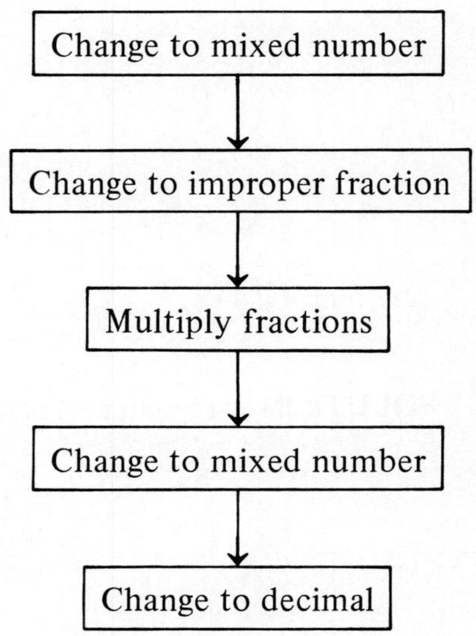

Notice how we multiply 39 X 273 and multiply the denominators. When we do this as a decimal problem, we add 1 + 2 = 3 decimal places. This stands for the denominators, 10 X 100 = 1000.

PROBLEM SET Multiply the following decimals.

See Examples 23 to 28

1. $\begin{array}{r} 1.2 \\ \times 3.7 \\ \hline \end{array}$	**2.** $\begin{array}{r} 99 \\ \times 0.8 \\ \hline \end{array}$	**3.** $\begin{array}{r} 1.25 \\ \times 6.66 \\ \hline \end{array}$	**4.** $\begin{array}{r} 0.85 \\ \times \ 6.3 \\ \hline \end{array}$
5. $\begin{array}{r} 73.9 \\ \times 9.85 \\ \hline \end{array}$	**6.** $\begin{array}{r} 11.8 \\ \times \ 8.3 \\ \hline \end{array}$	**7.** $\begin{array}{r} 0.951 \\ \times \ 8.8 \\ \hline \end{array}$	**8.** $\begin{array}{r} 65.4 \\ \times 80.1 \\ \hline \end{array}$
9. $\begin{array}{r} 24000 \\ \times \ 0.009 \\ \hline \end{array}$	**10.** $\begin{array}{r} 0.0091 \\ \times 0.0033 \\ \hline \end{array}$	**11.** $\begin{array}{r} 12550 \\ \times \ 6.3 \\ \hline \end{array}$	**12.** $\begin{array}{r} 0.0061 \\ \times \ 0.019 \\ \hline \end{array}$

13. Multiply 15.47×9.4.

14. Multiply 3500×0.0096.

15. Multiply $80{,}000 \times 5.9$.

16. Multiply 0.00046×0.0005.

17. Multiply 3.21×7.905.

18. Multiply 69.2×1.0009.

19. A school must buy 27 gallons of milk at \$1.69 a gallon. How much will this cost?

20. Lee buys 13.6 gallons of gasoline at \$1.44 a gallon. How much does the gas cost him?

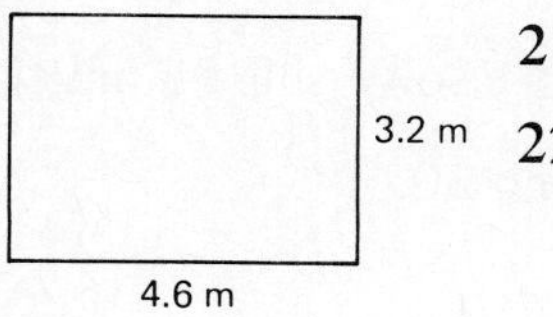

21. A room measures 3.2 meters by 4.6 meters. What is the area?

22. Fire insurance costs the Smedleys \$3.25 per thousand dollars of house value. Their house is worth \$28,400 (28.4 thousand). What will the insurance cost?

23. The standard car insurance is \$324.76. Because Ed has a bad driving record, he must pay 1.15 times this. What will his insurance cost?

24. In a certain state, the state income tax is 0.025 times your net income. The Arbuckles' net income is \$14,521.67. What is their state income tax? Round your answer to the nearest penny.

25. Jeanne makes \$4.27 a hour. Last week she worked 21.5 hours. How much did she earn?

ANSWERS 1. 4.44 2. 79.2 3. 8.325 4. 5.355 5. 727.915 6. 97.94 7. 8.3688 8. 5238.54 9. 216 10. 0.00003003 11. 79065 12. 0.0001159 13. 145.418 14. 33.6 15. 472,000 16. 0.00000023 17. 25.37505 18. 69.26228 19. 45.63 20. 19.584 (rounded to 19.58) 21. 14.72 22. 92.30 23. 373.474 (rounded to 373.47) 24. 363.04 25. 91.805 (rounded to 91.81)

In Chapters 1 and 2, we saw that *multiplication is the opposite (or inverse) of division.* As examples,

$$n \div 3 = 4 \quad \text{means} \quad n = 4 \times 3 = 12$$

$$n \div \frac{1}{2} = 10 \quad \text{means} \quad n = 10 \times \frac{1}{2} = 5$$

Similarly with decimals, multiplication is the opposite of division.

EXAMPLE 29 Find n: $n \div 2 = 4.6$.

SOLUTION Since multiplication is the opposite of division, we find n by multiplying both sides of the equation by 2.

$$n \div 2 = 4.6 \quad \text{means} \quad n = 4.6 \times 2 = 9.2$$

EXAMPLE 30 Find n: $n \div 3.9 = 7.1$.

SOLUTION We invert this division by multiplying by 3.9.

$$n \div 3.9 = 7.1 \qquad \text{means} \qquad n = 3.9 \times 7.1$$
$$= 27.69$$

EXAMPLE 31 Find n: $\dfrac{n}{16.5} = 2.4$.

SOLUTION Recall that $\dfrac{a}{b}$ means $a \div b$, so that $\dfrac{n}{16.5}$ means $n \div 16.5$. We solve this by multiplying both sides by 16.5 as we have done in Examples 29 and 30.

$$\frac{n}{16.5} = 2.4$$

Given

$$\cancel{16.5} \times \frac{n}{\cancel{16.5}} = 16.5 \times 2.4$$

Multiply both sides of the equation by 16.5

$$n = 39.6$$

Cancel and simplify

PROBLEM SET Find n or x in the following problems.

See Examples 29 to 31

1. $n \div 3 = 6.5$
2. $x \div 5 = 2.4$
3. $n \div 2.6 = 7.3$
4. $n \div 1.7 = 20.3$
5. $\dfrac{x}{7} = 3.4$
6. $\dfrac{n}{5.6} = 10$
7. $\dfrac{n}{0.02} = 8$
8. $\dfrac{x}{0.5} = 4.8$

ANSWERS **1.** 19.5 **2.** 12 **3.** 18.98 **4.** 34.51 **5.** 23.8 **6.** 56
7. 0.16 **8.** 2.4

Miscellaneous Exercises 3.3

Multiply the following decimals.

See Examples 21 and 22

1. 0.001×45.789
2. 0.01×5932
3. 0.1×594.38
4. 0.0001×672.85

See Examples 23 to 28

5. 2.3×5.8
6. 6.3×2.9
7. 4.38×6.1
8. 20×5.68
9. 0.0023×5.2
10. 0.0003×0.0045

11. $18{,}000 \times 0.0003$

12. 7200×0.0000005

See Examples 29 to 31

13. Find n: $n \div 6 = 4.3$.

14. Find n: $n \div 4.2 = 6.9$.

15. Find n: $\dfrac{n}{1.8} = 5.6$.

Business applications

16. A \$90 radio sells at a 25% markup. Compute this markup, 90×0.25.

17. A salesperson makes 7% commission on a \$23,000 sale. Compute this commission, $0.07 \times 23{,}000$.

18. The present value of \$1 (in 10 years, 8% interest) is 0.46319. Compute the present value of \$200,000, which is $0.46319 \times 200{,}000$.

Health applications

19. A patient received 0.25 of a 0.25-grain ephedrine sulfate tablet. How much medication did she receive?

20. A nurse is preparing 500 milliliters of a 3% cresol solution. Compute the amount of pure liquid cresol that is needed, 500×0.03.

Technical applications

21. Below is a sketch of a countersink head rivet. Dimension X is 0.425 times diameter D. Dimension Y is 1.85 times D. Find X and Y if $D = 0.6$.

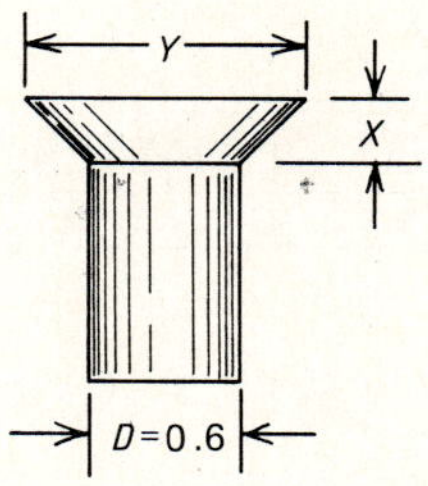

22. The tap-drill size for the screw shown below is given by $1.75 - (1.5 \times 0.13)$. Compute this. (Multiply first; then subtract.)

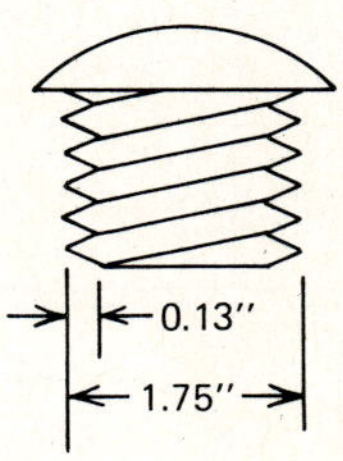

Science applications

23. The amount of heat (in Btu) needed to raise the temperature of a 2.5-pound silver bar by 7.5°F is given by $2.5 \times 7.5 \times 0.056$. (The number 0.056 is the specific heat capacity of silver.) Compute this.

24. The hydrostatic pressure exerted by water at a depth of 8.5 feet is given by 8.5×62.4. Compute this.

Psychology application

25. The average score on a learning test is 34.6. The 95% confidence range is the set of scores between
(a) $34.6 + (1.96 \times 6.1)$
(b) $34.6 - (1.96 \times 6.1)$
Compute these numbers. (Multiply first; then add or subtract.)

Consumer applications

26. The Hermans are buying fire insurance at \$2.08 per \$1000 value of the house. Their house is worth \$52,000 (52 thousands). How much is their insurance premium?

27. The Bakers are buying a house. Their mortgage payments are \$10.53 per thousand dollars borrowed. If they borrow \$47,000 (47 thousands), how much is their monthly mortgage payment?

28. The Gorgins put 15.8 gallons of gasoline in their car. At the time, gasoline was \$1.38 per gallon. How much did the gasoline cost them?

29. Because of inflation, food now costs 1.14 times what it did last year. If the Nelsons' food bill last year was \$78.34 per week, what is it this year?

30. Diane is tiling a rectangular room 5.2 meters by 3.1 meters. Find the area.

31. Distance in kilometers is 1.61 times the distance in miles. The distance from Chicago to Los Angeles is 2166 miles. What is this distance in kilometers?

32. Weight in kilograms is 0.454 times the equivalent weight in pounds. Fran weighs 132 pounds. How many kilograms is this?

33. Capacity in liters is 3.78 times the equivalent capacity in gallons. A certain Nova holds 22.6 gallons of gasoline. How many liters is this?

34. The amount that a person cannot deduct as a medical deduction on his or her income tax is 0.03 times the income. If the Gleasons' income is \$19,572, how much of their doctors' bills can they not deduct?

Now complete Post-Test 1 for Section 3.3 on page T45.

3.4 DIVISION OF DECIMALS

SECTION OBJECTIVE After completing this unit you will be able to divide decimals.

TYPICAL PROBLEMS You will be able to work problems like the ones below. Round (a) and (b) to two places; round (c) and (d) to three places.

(a) $4.8\overline{)87.21}$ **(b)** $9\overline{)43.87}$ **(c)** $0.01\overline{)32.6}$

(d) Divide $0.073 \div 58.3$. **(e)** Find n: $(2.5) \cdot n = 500$.

ANSWERS **(a)** 18.17 **(b)** 4.87 **(c)** 3260 **(d)** 0.001 **(e)** 200

Just as addition, subtraction, and multiplication of decimals are based on addition, subtraction, and multiplication of whole numbers, so is **division of decimals** based on division of whole numbers. We start with the divisor being a whole number.

> To divide a decimal by a whole number:
>
> 1. Divide the numbers as if they were both whole numbers.
> 2. The decimal point for the answer goes directly above the decimal point of the dividend.

EXAMPLE 32 Divide $38.29 \div 7$.

SOLUTION We set up the usual division process.

$$
\begin{array}{r}
5.47 \\
7\overline{)38.29} \\
\underline{35} \\
3\,2 \\
\underline{2\,8} \\
49 \\
\underline{49}
\end{array}
$$

Divide like whole numbers

↓

Put decimal point above the point in the dividend

EXAMPLE 33 Divide $56.94 \div 26$.

SOLUTION

$$
\begin{array}{r}
2.19 \\
26\overline{)56.94} \\
\underline{52} \\
4\,9 \\
\underline{2\,6} \\
2\,34 \\
\underline{2\,34}
\end{array}
$$

Divide like whole numbers

↓

Put decimal point above the point in the dividend

PROBLEM SET Divide the following decimals and whole numbers.

See Examples 32 and 33

1. $55.2 \div 8$ 2. $148.11 \div 3$ 3. $25.02 \div 9$ 4. $493.15 \div 5$
5. $102.9 \div 21$ 6. $304.87 \div 43$ 7. $3734.751 \div 159$

ANSWERS 1. 6.9 2. 49.37 3. 2.78 4. 98.63 5. 4.9 6. 7.09 7. 23.489

To divide decimals and round the answer:

1. Set up the problem like long division of whole numbers.
2. Move the decimal point of the divisor (outside number) to the right until it is a whole number.
3. Move the decimal point of the dividend (inside number) the *same* number of places to the right.
4. Place the decimal point for the quotient right above the new decimal point in the dividend.
5. Divide the numbers as if they were whole numbers.
6. Carry out the division to one place beyond that desired. Use rounding-off rules (round decimal place up or down, depending on the digit in the last place.)

EXAMPLE 34 Divide $34.86 \div 4.2$.

SOLUTION We move the decimal point one place to the right in both numbers to make 4.2 into a whole number, 42.

	Set up like long division
$4.2_{\wedge}\overline{)34.8_{\wedge}6}$	Move decimal point of 4.2 *one* place right
	Move decimal point of 34.86 *one* place right
$\begin{array}{r} 8.3 \\ 42_{\wedge}\overline{)348_{\wedge}6} \\ \underline{336\phantom{_{\wedge}6}} \\ 12\ 6 \\ \underline{12\ 6} \end{array}$	Divide like whole numbers
	Put decimal point above point in dividend

EXAMPLE 35 Divide 0.735 by 1.75.

SOLUTION We move the decimal point two places to the right in both numbers to make 1.75 into a whole number, 175.

	Set up like long division
$1.75_{\wedge}\overline{)0.73_{\wedge}5}$	Move decimal point of 1.75 *two* places right
	Move decimal point of 0.735 *two* places right
$\begin{array}{r} 0.42 \\ 175_{\wedge}\overline{)73_{\wedge}50} \\ \underline{70\ 0} \\ 3\ 50 \\ \underline{3\ 50} \end{array}$	Divide like whole numbers (fill in zero)
	Put decimal point above point in dividend

EXAMPLE 36 Divide $\dfrac{0.071656}{0.0104}$.

SOLUTION Recall that $\dfrac{a}{b}$ means $a \div b$.

$0.0104\overline{)0.071656}$	Set up like long division

```
0.0104⟌0.0716‸56
```

↓

Move decimal point of 0.0104 *four* places right

↓

Move decimal point of 0.071656 *four* places right

```
        6.89
104‸⟌716‸56
     624
      92 5
      83 2
       9 36
       9 36
```

↓

Divide like whole numbers

↓

Put decimal point above point in dividend

YES	NO
20 0.12‸⟌2.40‸	~~.20 0.12‸⟌2.40~~

We move the decimal point the *same* number of places in the divisor and dividend.

In these first three examples, the divisions all came out evenly; so we did not have to round off. In the next few examples we have to round off.

EXAMPLE 37 Divide 1.361 by 0.06 and round the answer to two decimal places.

SOLUTION Since we want two decimal places, we must first carry the division out to three places before we round off.

Set up like long division

↓

```
0.06‸⟌1.36‸100
```

Move decimal point of 0.06 *two* places right

↓

Move decimal point of 1.361 *two* places right

↓

Fill in zeros to get three places

↓

Divide like whole numbers

↓

```
     22.683
6‸⟌136‸100
   12
    16
    12
     4 1
     3 6
       50
       48
        20
        18
```

	Put decimal point above point in dividend
	↓
	Carry division out to three places
	↓
22.683	Determine place to be rounded
↑	↓
22.68[3]	Go one place right
↑	↓
22.68	Round down since digit is 3

EXAMPLE 38 Sherry drives 293.1 miles on 16.2 gallons of gas. How many miles per gallon is this (to one decimal place)?

SOLUTION To find miles *per* gallon, we divide the number of miles driven by the number of gallons of gas used. Here it is 293.1 ÷ 16.2. Since we want one decimal place, we must carry the division out to two places before we round off. (The word *per* usually indicates division.)

	Set up like long division
	↓
16.2) 293.1 00	Move decimal point of 16.2 *one* place right
	↓
	Move decimal point of 293.1 *one* place right
	↓
	Fill in zeros to get two places
	↓
	Divide like whole numbers
	↓
	Put decimal point above point in dividend
	↓
	Carry division out to two places

```
        18.09
162‸ ) 2931‸00
       162
       ----
       1311
       1296
       ----
         15 00
         14 58
         -----
            42
```

18.09

↑

18.0[9]

↑

18.1

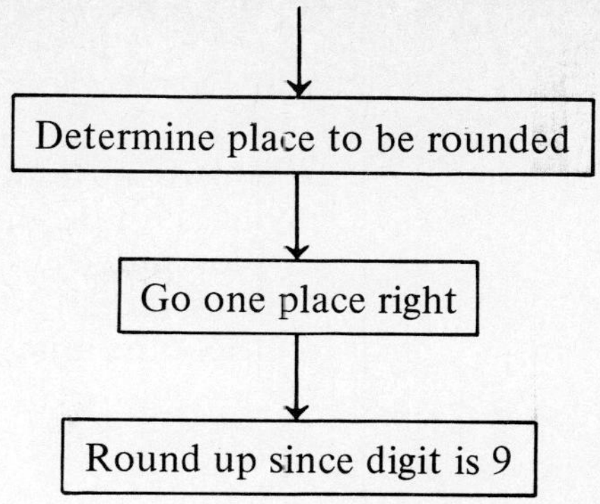

So Sherry gets about 18.1 miles per gallon.

EXAMPLE 39 Clyde Arbuckle leaves an estate of $1,239,543.71 to be split evenly among his wife and 6 children. What is each person's share (to the nearest penny)?

SOLUTION We divide the estate by 7 since there are 7 people. The nearest penny is the second decimal place, so we carry the division three places before we round off.

```
     177077.672
  7 ⟌1239543.710
     7
     53
     49
      49
      49
      054
       49
        53
        49
         47
         42
          51
          49
           20
           14
```

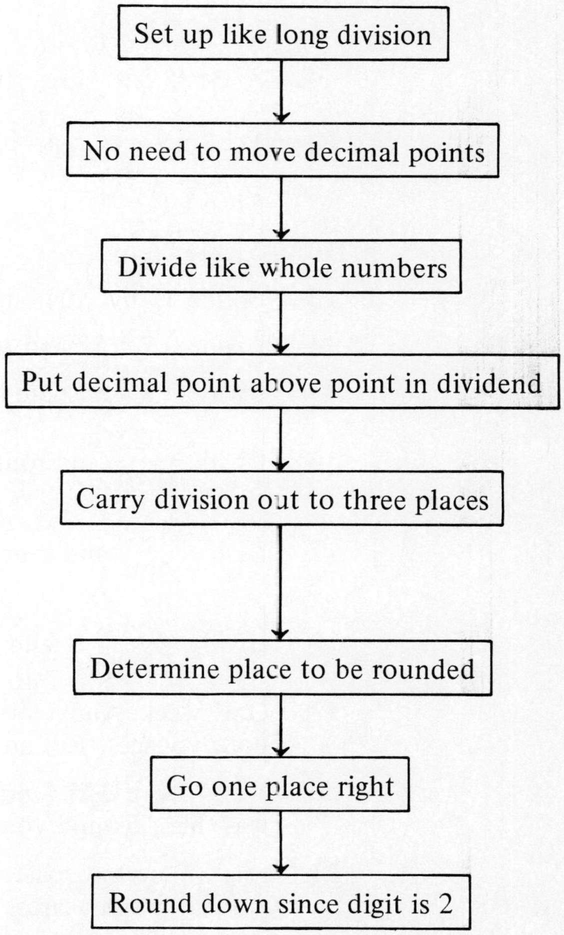

177077.672

↑

177077.67[2]

↑

177077.67

Thus each person's share of the estate is $177,077.67.

Why Does It Work?

When we move the decimal points in a division problem, we are really multiplying both the divisor and dividend by some power of 10 that will make the division simpler.

Consider $2.4 \div 0.12$. Suppose that we write this division as a fraction and multiply the numerator and denominator by 100.

$$\frac{2.4}{0.12} = \frac{2.4 \times 100}{0.12 \times 100} = \frac{240}{12} = 20$$

Now the problem is $240 \div 12$. This is the same result that we get when we move the decimal point two places:

$$\begin{array}{r} 20 \\ 0.12_{\wedge}\overline{)2.40_{\wedge}} \end{array}$$

PROBLEM SET Divide the following decimals.

See Examples 34 to 39

Round the answers in problems 1 to 6 to two decimal places.

1. $22\overline{)10.94}$
2. $1.3\overline{)7.467}$
3. $0.04\overline{)0.9368}$
4. $87.3\overline{)95.52}$
5. $0.005\overline{)5700}$
6. $9.5\overline{)660.00}$

Round the answers in problems 7 to 12 to three decimal places.

7. $57\overline{)60.04}$
8. $7\overline{)1}$
9. $0.25\overline{)7.014}$
10. $23.2\overline{)87.2}$
11. $52\overline{)51}$
12. $0.008\overline{)0.00039}$
13. Divide 58 by 0.04 and round your answer to one decimal place.
14. Divide $0.7 \div 58$ and round your answer to the nearest thousandth.
15. Divide $7.8 \div 52.6$ and round your answer to the nearest hundredth.
16. Divide $\dfrac{0.82}{0.39}$ and round your answer to the nearest tenth.
17. Divide $\dfrac{171}{309}$ and round your answer to three decimal places.
18. Divide $\dfrac{17}{0.00036}$ and round your answer to the nearest hundredth.
19. One week Andy worked 17.2 hours and earned \$69.37. What was his hourly wage? Round your answer to the nearest cent.
20. Mary drove 225.6 miles on 11.9 gallons of gas. How many miles per gallon was this? Round your answer to the nearest tenth.
21. If a 3.9-pound roast costs \$8.37, what is the price per pound? Round your answer to the nearest cent.
22. José runs 3.4 miles in 19.82 minutes. How many minutes per mile is this? Round your answer to two decimal places.

23. Dan Casey got 113 hits in 352 times at bat. What was his batting average (hits ÷ at bats, written as a three-place decimal)?

24. Five men have the following weights: 198.2, 182.6, 179.7, 158.4, and 216.4 pounds. Find the average weight. (Add the weights and divide by 5. Leave your answer as a one-place decimal.)

25. A disposable lighter selling for $1.39 lasted for 2984 lights. Find the cost per light as a decimal rounded to the nearest ten-thousandth of a cent. (*Hint:* First write $1.39 in cents.)

26. A clothes washing machine lasted exactly 5 years (one of them was a leap year). It cost $283.37 to buy new, plus $87.63 in repairs.
   (a) How many days was the machine owned?
   (b) How much did the machine cost to own (in cents)?
   (c) What was the average *daily* cost of owning the washer? Leave your answer in cents, rounded to the nearest tenth of a cent.

27. Lake Michigan covers 57,441 square kilometers. The U.S. portion of Lake Huron covers 23,245 square kilometers. Lake Michigan is how many times larger than the U.S. portion of Lake Huron? Leave your answer as a three-place decimal.

ANSWERS **1.** 0.50 **2.** 5.74 **3.** 23.42 **4.** 1.09 **5.** 1,140,000 **6.** 69.47
7. 1.053 **8.** 0.143 **9.** 28.056 **10.** 3.759 **11.** 0.981
12. 0.049 **13.** 1450 **14.** 0.012 **15.** 0.15 **16.** 2.1
17. 0.553 **18.** 47,222.22 **19.** 4.03 **20.** 19.0 **21.** $2.15
22. 5.83 **23.** 0.321 **24.** 187.1 **25.** 0.0466
26. **(a)** 1826 **(b)** 37100 **(c)** 20.3 **27.** 2.471

We saw in Chapters 1 and 2 that *division is the opposite of multiplication.* As examples (recall that dot · means multiplication),

$$6 \cdot n = 42 \quad \text{means} \quad n = 42 \div 6 = 7$$

$$\frac{2}{3} \cdot n = \frac{7}{5} \quad \text{means} \quad n = \frac{7}{5} \div \frac{2}{3} = \frac{7}{5} \cdot \frac{3}{2} = \frac{21}{10}$$

Similarly with decimals, we use division to invert multiplication.

EXAMPLE 40 Find n: $6 \cdot n = 0.84$.

SOLUTION Since division is the opposite of multiplication, we divide by 6 to find n.

$$6 \cdot n = 0.84 \quad \text{means} \quad n = \frac{0.84}{6}$$

$$\rightarrow 6\overline{)0.84} = 0.14$$

Thus $n = 0.14$.

EXAMPLE 41 Find n: $1.5 \cdot n = 750$.

SOLUTION We find n by dividing both sides of the equation by 1.5.

$$1.5 \cdot n = 750$$

Given

$$\frac{\cancel{1.5} \cdot n}{\cancel{1.5}} = \frac{750}{1.5}$$

Divide both sides by 1.5

$$\rightarrow 1.5_{\wedge}\overline{)750.0_{\wedge}} \quad 500.$$

Divide

Thus $n = 500$.

EXAMPLE 42 What number times 0.02 is 8.4?

SOLUTION We translate this as follows (we let n be the number we want):

What number times 0.02 is 8.4?

$$n \quad \cdot \quad 0.02 = 8.4$$

Translate

$$\frac{n \cdot \cancel{0.02}}{\cancel{0.02}} = \frac{8.4}{0.02}$$

Divide both sides by 0.02

$$\rightarrow 0.02_{\wedge}\overline{)8.40_{\wedge}} \quad 420.$$

Divide

Thus, $n = 420$.

PROBLEM SET Find n or x in the following problems.

See Examples 40 to 42

1. $3 \cdot n = 2.7$
2. $20 \cdot x = 4.8$
3. $0.2 \cdot n = 50$
4. $0.07 \cdot n = 210$
5. $0.14 \cdot x = 35$
6. $2.1 \cdot n = 0.714$
7. What number times 3.6 is 0.72?
8. What number times 0.25 is 13.2?

ANSWERS 1. 0.9 2. 0.24 3. 250 4. 3000 5. 250 6. 0.34
7. 0.2 8. 52.8

Miscellaneous Exercises 3.4

Divide the following decimals.

See Examples 32 and 33

1. $164.4 \div 3$
2. $13.44 \div 8$
3. $64.74 \div 13$
4. $0.9583 \div 37$

See Examples 34 to 39

5. $6.63 \div 1.7$
6. $0.8979 \div 7.3$
7. $3.8014 \div 0.83$
8. $2.3568 \div 0.03$
9. $\dfrac{0.7245}{20.7}$
10. $\dfrac{0.005559}{0.17}$

Divide the following decimals. Round the answers to problems 11 and 12 to the nearest tenth; problems 13 and 14 to the nearest hundredth; problems 15 and 16 to the nearest thousandth.

See Examples 37 to 39

11. $3.69 \div 0.7$
12. $\dfrac{56.8}{0.3}$
13. $0.46 \div 0.72$
14. $\dfrac{800}{1.2}$
15. $12.3 \div 0.0046$
16. $\dfrac{0.0057}{0.0009}$

Find n in the following problems.

See Examples 40 to 42

17. $4 \cdot n = 5.6$
18. $5 \cdot n = 87.2$
19. $0.03 \cdot n = 9$
20. $1.2 \cdot n = 0.0108$

Business applications

21. It costs a company $5.25 to produce 300 pencils. Find the cost per pencil to the nearest thousandth of a dollar.
22. A company pays its salespeople an 8% commission. The amount of sales needed to provide a $20,000 commission is given by $\dfrac{20{,}000}{0.08}$. Compute this.
23. An insurance company is studying the likelihood of men dying in their forty-fifth year. It studied 9049 men, of whom 48 died at age 45, and 9001 lived to be 46.
    (a) Compute the likelihood of a 45-year-old man living to be 46: $\dfrac{9001}{9049}$. Write the answer as a four-place decimal.
    (b) Compute the likelihood of a 45-year-old man dying before he is 46: $\dfrac{48}{9049}$. Write this as a four-place decimal.

Health applications

24. Six radiation workers reported the following radiation exposures for a certain time period: 1.3, 1.7, 0.9, 0.8. 0.7, and 1.1 rads. Find the average of these radiations (as a decimal accurate to the nearest hundredth).
25. The average adult dosage of atropine sulfate is 0.4 milligram. The dosage for a child with a body surface area of 0.56 square meter can be computed as $\dfrac{0.56}{1.73} \times 0.4$.
    (a) Compute this by dividing first, then multiplying.
    (b) Compute this by multiplying first, then dividing.

Electrical application

26. The effective current I through a 100-watt light bulb is given by

$$I = \frac{P}{V} = \frac{100 \text{ watts}}{115 \text{ volts}}$$

(a) Compute the effective current I.

(b) Compute the peak current by calculating $I_{peak} = \dfrac{I}{0.707}$.

Science applications

27. The molecular weight of oxygen (O_2) is 32. The weight of 1 liter of oxygen is given by $\dfrac{32}{22.4}$ grams. Compute this.

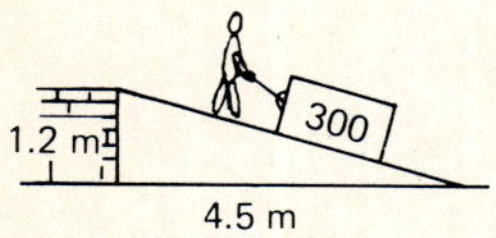

28. The ramp sketched at the left makes moving items easier. The *mechanical advantage MA* is given

$$MA = \frac{\text{length of ramp}}{\text{height of ramp}} = \frac{4.5}{1.2}$$

(a) Compute MA.

(b) For a 300-pound box, the *effort* is $\dfrac{300}{MA}$. Compute this.

Consumer applications

29. A gallon of low-fat milk costs about \$1.90. It contains about 128 grams of protein. Calculate the protein-per-dollar content of milk, $\dfrac{128}{1.90}$.

30. Frank's car gets 22.7 miles per gallon of gasohol. If gasohol costs \$1.45 per gallon, compute Frank's cost per mile (in cents per mile), $\dfrac{145}{22.7}$.

31. Denise borrows \$4552 to buy a car. The finance charge is 8% add-on for 3 years.

(a) Compute the finance charge = $4552 \times 0.08 \times 3$.

(b) Compute the total debt = 4552 + finance charge.

(c) Compute the monthly payments = $\dfrac{\text{total debt}}{36}$.

32. One month, Ken keeps the following records for gas mileage.

Miles	Gallons
289.2	14.6
312.7	16.1
257.9	12.3
298.6	14.9

(a) Find the total number of miles driven.

(b) Find the total number of gallons used.

(c) Compute the average miles per gallon for the month.

Now complete Post-Test 1 for Section 3.4 on page T47.

3.5 FRACTION-DECIMAL CONVERSIONS

SECTION OBJECTIVES

After completing this unit, you will be able to:

1. Change a fraction into a decimal of a stated number of places.
2. Change a decimal into a reduced fraction.

TYPICAL PROBLEMS You will be able to work problems like the ones below.

(a) Write $\frac{2}{7}$ as a three-place decimal.

(b) Write $5\frac{8}{9}$ as a two-place decimal.

(c) Write $\frac{100}{11}$ as a decimal accurate to the nearest hundredth.

(d) Write 0.083 as a reduced fraction.

(e) Write 9.25 as a reduced fraction.

(f) Write 6.4 as a reduced mixed number.

ANSWERS **(a)** 0.286 **(b)** 5.89 **(c)** 9.09 **(d)** $\frac{83}{1000}$ **(e)** $\frac{37}{4}$ **(f)** $6\frac{2}{5}$

Recall that the fraction $\frac{3}{4}$ means 3 divided by 4. We cannot do this division with whole numbers, but we can do it with decimals. We rewrite 3 as 3.00, and then we divide as decimals.

$$\begin{array}{r} 0.75 \\ 4\overline{)3.00} \\ \underline{2\,8} \\ 20 \\ \underline{20} \end{array}$$

We perform this division, and we get 0.75. We say that $\frac{3}{4}$ is 0.75 when written as a decimal.

> To change a fraction to a decimal:
>
> 1. Divide the denominator into the numerator.
> 2. Carry the division out to one extra place.
> 3. Use the last place to help round off.
> 4. If the number is a mixed number, leave the whole-number part alone and just change the fraction part to a decimal.

EXAMPLE 43 Write $\frac{5}{6}$ as a two-place decimal.

SOLUTION Since we want two decimal places, we will carry the division out to three places.

$$6\overline{)5.000}$$

Divide denominator into numerator

$6\overline{)5.000}$ = 0.833	Carry to three places
0.833 (arrow under the second 3)	Determine place to be rounded
0.83[3] (arrow under the first 3)	Go one place right
0.83	Round down since digit is 3

EXAMPLE 44 Write $\frac{1}{13}$ as a three-place decimal.

SOLUTION Since we want three decimal places, we carry the division out to four places.

$$\begin{array}{r} 0.0769 \\ 13\overline{)1.0000} \\ \underline{91} \\ 90 \\ \underline{78} \\ 120 \\ \underline{117} \\ 3 \end{array}$$	Divide denominator into numerator → Carry out to four places
0.0769 (arrow under the 6)	Determine place to be rounded
0.076[9] (arrow under the 6)	Go one place right
0.077	Round up since digit is 9

EXAMPLE 45 Write $\frac{7}{40}$ as a three-place decimal.

SOLUTION We carry this out to four places.

$$\begin{array}{r} 0.175 \\ 40\overline{)7.0000} \\ \underline{4\,0} \\ 3\,00 \\ \underline{2\,80} \\ 200 \\ \underline{200} \end{array}$$	Divide denominator into numerator → Division comes out evenly, so we stop

EXAMPLE 46 Write $10\frac{2}{3}$ as a decimal accurate to the nearest hundredth.

SOLUTION To do this, we forget the 10 for a moment and just change the $\frac{2}{3}$ to a decimal. We carry this out to three places. (Hundredth = two places.)

$3\overline{)2.000}$ with quotient 0.666	Divide denominator into numerator
	↓ Carry out to three places
0.666 (arrow under the second 6)	↓ Determine place to be rounded
$0.66\boxed{6}$ (arrow under the second 6)	↓ Go one place right
0.67	↓ Round up since digit is 6
So $10\frac{2}{3} = 10.67$	↓ Add whole number, 10

Why Does It Work?

As mentioned before, this rule works because the fraction $\frac{a}{b}$ means $a \div b$. When we divide, we get the decimal equivalent.

PROBLEM SET For problems 1 to 6, write each fraction as a decimal accurate to two decimal places.

See Examples 43 to 46

1. $\frac{8}{9}$ 2. $\frac{3}{7}$ 3. $\frac{7}{8}$ 4. $\frac{1}{17}$ 5. $\frac{2}{15}$ 6. $\frac{17}{3}$

For problems 7 to 12, write each fraction or mixed number as a decimal accurate to three places.

7. $\frac{5}{52}$ 8. $10\frac{5}{12}$ 9. $\frac{1}{75}$ 10. $\frac{114}{193}$ 11. $1\frac{1}{50}$ 12. $35\frac{6}{19}$

13. A board is $5\frac{7}{12}$ inches long. Write this as a decimal accurate to the nearest hundredth.

14. Ruth uses $\frac{11}{15}$ of a tank of gas on a trip. Write this as a decimal accurate to the nearest thousandth.

15. A certain lot is $\frac{1}{88}$ mile long. Write this as a decimal accurate to the nearest thousandth.

16. The Garcias' share of a town-house cooperative is $\frac{1}{120}$. Write this as a decimal accurate to the nearest ten-thousandth.

ANSWERS **1.** 0.89 **2.** 0.43 **3.** 0.88 **4.** 0.06 **5.** 0.13 **6.** 5.67 **7.** 0.096 **8.** 10.417 **9.** 0.013 **10.** 0.591 **11.** 1.02 **12.** 35.316 **13.** 5.58 **14.** 0.733 **15.** 0.011 **16.** 0.0083

Frequently, we wish to change a decimal into a fraction. This is the procedure that we use.

> To change a decimal to a reduced fraction:
> 1. Determine the column value of the last decimal place.
> 2. Put the original number (without any decimal point) over this column-value denominator.
> 3. Reduce the fraction, if necessary.

EXAMPLE 47 Write 0.93 as a reduced fraction.

SOLUTION The last column is the hundredths' column.

0.93 → $\frac{1}{100}$ — Determine last column value

$= \frac{93}{100}$ — Put number over this denominator

EXAMPLE 48 Write 0.025 as a reduced fraction.

SOLUTION The last column is the thousandths' column.

0.025 → $\frac{1}{1000}$ — Determine last column value

$$= \frac{25}{1000}$$

Put number over this denominator

$$= \frac{1}{40}$$

Reduce fraction

EXAMPLE 49 Write 6.7 as a reduced fraction.

SOLUTION The last column is the tenths' column.

$$6.7 \rightarrow \frac{1}{10}$$

Determine last place value

$$= \frac{67}{10}$$

Put number over this denominator

$$\text{or } 6\frac{7}{10}$$

Write as a mixed number

EXAMPLE 50 Write 0.0059 as a reduced fraction.

SOLUTION The last column is the ten-thousandths' column.

$$0.0059 \rightarrow \frac{1}{10{,}000}$$

Determine last column value

$$= \frac{59}{10{,}000}$$

Put number over this denominator

Why Does It Work?

The column values of decimals are fractions: $\frac{1}{10}$, $\frac{1}{100}$, $\frac{1}{1000}$, and so on. When we expand out a decimal, we see that the last place acts as a common denominator. As examples, consider

$$0.217 = \frac{2}{10} + \frac{1}{100} + \frac{7}{1000} = \frac{200}{1000} + \frac{10}{1000} + \frac{7}{1000} = \frac{217}{1000}$$

$$5.83 = 5 + \frac{8}{10} + \frac{3}{100} = \frac{500}{100} + \frac{80}{100} + \frac{3}{100} = \frac{583}{100} \quad \text{or} \quad 5\frac{83}{100}$$

EXAMPLE 51 The sponsors of a new TV show, Name That Clown, have computed that their share of the viewing market is 0.15. Write this as a reduced fraction.

SOLUTION The place value of the last digit is $\frac{1}{100}$. Thus we get

$$0.1\underset{\uparrow}{5} \quad \llcorner\!\!\longrightarrow \frac{1}{100}$$

$$= \frac{15}{100}$$

$$= \frac{3}{20}$$

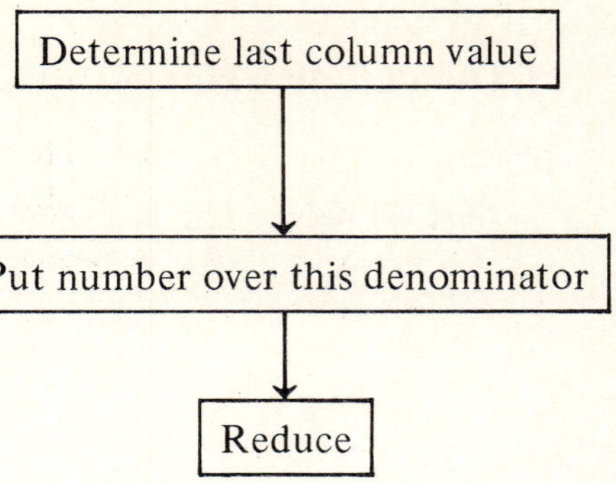

This says that 3 of every 20 viewers watch Name That Clown.

PROBLEM SET Write the following decimals as reduced fractions or mixed numbers.

See Examples 49 to 51

1. 0.7	2. 0.35	3. 0.231	4. 0.44	5. 2.1
6. 0.2	7. 7.25	8. 99.99	9. 0.98	10. 0.125
11. 6.14	12. 0.055	13. 1.12	14. 36.95	15. 0.76531

16. A car is 14.8 feet long. Write this as a mixed number.
17. Mary buys 0.64 pound of chocolate. Write this as a reduced fraction.
18. Reggie earns \$3.60 an hour. Write this as a reduced mixed number.

ANSWERS 1. $\frac{7}{10}$ 2. $\frac{7}{20}$ 3. $\frac{231}{1000}$ 4. $\frac{11}{25}$ 5. $2\frac{1}{10}$ 6. $\frac{1}{5}$ 7. $7\frac{1}{4}$ 8. $99\frac{99}{100}$ 9. $\frac{49}{50}$ 10. $\frac{1}{8}$ 11. $6\frac{7}{50}$ 12. $\frac{11}{200}$ 13. $1\frac{3}{25}$ 14. $36\frac{19}{20}$ 15. $\frac{76{,}531}{100{,}000}$ 16. $14\frac{4}{5}$ 17. $\frac{16}{25}$ 18. $3\frac{3}{5}$

Miscellaneous Exercises 3.5

Write the following fractions as three-place decimals.

See Examples 43 to 46

1. $\frac{6}{7}$ 2. $\frac{5}{9}$ 3. $\frac{11}{13}$ 4. $\frac{32}{29}$ 5. $3\frac{1}{4}$ 6. $23\frac{4}{11}$

Write the following decimals as reduced fractions (or mixed numbers).

See Examples 47 to 51

7. 0.74	8. 0.002	9. 0.35
10. 0.0005	11. 1.2	12. 5.25

Business applications

13. A marketing survey finds that $\frac{114}{349}$ of a sample population eats cereal in the morning. Write this as a three-place decimal.

14. According to life insurance tables, the chance of a 21-year-old man dying before he reaches 22 is 0.0025. Write this as a reduced fraction.

Health applications

15. A hospital stocks 0.375-grain ephedrine sulfate ampuls. Express this as a reduced fraction.

16. A patient receives $\frac{1}{4}$ of a $\frac{1}{2}$-grain codeine sulfate tablet.
 (a) How much medication did the patient receive (as a fraction of a grain)?
 (b) How much medication did the patient receive (as a decimal)?

Technical application

17. For the bolt sketched below, express each dimension as a decimal.

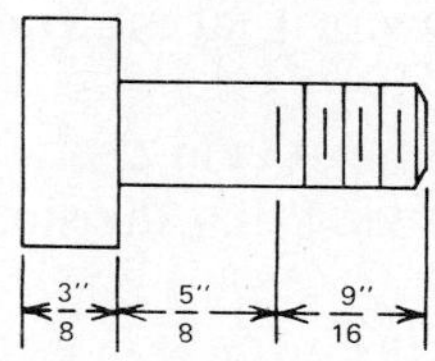

Chemistry application

18. One coulomb of electricity supplied to a solution of $AlCl_3$ (aluminum chloride) deposits $\frac{9}{96,500}$ gram of aluminum. If 19,300 coulombs are supplied to the solution, $19,300 \times \frac{9}{96,500}$ grams of aluminum are deposited.
 (a) Express this as a reduced fraction.
 (b) Express this as a decimal.

Consumer applications

19. At a sale, pants were $\frac{1}{4}$ off list price.
 (a) Express this as a decimal.
 (b) How much money was saved on a $23.75 purchase?

20. Mike calculates that 0.28 of his total income goes to taxes. Express this as a reduced fraction.

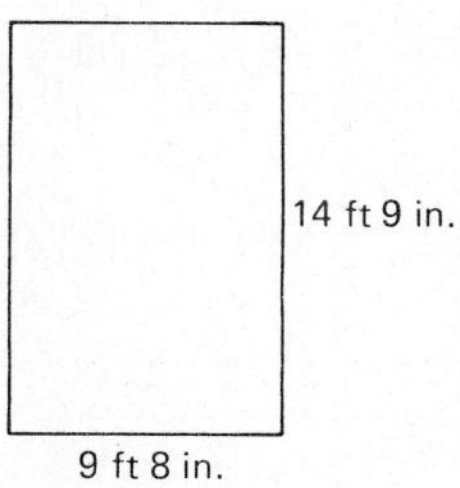

21. At the left is a sketch of a room that the Rossis are carpeting.
 (a) Express each dimension as a mixed number of feet. (For example, 2 feet 7 inches is $2\frac{7}{12}$ feet.)
 (b) Express each dimension as a two-place decimal.
 (c) Compute the area as a mixed number and as a decimal. [Use the numbers from parts (a) and (b).]

Now complete Post-Test 1 for Section 3.5 on page T49.

CHAPTER 3 SUMMARY

Important Words and Phrases

addition (3.2)
decimal point (3.1)
decimal powers of 10 (3.3)
decimals (3.1)

division (3.4)
fractional place value (3.1)
multiplication (3.3)
rounding off (3.1)
subtraction (3.2)

Important Ideas

(3.1) Knowing the place value of decimals.
Rounding off numbers.
Comparing the size of decimals.

(3.2) Adding decimals.
Subtracting decimals.
Knowing that addition and subtraction are opposite operations.

(3.3) Multiplying decimals.
Knowing that multiplication is the opposite of division.

(3.4) Dividing decimals.
Knowing that division is the opposite of multiplication.

(3.5) Writing a fraction as a decimal.
Writing a decimal as a fraction.

CHAPTER 3 REVIEW TEST

Choose the correct answer.

1. In 2.4512, what is the value of the 5?
   (a) $\frac{5}{10}$ (b) $\frac{5}{100}$ (c) $\frac{5}{1000}$ (d) $\frac{512}{10{,}000}$ (e) $\frac{45}{100}$

2. Round 16.7169 to the nearest hundredth.
   (a) 16.72 (b) 16.71 (c) 16.717 (d) 17 (e) 16.8

3. Round 25289.3 to the nearest thousand.
   (a) 25300 (b) 26000 (c) 25000 (d) 2500 (e) 2520

4. Which of the following decimals is the largest?
   (a) 0.02 (b) 0.022 (c) 0.0202 (d) 0.0222 (e) 0.00222

5. Add
   $$\begin{array}{r} 3.14 \\ 5.8 \\ +13.982 \\ \hline \end{array}$$
   (a) 22.922 (b) 21.882 (c) 21.822 (d) 11.82 (e) 11.92

6. Subtract
   $$\begin{array}{r} 325.6 \\ -144.73 \\ \hline \end{array}$$
   (a) 279.3 (b) 281.13 (c) 180.93 (d) 181.33 (e) 180.87

7. Add 2.3 + 8.17 + 14 + 6.397.
   (a) 22.767 (b) 725.1 (c) 30.867
   (d) 72.51 (e) 7.251

8. Subtract 1000 − 231.8.
(a) 178.2 (b) 879.2 (c) 779.2 (d) 868.2 (e) 768.2

9. Find n: $n - 0.6 = 3.1$.
(a) 3.7 (b) 2.5 (c) 18.6 (d) 9.1 (e) 0.51

10. Multiply 2.6×1.9.
(a) 4.94 (b) 49.4 (c) 494 (d) 4.84 (e) 48.4

11. Multiply 31.2×1.003.
(a) 4.056 (b) 0.456 (c) 31.2936
(d) 312.936 (e) 3129.36

12. Multiply 0.004×0.07.
(a) 0.28 (b) 0.028 (c) 0.0028
(d) 0.00028 (e) 0.000028

13. Divide $6.444 \div 1.8$.
(a) 0.347 (b) 3.47 (c) 0.358 (d) 35.8 (e) 3.58

14. Divide $0.07\overline{)4.531}$.
(a) 64.73 (b) 6.47 (c) 0.65 (d) 63.47 (e) 6.48

15. Divide $0.397 \div 0.51$. (Round your answer to the nearest hundredth.)
(a) 1.28 (b) 0.77 (c) 7.78 (d) 0.78 (e) 0.08

16. Find n: $(0.8) \cdot n = 6.4$.
(a) 7.2 (b) 5.6 (c) 5.12 (d) 8 (e) 0.08

17. Write $\frac{7}{20}$ as a decimal.
(a) 0.286 (b) 2.86 (c) 0.35 (d) 3.5 (e) 0.035

18. Write $6\frac{1}{8}$ as a decimal.
(a) 6.120 (b) 6.180 (c) 6.125 (d) 0.613 (e) 0.612

19. Write 0.97 as a fraction.
(a) $\frac{100}{97}$ (b) $\frac{10}{97}$ (c) $\frac{97}{10}$ (d) $\frac{97}{100}$ (e) $\frac{97}{1000}$

20. Write 4.8 as a mixed number.
(a) $4\frac{8}{100}$ (b) $4\frac{4}{5}$ (c) $4\frac{4}{50}$ (d) $\frac{48}{50}$ (e) $4\frac{48}{100}$

Now complete Post-Test 1 for Chapter 3 on page T51.

4

PERCENTS AND PROPORTIONS

Before you begin this chapter, complete the Pre-Test for Chapter 4 on page T55.

4.1 CHANGING PERCENTS TO FRACTIONS AND DECIMALS

SECTION OBJECTIVES After completing this unit, you will be able to:

1. Change a percent to a reduced fraction.
2. Change a percent to a decimal.

TYPICAL PROBLEMS You will be able to work problems like the ones below.

(a) Write 45% as a reduced fraction.
(b) Write 0.002% as a reduced fraction.
(c) Write 310% as a reduced fraction.
(d) Write 67% as a decimal.
(e) Write 4.5% as a decimal.
(f) Write 340% as a decimal.

ANSWERS **(a)** $\frac{9}{20}$ **(b)** $\frac{1}{50,000}$ **(c)** $\frac{31}{10}$ or $3\frac{1}{10}$ **(d)** 0.67 **(e)** 0.045 **(f)** 3.4

We have already studied fractions and decimals. A third type of numeral is the **percent**. It is virtually impossible to go through a day without hearing statements such as:

"There is a 60% chance of rain today."

"Our bank pays $6\frac{1}{2}\%$ interest on deposits."

"Savings up to 50% in every department of our store."

Percent means "out of 100." If we have 67%, we mean 67 out of 100, or $\frac{67}{100}$.

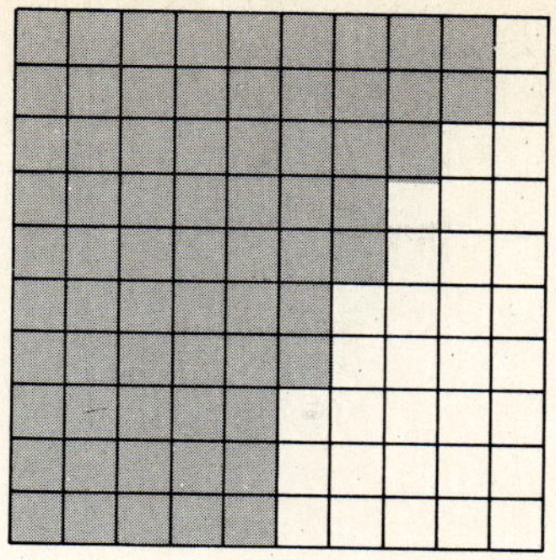

67%

$$N\% \quad \text{means} \quad \frac{N}{100} \quad \text{or} \quad N \times \frac{1}{100} \quad \text{or} \quad N \div 100$$

To write $N\%$ as a reduced fraction:

1. Write N as a whole number or fraction.
2. Replace the % symbol by $\left(\times \frac{1}{100}\right)$.
3. Multiply and reduce.

EXAMPLE 1

$39\% = 39 \times \frac{1}{100}$	Replace % by $\left(\times \frac{1}{100}\right)$ ↓
$= \frac{39}{100}$	Multiply

EXAMPLE 2

$75\% = 75 \times \frac{1}{100}$	Replace % by $\left(\times \frac{1}{100}\right)$ ↓
$= \frac{75}{100}$	Multiply ↓
$= \frac{15}{20} = \frac{3}{4}$	Reduce

EXAMPLE 3

$32.4\% = \frac{324}{10}\%$	Write 32.4 as a fraction ↓
$= \frac{324}{10} \times \frac{1}{100}$	Replace % by $\left(\times \frac{1}{100}\right)$ ↓

$= \frac{324}{1000}$	Multiply
$= \frac{162}{500} = \frac{81}{250}$	Reduce

EXAMPLE 4

$0.005\% = \frac{5}{1000}\%$	Write 0.005 as a fraction
$= \frac{5}{1000} \times \frac{1}{100}$	Replace % by $\left(\times \frac{1}{100}\right)$
$= \frac{5}{100{,}000}$	Multiply
$= \frac{1}{20{,}000}$	Reduce

EXAMPLE 5

$250\% = 250 \times \frac{1}{100}$	Replace % by $\left(\times \frac{1}{100}\right)$
$= \frac{250}{100}$	Multiply
$= \frac{25}{10} = \frac{5}{2}$	Reduce
or $2\frac{1}{2}$	Write as a mixed number

EXAMPLE 6

$2\frac{1}{4}\% = \frac{9}{4}\%$	Write $2\frac{1}{4}$ as a fraction
$= \frac{9}{4} \times \frac{1}{100}$	Replace % by $\left(\times \frac{1}{100}\right)$
$= \frac{9}{400}$	Multiply

Why Does It Work?

$N\%$ means N out of 100, or $\frac{N}{100}$, or $N \div 100$. Recall from Section 2.4 that we invert and multiply to divide fractions.

$$N \div 100 = N \times \frac{1}{100}$$

In these problems, it is usually easier to multiply by $\frac{1}{100}$.

Since $100\% = 100 \times \frac{1}{100} = 1$, we have the following useful fact.

$$100\% = 1$$

In words, 100% is the whole quantity.

PROBLEM SET Write the following percents as reduced fractions or mixed numbers.

See Examples 1 to 6

1. 57% 2. 89% 3. 40% 4. 22% 5. 8%
6. 2% 7. 45.1% 8. 67.4% 9. 4.25% 10. 0.003%
11. 0.6% 12. $4\frac{1}{3}\%$ 13. $10\frac{1}{2}\%$ 14. 240% 15. 175%
16. There is a 70% chance of rain today. Write this as a reduced fraction.
17. A certain math class is 35% female. Write this as a reduced fraction.
18. A bank pays $6\frac{1}{4}\%$ interest on a certain account. Write this as a reduced fraction.
19. A nurse uses a 0.2% solution. Write this as a reduced fraction.
20. The price of gasoline is 210% of what it was a year ago. Write this as a reduced mixed number.

ANSWERS 1. $\frac{57}{100}$ 2. $\frac{89}{100}$ 3. $\frac{2}{5}$ 4. $\frac{11}{50}$ 5. $\frac{2}{25}$ 6. $\frac{1}{50}$
7. $\frac{451}{1000}$ 8. $\frac{337}{500}$ 9. $\frac{17}{400}$ 10. $\frac{3}{100{,}000}$ 11. $\frac{3}{500}$ 12. $\frac{13}{300}$
13. $\frac{21}{200}$ 14. $2\frac{2}{5}$ 15. $1\frac{3}{4}$ 16. $\frac{7}{10}$ 17. $\frac{7}{20}$ 18. $\frac{1}{16}$
19. $\frac{1}{500}$ 20. $2\frac{1}{10}$

Changing percents into decimals is very similar to changing percents into fractions. Again, we replace the % symbol by division by 100. Here it is easier to use ($\div$ 100) instead of $\left(\times \frac{1}{100}\right)$.

To write $N\%$ as a decimal:

1. Replace % by ($\div$ 100) or $\left(\times \frac{1}{100}\right)$.
2. Divide as decimals.

EXAMPLE 7

$$34\% = 34 \div 100$$
$$= 100\overline{)34.00} \quad (\text{quotient } 0.34)$$
$$= 0.34$$

Replace % by ($\div$ 100) → Divide

EXAMPLE 8

$$4.5\% = 4.5 \div 100$$
$$= 100\overline{)4.500} \quad (\text{quotient } 0.045)$$
$$= 0.045$$

Replace % by ($\div$ 100) → Divide

EXAMPLE 9

$$6\frac{1}{4}\% = 6.25\%$$
$$= 6.25 \div 100$$
$$= 100\overline{)6.2500} \quad (\text{quotient } 0.0625)$$
$$= 0.0625$$

Write $6\frac{1}{4}$ as 6.25 → Replace % by ($\div$ 100) → Divide

EXAMPLE 10

$$0.03\% = 0.03 \div 100$$
$$= 100\overline{)0.0300} \quad (\text{quotient } 0.0003)$$
$$= 0.0003$$

Replace % by ($\div$ 100) → Divide

From these examples, we can see a shortcut.

To change a percent to a decimal:

1. Move the decimal point two places to the *left*.
2. Drop the % symbol.

EXAMPLE 11 275% = 2.75%   [Move decimal point two places left] → [Drop % sign]

= 2.75

EXAMPLE 12 A recent poll showed Thomas Righteous getting 48.6% of the vote. Write this as a decimal.

SOLUTION We move the decimal point two places left.

.48.6%   [Move decimal point two places left] → [Drop % sign]

= 0.486

PROBLEM SET Write the following percents as decimals.

See Examples 7 to 12

1. 54% 2. 68% 3. 5% 4. 12.3% 5. 4.25%
6. 18.3% 7. 230% 8. 99.99% 9. 0.001% 10. 80%
11. 101% 12. $6\frac{1}{2}\%$ 13. $\frac{1}{10}\%$ 14. 1% 15. 0.5%
16. A recent poll showed Senator Goodman getting 50.3% of the vote. Write this as a decimal.
17. Barry and Betty spend 19% of their income on food. Write this as a decimal.
18. Roosevelt Federal Savings pays $6\frac{3}{4}\%$ interest. Write this as a decimal.
19. Greg gets a 15% discount on a stereo set. Write this a decimal.
20. A nurse prepares a 12.5% iodine solution. Write this as a decimal.

ANSWERS 1. 0.54 2. 0.68 3. 0.05 4. 0.123 5. 0.0425 6. 0.183 7. 2.3 8. 0.9999 9. 0.00001 10. 0.8 11. 1.01 12. 0.065 13. 0.001 14. 0.01 15. 0.005 16. 0.503 17. 0.19 18. 0.0675 19. 0.15 20. 0.125

Miscellaneous Exercises 4.1

Write the following percents as reduced fractions.

See Examples 1 to 6

1. 37% 2. 49% 3. 15%
4. $5\frac{1}{2}\%$ 5. 350% 6. 0.04%

Write the following percents as decimals.

See Examples 7 to 12

7. 29%   8. 3.8%   9. $2\frac{1}{2}\%$

10. 0.007%   11. 125%   12. 30%

Complete the following table.

See Examples 1 to 12

	Percent	Fraction	Decimal
13.	24%	?	?
14.	0.005%	?	?
15.	450%	?	?
16.	$1\frac{1}{4}\%$	?	?

Business applications

17. A retail store works on a $12\frac{1}{2}\%$ markup. Write this as a reduced fraction.
18. A certain manufacturer finds its sales down 27%. Write that as a decimal.
19. Bigfish Conglomerates buys 51% of the stock in Littlefish Industries. Write this as a decimal.

Health applications

20. A nurse is to administer 75% of an ampul (a sealed vessel of liquid medicine) to a patient. Write this as a reduced fraction.
21. A hospital is preparing a 0.9% saline solution. Write this as a decimal.

Technical applications

22. The efficiency of a certain boiler is 63%. Write this as a decimal.
23. A mechanic was able to increase his output by 125%. Write this as a reduced mixed number.

Ecology application

24. Under natural conditions, the Shrimp-Hydra-Man food chain has a gross ecological efficiency of 10%. Write this as a reduced fraction.

Consumer applications

25. A certain bank offers $6\frac{1}{4}\%$ interest. Write this as a decimal.
26. Judy is going to get a 7.8% raise. Write this as a decimal.
27. Nick receives a 20% discount at the store where he works. Write this as a fraction.
28. The Filmores are in the 31% tax bracket. Write this as a decimal.
29. The price of gasoline went up 80% one year. Write this as a reduced fraction.

Now complete Post-Test 1 for Section 4.1 on page T57.

4.2 CHANGING FRACTIONS AND DECIMALS TO PERCENTS

SECTION OBJECTIVES

After completing this unit, you will be able to:

1. Change a fraction to a percent.
2. Change a decimal to a percent.

TYPICAL PROBLEMS You will be able to work problems like the ones below.

(a) Write $\frac{3}{11}$ as a percent. **(b)** Write $\frac{1}{700}$ as a percent.

(c) Write $1\frac{1}{5}$ as a percent. **(d)** Write 0.531 as a percent.

(e) Write 0.002 as a percent. **(f)** Write 2.7 as a percent.

ANSWERS **(a)** 27.3% **(b)** $\frac{1}{7}$% or 0.14% **(c)** 120% **(d)** 53.1% **(e)** 0.2% **(f)** 270%

In Section 4.1, we saw that 100% = 1. We can use this fact to convert numbers to percent form: we multiply by 100%. Since 100% = 1, this is the same as multiplying by 1, which does not change the number's value.

> To write a decimal as a percent:
> 1. Multiply the decimal by 100% (which is 1).
> 2. Simplify.

EXAMPLE 13

0.31 = 0.31 X 100% — Multiply by 100% (= 1)

$$= \begin{array}{r} 100\% \\ \times\ 0.31 \\ \hline 31.00\% \end{array}$$ — Multiply

= 31%

EXAMPLE 14

0.007 = 0.007 X 100% — Multiply by 100% (= 1)

$$= \begin{array}{r} 100\% \\ \times\ 0.007 \\ \hline 0.700\% \end{array}$$ — Multiply

= 0.7%

EXAMPLE 15

3.5 = 3.5 X 100% — Multiply by 100% (= 1)

$$= \begin{array}{r} 100\% \\ \times\ 3.5 \\ \hline 350.0\% \end{array}$$ — Multiply

= 350%

Perhaps from these examples you can see a shortcut.

> To write a decimal as a percent:
> 1. Move the decimal point two places to the *right*.
> 2. Attach a % symbol.

EXAMPLE 16

0.0015 = 0.0015 (Move two places right)

= 0.15% (Attach % sign)

EXAMPLE 17 In a certain church, the proportion of men is 0.459. Write this as a percent.

SOLUTION We move the decimal point two places to the right.

0.459 (Move decimal point two places right)

= 45.9% (Attach % sign)

EXAMPLE 18 A roller skate company reports that its sales are up 2.25 times. Write this as a percent.

SOLUTION We move the decimal point two places to the right.

2.25 = 225%

Thus sales are up 225%.

PROBLEM SET Write the following decimals as percents.

See Examples 13 to 18

1. 0.79	**2.** 0.26	**3.** 0.06	**4.** 0.015	**5.** 0.98
6. 0.876	**7.** 0.002	**8.** 0.1234	**9.** 0.0004	**10.** 0.1
11. 1.3	**12.** 0.00007	**13.** 1	**14.** 2.56	**15.** 0.0543

16. The new cars are 0.88 times as long as they used to be. Write this as a percent.

17. John's salary goes up by a factor of 0.072. Write this as a percent.

18. Green peppers cost 1.25 times what they did two years ago. Write this as a percent.

19. Judy buys a car and computes the interest rate to be 0.157. Write this as a percent.

20. A nurse measures the strength of a solution to be 0.0025. Write this as a percent.

ANSWERS **1.** 79% **2.** 26% **3.** 6% **4.** 1.5% **5.** 98% **6.** 87.6% **7.** 0.2% **8.** 12.34% **9.** 0.04% **10.** 10% **11.** 130% **12.** 0.007% **13.** 100% **14.** 256% **15.** 5.43% **16.** 88% **17.** 7.2% **18.** 125% **19.** 15.7% **20.** 0.25%

> To write a fraction as a percent:
>
> 1. Multiply the fraction by 100% (which is 1).
> 2. Simplify. (Convert to a decimal, if necessary.)

EXAMPLE 19

$$\frac{2}{5} = \frac{2}{5} \times 100\%$$

$$= \frac{200}{5}\%$$

$$= 40\%$$

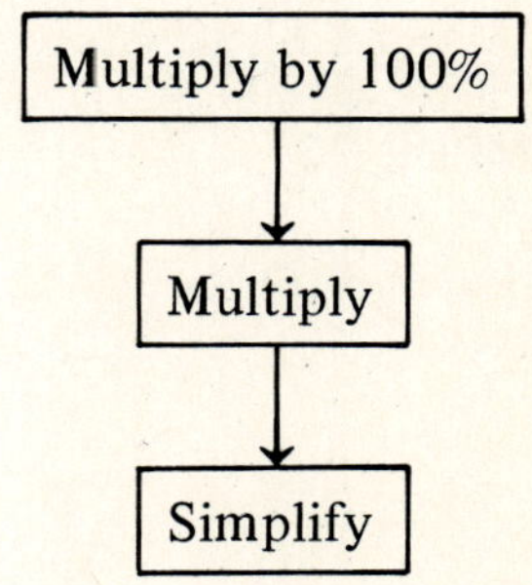

EXAMPLE 20

$$\frac{5}{13} = \frac{5}{13} \times 100\%$$

$$= \frac{500}{13}\%$$

$$= 38\frac{6}{13}\%$$

or 38.46%

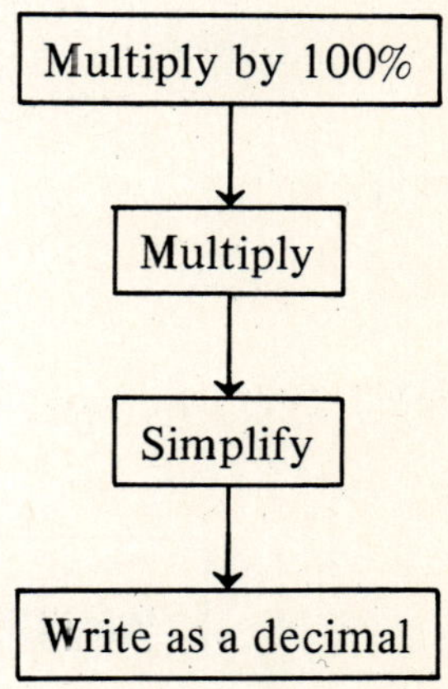

EXAMPLE 21

$$\frac{1}{40} = \frac{1}{40} \times 100\%$$

$$= \frac{100}{40}\%$$

$$= \frac{5}{2}\%$$

$$= 2\frac{1}{2}\%$$

or 2.5%

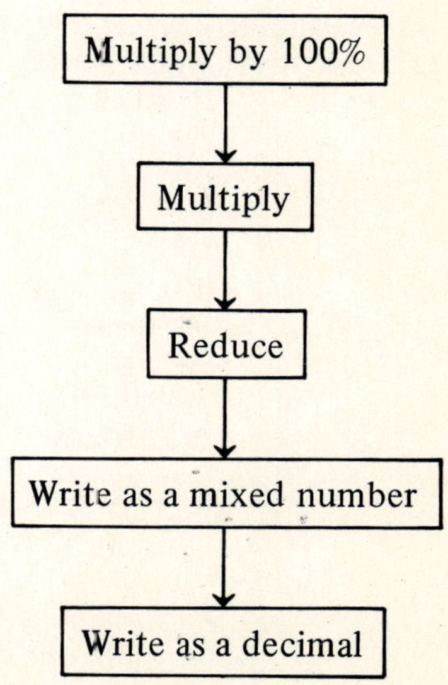

EXAMPLE 22

$1\frac{1}{4} = 1\frac{1}{4} \times 100\%$	Multiply by 100%
$= \frac{5}{4} \times 100\%$	Change $1\frac{1}{4}$ to a fraction
$= \frac{500}{4}\%$	Multiply
$= 125\%$	Simplify

EXAMPLE 23

$\frac{1}{20{,}000} = \frac{1}{20{,}000} \times 100\%$	Multiply by 100%
$= \frac{100}{20{,}000}\%$	Multiply
$= \frac{1}{200}\%$	Reduce
or 0.005%	Write $\frac{1}{200}$ as a decimal

In Example 23, note that *both* answers are correct: $\frac{1}{200}\%$ *and* 0.005%.

> **Why Does It Work?**
>
> Suppose that we want to write any number (fraction or decimal) A as a percent. We have to write this as something over 100. To do this, we multiply A by $\frac{100}{100}$, which is just 1.
>
> $$A = A \times \frac{100}{100} = \frac{A \times 100}{100} = (A \times 100)\%$$
>
> This is why we use the rule $A = A \times 100\%$.

Another method that many students prefer for changing a fraction to a decimal is as follows:

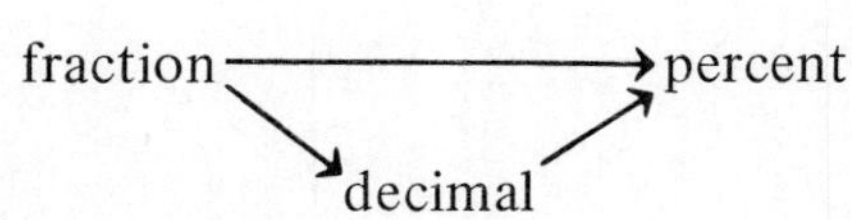

> To change a fraction to a percent:
>
> 1. Change the fraction to a decimal (by dividing).
> 2. Change the decimal to percent (by moving the decimal point two places to the right).

EXAMPLE 24 Write $\frac{2}{17}$ as a percent.

SOLUTION We first write this as a decimal, then as a percent.

$$17\overline{)2.000}\quad 0.117$$

Divide to convert to a decimal

$$\rightarrow 0.12$$

Round off

$$= 0.12\% = 12\%$$

Convert to a percent

EXAMPLE 25 The average human brain is $\frac{1}{47}$ of the body weight. The average beaver's brain is $\frac{1}{197}$ of its body weight. Write these fractions as percents.

SOLUTION We first write these as decimals and then as percents.

Human	*Beaver*
$\frac{1}{47} \rightarrow 47\overline{)1.0000}$ (quotient 0.0212)	$\frac{1}{197} \rightarrow 197\overline{)1.0000}$ (quotient 0.0050)
$\rightarrow 0.021$	$\rightarrow 0.005$
$= 0.021\% = 2.1\%$	$= 0.005\% = 0.5\%$

PROBLEM SET Write the following fractions as percents. Leave the answers to problems 1 to 8 as fraction–percents, such as $2\frac{1}{2}\%$. Leave the answers to problems 9 to 15 as decimal–percents, such as 2.5%.

See Examples 19 to 25

1. $\frac{1}{25}$ 2. $\frac{7}{40}$ 3. $\frac{119}{150}$ 4. $\frac{5}{12}$ 5. $\frac{1}{75}$

6. $\frac{8}{300}$ 7. $6\frac{4}{5}$ 8. $\frac{3}{400}$ 9. $\frac{4}{19}$ 10. $\frac{5}{24}$

11. $\frac{2}{750}$ 12. $\frac{3}{5000}$ 13. $\frac{137}{259}$ 14. $5\frac{3}{7}$ 15. $1\frac{2}{17}$

16. In one class, $\frac{12}{29}$ are men. Write this as a percent (with a fraction).

17. The number of cars on the road is $1\frac{1}{4}$ times what it was a few years ago. Write this as a percent.

18. The chance of rolling an eight with a pair of dice is $\frac{5}{36}$. Write this as a percent (with decimals).

19. In a survey, $\frac{178}{298}$ of the people favored Senator Higgins for reelection. Write this as a percent (with a decimal).

20. Monica's pay went up by $\frac{1}{15}$. Write this as a percent (with a fraction).

ANSWERS 1. 4% 2. $17\frac{1}{2}\%$ 3. $79\frac{1}{3}\%$ 4. $41\frac{2}{3}\%$ 5. $1\frac{1}{3}\%$ 6. $2\frac{2}{3}\%$

7. 680% 8. $\frac{3}{4}\%$ 9. 21.1% 10. 20.8% 11. 0.27%

12. 0.06% 13. 52.9% 14. 542.9% 15. 111.8% 16. $41\frac{11}{29}\%$

17. 125% 18. 13.9% 19. 59.7% 20. $6\frac{2}{3}\%$

Miscellaneous Exercises 4.2

Write the following fractions and decimals as percents.

See Examples 13 to 18

1. 0.43 2. 0.0023 3. 4.3

4. 0.0362 5. 0.4 6. 0.15

See Examples 19 to 25

7. $\frac{5}{9}$ 8. $\frac{5}{14}$ 9. $\frac{1}{50}$

10. $\frac{3}{4000}$ 11. $3\frac{3}{4}$ 12. $\frac{23}{82}$

Complete the following table.

See Examples 1 to 25

	Percent	Fraction	Decimal
13.	42%	?	?
14.	0.02%	?	?
15.	?	$\frac{2}{7}$	?

	Percent	Fraction	Decimal
16.	?	$\frac{3}{1700}$	?
17.	?	?	0.6
18.	?	?	2.2

Chemistry application

19. Sulfuric acid (H_2SO_4) has a molecular weight of 98. The elements hydrogen (H), oxygen (O), and sulfur (S) contribute as follows:

(number of atoms of H) × (atomic weight of H) = 2 × 1	= 2
(number of atoms of S) × (atomic weight of S) = 1 × 32	= 32
(number of atoms of O) × (atomic weight of O) = 4 × 16	= 64
Total molecular weight	= 98

(a) Find the percent of hydrogen = $\frac{2}{98}$.

(b) Find the percent of sulfur = $\frac{32}{98}$.

(c) Find the percent of oxygen = $\frac{64}{98}$.

Technical application

20. An engine is rated at 160 horsepower but only delivers 110 horsepower. Find the percent of efficiency = $\frac{110}{160}$.

Business applications

21. Wheeldeal Industries controls 0.09835 of the proxies of another company. Write this as a percent.

22. A marketing survey finds that $\frac{97}{145}$ of a sample are familiar with Rosehips soft drinks. Write this as a percent.

23. A company votes to pay a \$6 dividend on each \$50 share of stock. Find the percent of this dividend = $\frac{6}{50}$.

Health applications

24. An epinephrine solution is made from 2 grams of solid epinephrine in 800 milliliters (grams) of water. Find the percent strength of this solution = $\frac{2}{800}$.

25. The proportion of protein in a certain food is 0.025. Write this as a percent.

Consumer applications

26. The price of bacon went up by a third one year. Write this fraction as a percent.

27. The Wentzels are told that their share in a condominium town house is 0.09645. Write this as a percent.

28. Sherry earns \$12,000. Then she gets a \$1000 raise. Find the percent of her raise = $\frac{1000}{12{,}000}$.

Now complete Post-Test 1 for Section 4.2 on page T59.

4.3 PERCENTAGE PROBLEMS

SECTION OBJECTIVES

After completing this unit, you will be able to:

1. Translate and solve simple percentage problems.
2. Set up and solve verbal percentage problems.

TYPICAL PROBLEMS

You will be able to solve problems like the ones below.

(a) What is 14% of 200?

(b) One hundred fifty is what percent of 500?

(c) Twenty-five is 5% of what?

(d) Helen earns $11,000 per year. She receives a 5.6% raise. How much is the raise and her new salary?

(e) Andy surveys 150 people on the street. Seventy-nine say that they will vote for Senator Washington. What percent is this?

(f) A person should spend about 25% of his or her income on rent. Janet is interested in an apartment that rents for $265 a month. What must her monthly income be to afford the apartment?

ANSWERS

(a) 28 **(b)** 30% **(c)** 500 **(d)** Raise = $616; new salary = $11,616 **(e)** 52.7% **(f)** $1060

Many problems in everyday life involve percents. But when we do the mathematics of the problem, we must use decimals (and sometimes fractions). Recall how we change percents and decimals:

$$34\% = 34 \div 100 = 0.34 \quad \text{(percent to decimal)}$$

$$0.074 = 0.074 \times 100\% = 7.4\% \quad \text{(decimal to percent)}$$

Many percentage problems can be solved by **translating** from an English sentence into a mathematical equation using the following table.

English	Mathematics
of ⟶	$\cdot$ (multiplication)
what ⟶	x (unknown)
out of ⟶	$\div$ (division)
is ⟶	$=$ (equals)

EXAMPLE 26 The following examples show how we translate English sentences (or questions) into mathematical equations. Notice that the word *is* translates into =, which then acts like the verb of the equation.

(a) Forty is what percent of 75?

$$40 = x \cdot 75$$

(b) What is 15% of 5000?

$$x = 0.15 \cdot 5000$$

(c) Two hundred is 30% of what number?

$$200 = 0.30 \cdot x$$

(d) Five out of 19 is what percent?

$$5 \div 19 = x$$

(e) Ten percent of what number is 350?

$$0.10 \cdot x = 350$$

PROBLEM SET Translate the following English sentences into mathematics.

See Example 26

1. Fifty is 2% of what number?
2. What is 28% of 5900?
3. One thousand is what percent of 3500?
4. Ten out of 25 is what percent?
5. What is 125% of 400?
6. Twenty out of 30 is what percent?
7. Nine hundred is what percent of 3200?
8. Five percent of what number is 235?

ANSWERS 1. $50 = 0.02 \cdot x$ 2. $x = 0.28 \cdot 5900$ 3. $1000 = x \cdot 3500$ 4. $10 \div 25 = x$ 5. $x = 1.25 \cdot 400$ 6. $20 \div 30 = x$ 7. $900 = x \cdot 3200$ 8. $0.05 \cdot x = 235$

Before we begin solving percentage problems, let us review the fact that division is the opposite of multiplication. We use this idea to find an unknown in an equation.

EXAMPLE 27 Find x: $20 \cdot x = 100$.

SOLUTION Since division is the opposite of multiplication, we divide both sides by 20 to find x.

$$20 \cdot x = 100$$

$$\frac{\cancel{20} \cdot x}{\cancel{20}} = \frac{100}{20}$$

$$x = 5$$

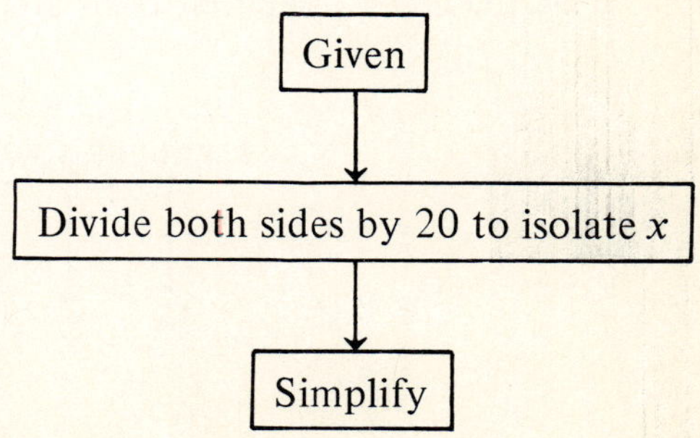

EXAMPLE 28 Find x: $0.15 \cdot x = 3000$.

SOLUTION We isolate x by dividing both sides by 0.15. Recall that we move the decimal point two places to the right in both numbers when we divide.

$$0.15 \cdot x = 3000$$

$$\frac{\cancel{0.15} \cdot x}{\cancel{0.15}} = \frac{3000}{0.15}$$

$$x = 0.15\overline{)3000.00} \quad \begin{array}{r} 200\,00 \\ 3000\,00 \\ 30 \\ \hline 00\,00 \end{array}$$

$$x = 20{,}000$$

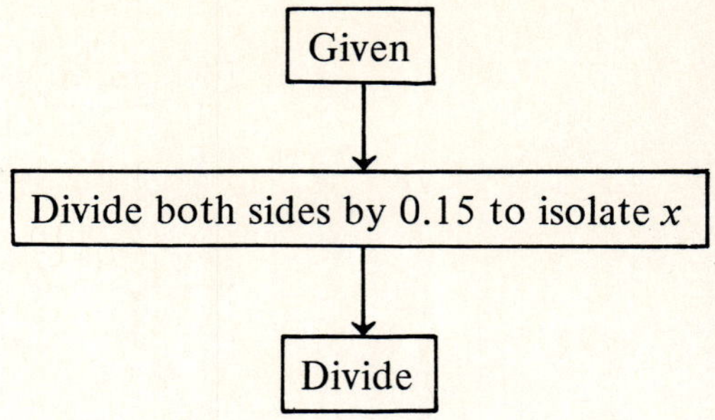

PROBLEM SET Find x in the following problems.

See Examples 27 and 28 (See also Examples 40 to 42 of Chapter 3)

1. $2 \cdot x = 12$
2. $5 \cdot x = 50$
3. $6 \cdot x = 2.4$
4. $12 \cdot x = 0.48$
5. $0.12 \cdot x = 8.4$
6. $0.4 \cdot x = 0.08$
7. $3.5 \cdot x = 700$
8. $50 \cdot x = 2.4$
9. $0.03 \cdot x = 0.009$
10. $0.45 \cdot x = 6.75$

ANSWERS **1.** 6 **2.** 10 **3.** 0.4 **4.** 0.04 **5.** 70 **6.** 0.2 **7.** 200 **8.** 0.048 **9.** 0.3 **10.** 15

> To solve simple percentage problems:
>
> 1. Translate the English into mathematics.
> 2. Solve the equation for the unknown.
> 3. If the unknown is a percent, change the decimal to a percent.

EXAMPLE 29 Eighteen is what percent of 42?

SOLUTION We first translate, and then we solve.

Eighteen is what percent of 42?

$$18 = x \cdot 42$$

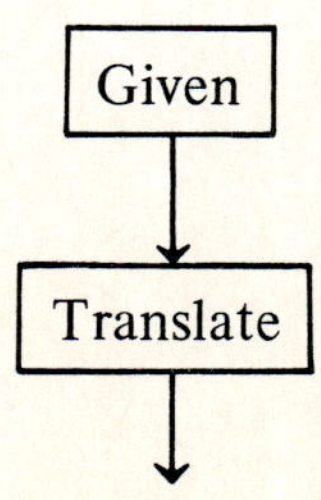

$$\frac{18}{42} = \frac{x \cdot \cancel{42}}{\cancel{42}}$$

Divide both sides by 42 to isolate x

$$x = 42\overline{)18.000} = 0.428$$

```
        0.428
x = 42 ⟌18.000
        16 8
         1 20
           84
           360
           336
            24
```

Divide

$$x = 0.43$$

Round off

$$x = 43\%$$

Convert to percent

EXAMPLE 30 What is 12% of 145?

SOLUTION We translate and then solve.

What is 12% of 145?

$$x = 0.12 \cdot 145$$

Given → Translate

```
      145
    ×0.12
     2 90
    14 5
x = 17.40
```

Multiply (x is already alone)

EXAMPLE 31 Ninety-five is 26% of what?

SOLUTION Again, we translate and solve.

Ninety-five is 26% of what?

$$95 = 0.26 \cdot x$$

Given → Translate

$$\frac{95}{0.26} = \frac{\cancel{0.26} \cdot x}{\cancel{0.26}}$$

Divide both sides by 0.26 to isolate x

```
       3 65 38
0.26 ) 95.00 00  = x
       78
       17 0
       15 6
        1 40
        1 30
          10 0
           7 8
           2 20
           2 08
             12
```

$365.4 = x$

Divide → Round off

EXAMPLE 32 Fourteen out of 20 is what percent?

SOLUTION We translate and solve.

Fourteen out of 20 is what percent?

$$14 \div 20 = x$$

```
      0.70
20 ) 14.00  = x
     14 0
        0
        0
```

$70\% = x$

Given → Translate → Divide (*x* is already alone) → Convert to percent

EXAMPLE 33 Forty is $1\frac{1}{4}\%$ of what number?

SOLUTION We must be careful when we translate $1\frac{1}{4}\%$ into a decimal. $1\frac{1}{4}\% = 1.25\% = 1.25 \div 100 = 0.0125$.

Forty is $1\frac{1}{4}\%$ of what?

$$40 = 0.0125 \cdot x$$

Given → Translate →

$$\frac{40}{0.0125} = \frac{\cancel{0.0125} \cdot x}{\cancel{0.0125}}$$

Divide by 0.0125 to isolate x

$$0.0125\overline{)40.0000} = x$$

$$\begin{array}{r} 3200 \\ 0.0125\overline{)40.0000} \\ 37\,5 \\ \hline 2\,50 \\ 2\,50 \\ \hline 00 \\ 00 \\ \hline \end{array}$$

Divide

$$3200 = x$$

Answer

PROBLEM SET Solve the following percentage problems.

See Examples 29 to 33

1. What is 60% of 500?
2. Seventy is what percent of 200?
3. Ninety-two is 46% of what?
4. What is $5\frac{1}{2}\%$ of 300?
5. Forty-five is what percent of 300?
6. One hundred thirty is 65% of what number?
7. Fifteen is what percent of 2250?
8. Two hundred sixty-one out of 700 is what percent?
9. What is 0.3% of 7500?
10. Five thousand is what percent of 1500?
11. Thirteen thousand is 65% of what?
12. What is 350% of 70,000?
13. Twenty-four is what percent of 850?
14. Ninety-three out of 98 is what percent?
15. What is $\frac{1}{4}\%$ of 68,000?
16. Fifty-five percent of what is 330?
17. What percent of 980 is 320?
18. Eighty-two percent of 4500 is what?
19. What percent is 68 out of 360?
20. What is 7% of 600,000?

ANSWERS **1.** 300 **2.** 35% **3.** 200 **4.** 16.5 **5.** 15% **6.** 200
7. 0.67% **8.** 37.3% **9.** 22.5 **10.** 333% **11.** 20,000

12. 245,000 13. 2.8% 14. 94.9% 15. 170 16. 600
17. 32.7% 18. 3690 19. 18.9% 20. 42,000

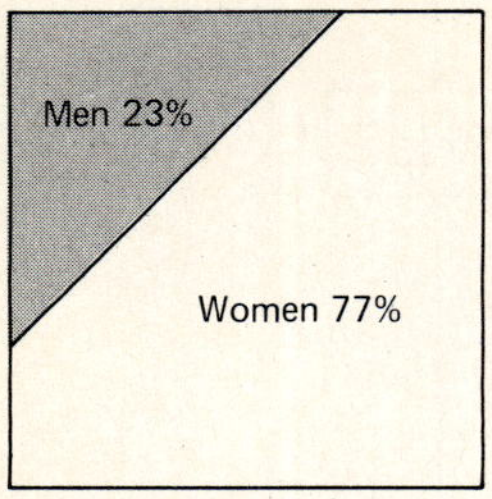

Whole=100%

The following fact is very important when working percentage problems.

> The whole is 100%:
> If a part of the whole is $N\%$, the remainder is $(100 - N)\%$.

EXAMPLE 34 (a) Of a group, 23% are men; then 77% of the group are women.
(b) There is an 80% chance of rain; there is a 20% chance of no rain.
(c) A man got a 30% discount; he paid 70% of the price.

PROBLEM SET

See Example 34

1. If 12% of a class wear gym shoes, what percent do not?
2. If a man pays 22% of his income for tax, what percent does he keep?
3. If $2\frac{1}{2}\%$ of a product is sugar, what percent is not sugar?
4. If there is a 50% chance of flipping a head on a fair coin, what is the chance of a tail?
5. If 94.3% of American households own TV sets, what percent do not?

ANSWERS 1. 88% 2. 78% 3. $97\frac{1}{2}\%$ 4. 50% 5. 5.7%

Some percentage problems are not simple translations. But we do notice that most of the percentage problems in this section had the form, "*A* is some percent of *B*." Every percentage problem has three parts:

> 1. *Amount* (A): This is a *part* of the whole; something *new*; a *change.*
> 2. *Percent* (P): This is the *percent* in the problem; for example, 25%, $5\frac{1}{2}\%$, 93.8%, and so on.
> 3. *Base* (B): This is the *whole*, the *original* size, the *list* price, and so on.

For example, suppose that there are 200 people in a room, and 31% (or 62 people) smoke. Then

$$A = 62 \textit{ smokers}$$

$$P = 31\% \textit{ smokers}$$

$$B = 200 \text{ total}$$

In every percentage problem, you will know two of these three parts, and the third will be the unknown. The following is a very important fact to remember.

> The amount (A) and the percent (P) must represent the same quantity in the problem.

For example,

$$A = \text{amount of } \textit{increase}, \quad P = \text{percent of } \textit{increase}$$

$$A = \text{amount of } \textit{bank interest}, \quad P = \text{percent of } \textit{bank interest}$$

These three terms are connected as follows. The *amount* is some *percent* of the *base.* We can translate this into a formula.

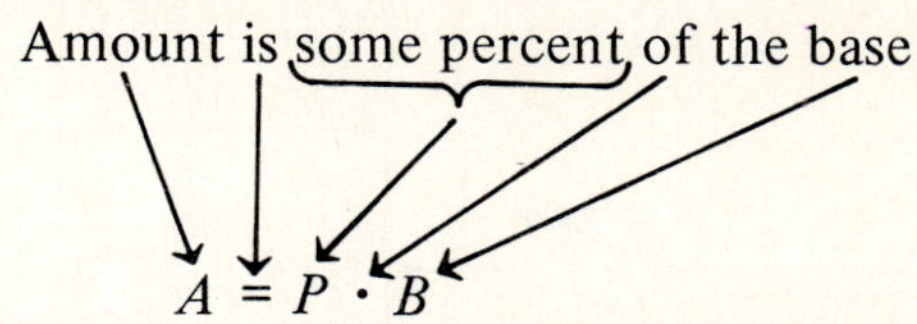

> To solve percentage problems:
>
> 1. Identify the terms A, P, and B. (Be sure that A and P represent the same quantity.)
> 2. Write down the basic formula, $A = P \cdot B$.
> 3. Substitute the known quantities into the basic equation.
> 4. Solve for the unknown.
> 5. If P is the unknown, be sure to convert it to a percentage.

EXAMPLE 35 City Bank pays 6% interest on deposits. If Jan and Wayne deposit \$1300, how much interest will they earn in 1 year?

SOLUTION We must first identify the terms before we can substitute into the basic equation and solve. Here we know the percent and the base.

$$\begin{cases} A = \text{amount of } \textit{interest} = ? \\ P = \text{percent of } \textit{interest} = 6\% = 0.06 \\ B = \text{base} = 1300 \end{cases}$$

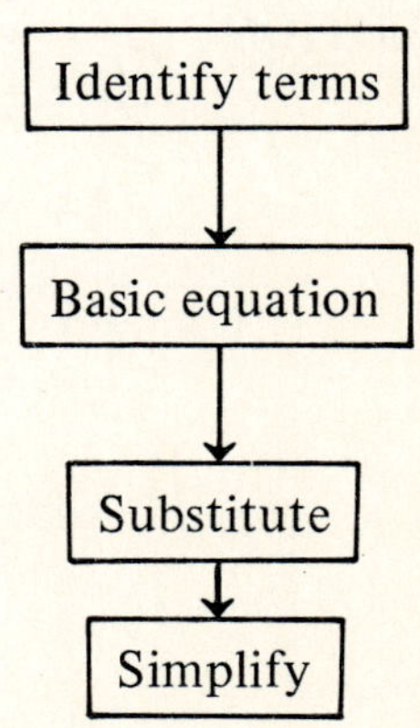

$A = P \cdot B$

$A = (0.06) \cdot (1300) \longrightarrow$

$$\begin{array}{r} 1300 \\ \times\ 0.06 \\ \hline 78.00 \end{array}$$

$A = 78 \longleftarrow$

Thus they will earn \$78 interest. Notice that A and P were the same quantity, *interest.*

EXAMPLE 36 There are 217 employees in a certain company, and 52 of them are joggers. What percent is this?

SOLUTION We are given the total (base) and the part (amount), and the unknown is the percent.

$$\begin{cases} A = \text{amount of } \textit{joggers} = 52 \\ P = \text{percent of } \textit{joggers} = ? \\ B = \text{base, total} = 217 \end{cases}$$

$$A = P \cdot B$$

$$52 = P \cdot 217$$

$$\frac{52}{217} = \frac{P \cdot \cancel{217}}{\cancel{217}} \longrightarrow$$

$$\begin{array}{r} 0.239 \\ 217\overline{)52.000} \\ \underline{43\ 4} \\ 8\ 60 \\ \underline{6\ 51} \\ 2\ 090 \\ \underline{1\ 953} \\ 137 \end{array}$$

$$0.239 = P \longleftarrow$$

$$0.24 = P$$

$$24\% = P$$

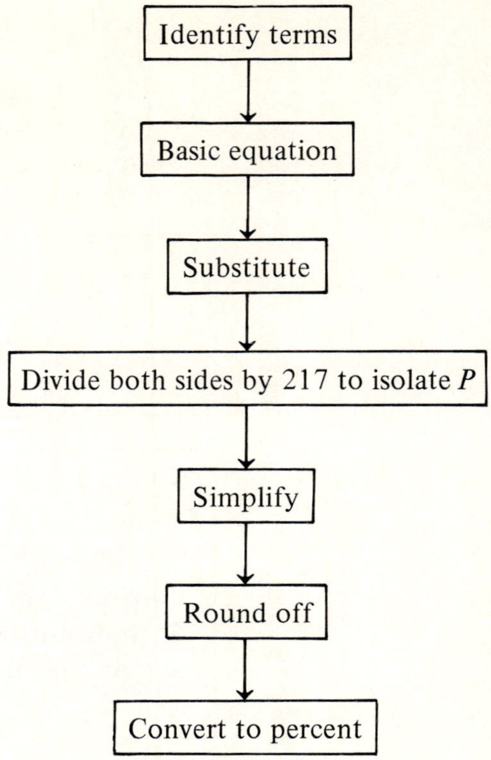

So 24% of the employees are joggers. Notice again that the quantity *joggers* was the same for A and P.

EXAMPLE 37 Eddie spends 15% of his salary on his car. He spends \$210 a month on his car. What is his monthly income?

SOLUTION Here we are given the amount spent on the car, and the percent spent on the car. The unknown is the base or the total monthly salary.

$$\begin{cases} A = \text{amount } \textit{spent on car} = 210 \\ P = \text{percent } \textit{spent on car} = 15\% = 0.15 \\ B = \text{base, total} = ? \end{cases}$$

$$A = P \cdot B$$

$$210 = (0.15) \cdot B$$

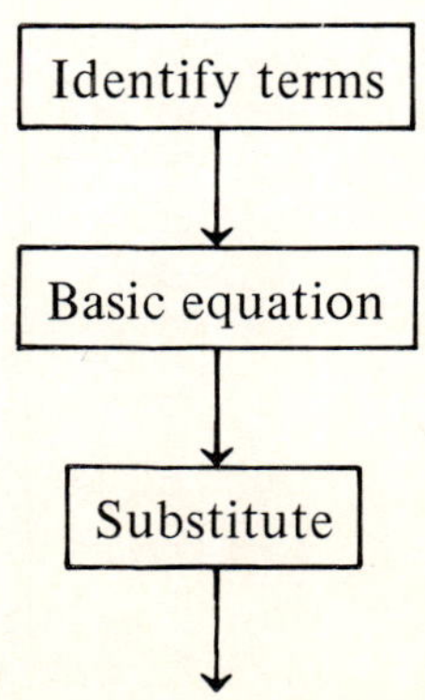

$$\frac{210}{0.15} = \frac{(0.15) \cdot B}{0.15}$$

$$0.15\overline{)210.00} = 1400$$

14 00
0.15) 210.00
15
60
60
00
00

$$1400 = B$$

Divide both sides by 0.15 to get B alone → Simplify

Therefore, Eddie earns \$1400 per month. Here we divided by 0.15 to get B by itself. We also have to be careful to divide the decimals properly.

EXAMPLE 38 Bob is interested in a \$360 stereo set. There is a 30% discount. What will the set cost?

SOLUTION The list price, \$360, is the base. Since we are given the percent of discount, we must first solve for the amount of the discount.

$$\begin{cases} A = \text{amount of } discount = ? \\ P = \text{percent of } discount = 30\% = 0.30 \\ B = \text{base, list price} = 360 \end{cases}$$

Identify terms

$$A = P \cdot B$$

Basic equation

$$A = (0.30) \cdot (360)$$

$$\begin{array}{r} 360 \\ \times \quad 0.30 \\ \hline 108.00 \end{array}$$

Substitute

$$A = 108$$

Simplify

So the discount is \$108. Therefore, he will pay \$360 − \$108 = \$252.

Another way to solve this is as follows. Since he will get a 30% *discount*, he will *pay* 70% (= 100% − 30%).

$$\begin{cases} A = \text{amount } paid = ? \\ P = \text{percent } paid = 70\% = 0.70 \\ B = \text{base, list price} = 360 \end{cases}$$

Identify terms

$$A = P \cdot B$$

Basic equation

$$A = (0.70) \cdot (360)$$

$$\begin{array}{r} 360 \\ \times \quad 0.70 \\ \hline 252.00 \end{array}$$

Substitute

$$A = 252$$

Simplify

We get the same answer for the price, \$252. Notice that in either case, the amount and the percent had to be the same quantity.

EXAMPLE 39 Sue earned \$12,200 last year. This year, she earned \$13,100. What was the percent of increase?

SOLUTION We must be careful here. We are looking for the percent of *increase*; therefore, we must first find the amount of *increase*. This is 13,100 − 12,200 = 900. Sometimes we have to do a little work like this to make the A and P quantities the same.

$$\begin{cases} A = \text{amount of } increase = 900 \\ P = \text{percent of } increase = \, ? \\ B = \text{base, original salary} = 12{,}200 \end{cases}$$ Identify terms

$$A = P \cdot B$$ Basic equation

$$900 = P \cdot 12{,}200$$ Substitute

$$\frac{900}{12{,}200} = \frac{P \cdot \cancel{12{,}200}}{\cancel{12{,}200}}$$ Divide both sides by 12,200 to get P alone

$$\begin{array}{r} 0.0737 \\ 12200 \overline{)\,900.0000} \\ 854\;00 \\ \hline 46\;000 \\ 36\;600 \\ \hline 9\;4000 \\ 8\;5400 \\ \hline 8600 \end{array}$$

$$0.0737 = P$$ Simplify

$$0.074 = P$$ Round off

$$7.4\% = P$$ Convert to percent

YES	NO
$P = \dfrac{900}{12{,}200}$	~~$P = \dfrac{13{,}100}{12{,}200}$~~

If we want percent of *increase*, we must use amount of *increase*.

EXAMPLE 40 A certain church membership is 62% women. If there are 145 men in the church, what is the total membership?

SOLUTION We have a slight problem here. We are given the amount (number) of men and the percent of women. Can we find the amount (number) of women? No. Can we find the percent of men? Yes. If there are 62% women, then there are 100% − 62% = 38% men. Now we can work the problem.

$$\begin{cases} A = \text{amount (number) of } men = 145 \\ P = \text{percent of } men = 38\% = 0.38 \\ B = \text{base, total} = ? \end{cases}$$

$$A = P \cdot B$$

$$145 = (0.38) \cdot B$$

$$\frac{145}{0.38} = \frac{\cancel{(0.38)} \cdot B}{\cancel{0.38}} \longrightarrow$$

$$\begin{array}{r} 381.5 \\ 0.38\overline{)\,145.000} \\ \underline{114} \\ 31\,0 \\ \underline{30\,4} \\ 60 \\ \underline{38} \\ 22\,0 \\ \underline{19\,0} \\ 3\,0 \end{array}$$

$$381.5 = B \longleftarrow$$

$$382 = B$$

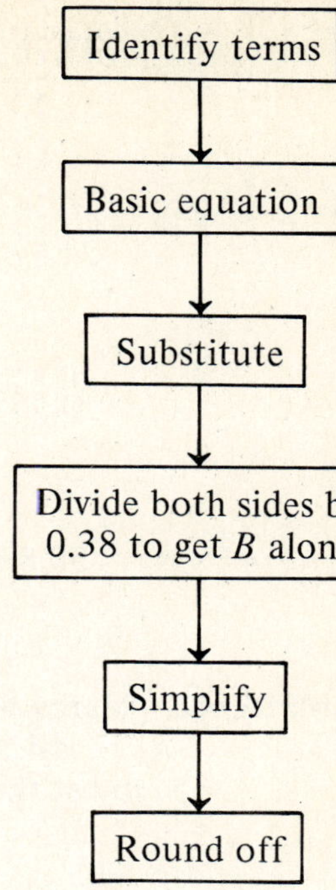

YES	NO
A = amount of *men* = 145	~~A = amount of *men* = 145~~
P = percent of *men* = 38%	~~P = percent of *women* = 62%~~
B = base, total = ?	~~B = base, total = ?~~

A and P must be the *same* quantity.

EXAMPLE 41 In a recent election, Ken Kleen defeated Neal Niceguy 123,965 to 98,344. What percent of the total did Kleen get?

SOLUTION Here we must first find the base or total. To do this, we add the two vote tallies: 123,965 + 98,344 = 222,309.

$$\begin{cases} A = \text{amount } Kleen\ got = 123{,}965 \\ P = \text{percent } Kleen\ got = ? \\ B = \text{base, total} = 222{,}309 \end{cases}$$

$$A = P \cdot B$$

$$123{,}965 = P \cdot 222{,}309$$

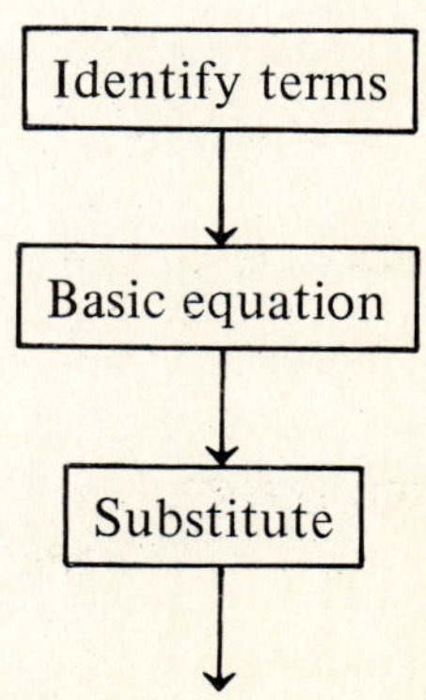

$\frac{123,965}{222,309} = P$	Divide both sides by 222,309 to get P alone
$0.5576 = P$	Simplify
$0.558 = P$	Round off
$55.8\% = P$	Convert to percent

```
             0.5576
222309 | 123965.0000
          111154 5
          12810 50
          11115 45
           1695 050
           1556 163
            138 8870
            133 3854
              5 5016
```

EXAMPLE 42 Marcy and Lee are selling their house through a real estate broker who charges a 7% sales commission. If they wish to clear $32,000 *after the commission*, what must the selling price be?

SOLUTION Again, we must be careful. The 7% commission is *not* charged on the $32,000, but on the actual sale price. The $32,000 is the amount left *after commission*. To be the same, the percent must be 100% − 7% = 93% *after commission.*

A = amount *after commission* = 32,000 P = percent *after commission* = 93% = 0.93 B = base, total price = ?	Identify terms
$A = P \cdot B$	Basic equation
$32,000 = (0.93) \cdot B$	Substitute
$\frac{32,000}{0.93} = B$	Divide both sides by 0.93 to get B alone
$34,408.6 = B$	Simplify
$34,409 = B$	Round off

```
          344 08 6
0.93 | 32000.00 0
       279
        410
        372
         380
         372
           8 00
           7 44
             56 0
             55 8
                2
```

So, in order to make $32,000 after their sale, they must actually sell the house for at least $34,409. Also, in problems like this it is easy to make a mistake dividing the decimals. Be careful when you divide; be sure to move the proper number of places.

PROBLEM SET Work the following percentage problems.

See Examples 35 to 42

1. In one town, 2000 children take an IQ test. Sixty-eight percent of the scores are between 85 and 115. How many children is this?
2. The price of hamburger was \$1.10 per pound. The price went up 12%. What was the price increase and the new price?
3. Jerry earns \$1100 per month. He spends \$220 per month on his car. What percent is this?
4. In a recent election, Senator Arbuckle got 1,100,000 votes out of 2,000,000 cast. What percent was this?
5. Angie pays \$270 for a \$400 stereo set. What percent discount is this?
6. Fran earns \$1400 per month. Next year she will earn \$1470 per month. What percent increase is this?
7. A certain housing development requires a 10% down payment. What down payment is needed to buy a \$48,000 house in the development?
8. A group is 80% men. If there are 180 women, how large is the group?
9. Sid pays \$5200 for a new car. A year later it is worth only \$3900. What percent has the value decreased?
10. It takes a 20% down payment to buy a used house. Jeff and Debby have saved \$7300. How expensive a used house can they buy?
11. In a recent election, Maura Media defeated Ian Image by 780,000 to 700,000. There were 20,000 votes cast for other candidates. What percent of the vote did Media get?
12. Jeni buys an antique chair for \$55. She later sells it for \$135. What percent profit is this?
13. Terry pays 22% income tax on every extra dollar he earns. He needs \$1200 (after taxes) to buy a camera and dark-room equipment. How much does he need to earn so that he will have \$1200 after taxes?
14. A box of oatmeal claims that it provides 5 grams of protein per ounce. It also claims that each ounce provides 6% of an adult's daily recommended need for protein. Using this, what is an adult's daily recommended need for protein?
15. Experts say that the monthly cost of a house is about $1\frac{1}{4}\%$ of the total cost. Danny and Diane are interested in a \$60,000 house.
    (a) What will their monthly cost be (by this rule)?
    (b) If this cost per month is 25% of their income, what must their monthly income be to afford the house?
    (c) If the bank requires a 20% down payment, how large a down payment do they need?
16. Judy and Mike want to make \$35,000 (clear) on the sale of their home.
    (a) How much should the sale price be so that they clear \$35,000 after a 6% commission?

(b) What should their asking price be so that they can appear to discount their asking price by 10% and get the needed sale price in part (a)?

17. In 1960, Harbor State College had 1500 students, 200 of whom were women. In 1980, there were 2400 students, 900 of whom were women.
(a) What was the percent of women in 1960?
(b) What was the percent of women in 1980?
(c) What was the percent of increase in students in the 20 years?
(d) What was the percent of increase in women students in the 20 years?

18. Harold Chace owns a store. He is trying to sell a \$300 washing machine. He first raises the price by 20%. Then he offers a 20% discount. What is the final price?

19. An item costs \$250. Is it cheaper to charge 5% tax and then take off a 20% discount, or is it cheaper to take off the 20% discount first and then add the 5% tax? Do it both ways.

20. Sally pays \$6600 for a car. She is told that this is 10% over cost. What is the list price? (*Hint:* Her cost is 110% of list.)

21. The Rogers own a \$55,000 house. The real estate taxes are \$1430 per year. What percent is the tax rate?

22. If the population of the world is about 4,000,000,000 and it increases about 2.1% per year, how much will the earth's population increase in 1 year?

23. Fred can earn 7% interest on his money. How much money must he invest in order to earn \$5000 a year interest?

ANSWERS 1. 1360 2. Increase = 13 cents; new price = \$1.23 3. 20% 4. 55% 5. 32.5% 6. 5% 7. 4800 8. 900 9. 25% 10. 36500 11. 52% 12. 145.5% 13. 1538.46 14. 83.3 15. (a) 750 (b) 3000 (c) 12,000 16. (a) 37,234 (b) 41,371 17. (a) 13.3% (b) 37.5% (c) 60% (d) 350% 18. 288 19. Same: 210 both ways 20. 6000 21. 2.6% 22. 84,000,000 23. 71,428.57

Miscellaneous Exercises 4.3

Translate the following into mathematical equations.

See Example 26

1. What is 54% of 490?
2. Twenty-one is what percent of 460?
3. Fifty is 3% of what number?

Find x in the following problems.

See Examples 27 and 28

4. $5 \cdot x = 40$
5. $12 \cdot x = 68$
6. $0.25 \cdot x = 48$
7. $0.02 \cdot x = 350$

Translate and solve the following.

See Examples 29 to 33

8. What is 30% of 2500?
9. Thirty is what percent of 500?
10. One hundred is 5% of what number?

11. Two out of 8 is what percent?

12. Forty percent of 25 is what?

13. What number is 45% of 3000?

14. Fifteen percent of what number is 45?

See Examples 35 to 42 The following applications are typical percentage word problems. Solve these using the methods shown in Examples 35 to 42.

Business applications

15. A certain company pays its salespeople's commissions as follows: 5% on the first $100,000 sales and 6% on all sales over $100,000. Compute the following commissions. (*Hint:* For sales over $100,000, compute the commission in two parts—on the first $100,000, then on the sales minus $100,000.)
    (a) Wilson: $85,000
    (b) Haken: $120,000
    (c) Dimbrowskas: $95,000
    (d) Ruben: $105,000

16. There are about 70,000,000 households in the United States. An advertiser estimates that 34% of those households were watching a TV show that it sponsored. How many households is this?

17. The ABC YoYo Company pays a $7 dividend on each $80 share of common stock. Find the percent of this dividend.

Health applications

18. A patient's systolic blood pressure increased from 150 to 180. Find the percent of increase.

19. A $2\frac{1}{2}$% dextrose solution contains 10 ml of pure dextrose. How large is the total solution?

Technical applications

20. A certain automobile engine has 48 horsepower. The transmission and rear-end drive deliver only 88% of this power to the rear wheels. How much power is delivered to the rear wheels?

21. A blanking die cuts pieces from sheets of aluminum. Each piece is 0.815 inch, with a possible error of 0.015 inch. What is the percent of error?

Psychology application

22. In an experiment, some mice were raised by their natural mothers and others by foster mothers. They were then tested for aggressive fighting behavior.
    (a) Of those with their natural mothers, 27 of 167 were "fighters." What percent is this?
    (b) Of those with a foster mother, 47 of 140 were "fighters." What percent is this?

Literature application

23. In testing a piece of writing for its true authorship, it is common to compare word-length frequency with a known work. For example, to test a writing to determine if it was written by Mark Twain, his percent usage of words was studied. Find the percent of one-letter words, two-letter words, and so on, in a 1885-word writing.

Word Length	Number	Percent
1	74	?
2	349	?
3	456	?
4	374	?
5	212	?

Word Length	Number	Percent
6	127	?
7	107	?
8	84	?
9	45	?
10	27	?
11	13	?
12	8	?
Over 12	9	?
Total	1885	100%

Chemistry application

24. Salt is sodium chloride (NaCl). Sodium has an atomic weight of 23, and chlorine has an atomic weight of 35.5.
(a) What is the total atomic weight of NaCl?
(b) What percent of the weight is sodium?

Ecology application

25. The food chain only has an efficiency of 10%. Suppose that a hamburger has 300 calories of energy. This is 10% of the energy that the cow consumed to produce that hamburger. How many calories of energy did the cow consume to produce that 300-calorie hamburger?

Consumer applications

26. The Gomez family invests $2500 at 8% interest. How much interest will they earn in 1 year?

27. Chuck and Donna have saved $5500 for the down payment on a house. If this is a 20% down payment, how expensive a house can they afford?

28. Greg buys a $6400 car. Two years later, it is worth $3600. What percent has the car depreciated?

29. A certain credit card charges $1\frac{1}{2}$% interest per month on the unpaid balance. If the Daleys have an unpaid balance of $650, what will their finance charge be?

30. Jerry buys a $250 guitar. He later sells it for $400. What is the percent of profit?

Now complete Post-Test 1 for Section 4.3 on page T61.

4.4 RATIO AND PROPORTION

SECTION OBJECTIVES

After completing this unit, you will be able to:

1. Translate and solve simple proportion problems.

2. Set up and solve verbal proportion problems.

TYPICAL PROBLEMS

You will be able to work problems like the ones below. Solve the following proportion problems.

(a) x is to 5 as 7 is to 13; solve for x.

(b) Eight is to 7 as 17 is to what?

(c) A bag of 110 nails costs 63 cents. How many nails can one get for $1.50?

(d) Donna is 5 feet, 4 inches tall (or 64 inches). One inch is 2.54 centimeters. How tall is Donna in centimeters?

(e) Rudy can go 193 miles on 18.3 gallons of gasoline. How many gallons will it take to go 1500 miles?

ANSWERS **(a)** $2\frac{9}{13}$ **(b)** $14\frac{7}{8}$ **(c)** 262 **(d)** 162.56 **(e)** 142.2

The first thing that we study is a ratio. **A ratio** is a comparison of two quantities by division. A ratio may be written as a fraction or a decimal, whichever is more convenient. As examples of ratios, we have

(a) $\frac{7}{9}$ (also written 7 to 9 or 7:9)

(b) $\frac{240\text{ miles}}{15\text{ gallons}} = 16\,\frac{\text{miles}}{\text{gallon}}$

(c) $\frac{2200\text{ TV sets}}{1000\text{ families}} = 2.2\,\frac{\text{TV sets}}{\text{family}}$

(d) $\frac{1\text{ inch}}{2.54\text{ centimeters}} = 0.39\,\frac{\text{inch}}{\text{centimeter}}$

Notice that the ratios with units retained these units, such as $\frac{\text{miles}}{\text{gallon}}$ or $\frac{\text{TV sets}}{\text{family}}$.

EXAMPLE 43 A certain university has 7000 students and 280 faculty members. Write this as a ratio with its units.

SOLUTION A ratio is just a comparison of two quantities:

$$\frac{7000\text{ students}}{280\text{ faculty}} = 25\,\frac{\text{students}}{\text{faculty}}$$

EXAMPLE 44 In a certain church, there are 230 men and 415 women. Write this as a ratio with its units.

SOLUTION We again write this as a quotient:

$$\frac{230\text{ men}}{415\text{ women}} \quad \text{or} \quad 0.55\,\frac{\text{man}}{\text{woman}}$$

PROBLEM SET Write the following statements as ratios with their units.

See Examples 43 and 44

1. Two pounds of steak cost \$4.75.
2. Ron drove 450 miles on 27.1 gallons of gasoline.
3. One foot is 0.305 meter.
4. Thirteen cars for nine families.
5. A basket of food cost \$35.15 in 1981 and \$39.42 in 1982.

ANSWERS 1. $\dfrac{2 \text{ pounds}}{4.75 \text{ dollars}}$ 2. $\dfrac{450 \text{ miles}}{27.1 \text{ gallons}}$ 3. $\dfrac{1 \text{ foot}}{0.305 \text{ meter}}$ 4. $\dfrac{13 \text{ cars}}{9 \text{ families}}$

5. $\dfrac{35.15 \text{ (1981 dollars)}}{39.42 \text{ (1982 dollars)}}$

A **proportion** is an equality of two ratios. Let us look at this statement. A ratio is a comparison of two quantities, and a proportion tells us when two such comparisons are equal. For example, if hot dogs cost 8 for \$1.50 or 16 for \$3.00, this is a proportion since the ratios are equal:

$$\frac{8 \text{ hot dogs}}{1.50 \text{ dollars}} = \frac{16 \text{ hot dogs}}{3.00 \text{ dollars}}$$

Notice that there are four numbers (since there are two fractions). Most proportions are like this. Usually, *you will be given three of the numbers and will have to find the fourth*.

Just as with percentage problems, one type of ratio and proportion problem needs only to be translated and then solved. The code is as follows.

English	Mathematics
what ⟶	x (unknown)
a is to b ⟶	$\dfrac{a}{b}$ (ratio)
as ⟶	= (equals)

EXAMPLE 45 The following examples show how to translate proportion-type sentences into mathematical equations using the code that *what* means x, *is to* indicates a ratio, and *as* becomes =.

(a) a is to b as 2 is to 3

$$\frac{a}{b} = \frac{2}{3}$$

(b) What is to 3 as 5 is to 7?

$$\frac{x}{3} = \frac{5}{7}$$

(c) Five is to what, as 8 is to 11?

$$\frac{5}{x} = \frac{8}{11}$$

(d) Three is to 10, as what is to 30?

$$\frac{3}{10} = \frac{x}{30}$$

PROBLEM SET Translate the following statements into mathematical equations.

See Example 45

1. x is to y as 3 is to 4.
2. m is to n as p is to q.
3. What is to 4 as 5 is to 8?
4. Ten is to what as 3 is to 50?
5. Eight is to 7 as 50 is to what?

ANSWERS 1. $\frac{x}{y} = \frac{3}{4}$ 2. $\frac{m}{n} = \frac{p}{q}$ 3. $\frac{x}{4} = \frac{5}{8}$ 4. $\frac{10}{x} = \frac{3}{50}$ 5. $\frac{8}{7} = \frac{50}{x}$

Before we begin solving proportion problems, let us review the fact that multiplication is the opposite of division. We will use this fact in solving equations for an unknown.

EXAMPLE 46 Find n: $n \div 2 = 10$.

SOLUTION Since multiplication is the opposite of division, we find n by multiplying by 2.

$$n \div 2 = 10 \qquad \text{means} \qquad n = 2 \cdot 10 = 20$$

EXAMPLE 47 Find x: $\frac{x}{7} = 3$.

SOLUTION We undo this division by multiplying both sides by 7.

$$\frac{x}{7} = 3$$

$$\cancel{7} \cdot \frac{x}{\cancel{7}} = 7 \cdot 3$$

$$x = 21$$

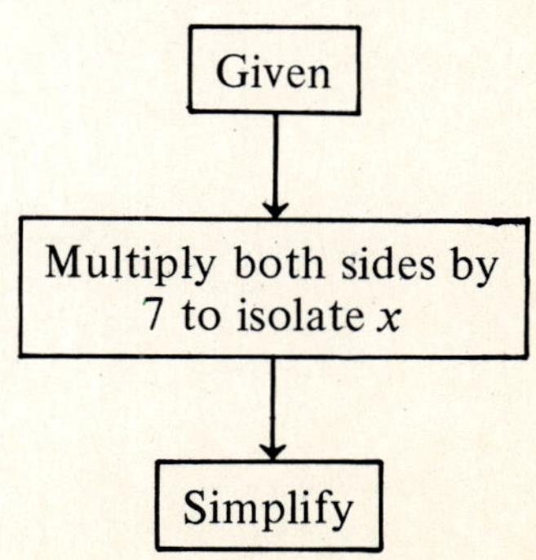

EXAMPLE 48 Find x: $\dfrac{x}{1.5} = \dfrac{100}{75}$.

SOLUTION We multiply both sides by 1.5 to get x by itself.

$$\frac{x}{1.5} = \frac{100}{75}$$

Given

$$\cancel{1.5} \cdot \frac{x}{\cancel{1.5}} = 1.5 \cdot \frac{100}{75}$$

Multiply both sides by 1.5 to isolate x

$$x = \frac{150}{75} = 2$$

Simplify

PROBLEM SET Find the unknown letter (x or n) in the following problems.

See Examples 46 to 48

1. $n \div 3 = 5$.
2. $x \div 4 = 12$.
3. $\dfrac{n}{5} = 8$.
4. $\dfrac{x}{9} = 10$.
5. $\dfrac{x}{1.2} = 50$.
6. $\dfrac{x}{0.6} = 40$.
7. $\dfrac{x}{12} = \dfrac{6}{9}$.
8. $\dfrac{n}{15} = \dfrac{16}{20}$.

ANSWERS **1.** 15 **2.** 48 **3.** 40 **4.** 90 **5.** 60 **6.** 24 **7.** 8 **8.** 12

We are now ready to put these skills (translating and solving) together to solve problems. The key to these (and all math problems) is to break down the problem into smaller simpler steps.

> To solve simple proportion problems:
>
> 1. Translate the sentence into an equation using the table on page 212.
> 2. Solve for the unknown. (If necessary, both fractions may be inverted to get x into the numerator.)

EXAMPLE 49 Solve for x: x is to 14 as 5 is to 7.

SOLUTION As with percentage problems, we first translate the sentence into an equation and then solve.

x is to 14, as 5 is to 7

$$\frac{x}{14} = \frac{5}{7}$$

Given

Translate

$\cancel{14} \cdot \dfrac{x}{\cancel{14}} = 14 \cdot \dfrac{5}{7}$	Multiply both sides by 14 to get x alone
$x = \dfrac{70}{7} = 10$	Simplify

We could also have canceled when we multiplied:

$$\overset{2}{\cancel{14}} \cdot \frac{5}{\underset{1}{\cancel{7}}} = 10$$

EXAMPLE 50 Eight is to 15 as 20 is to what?

SOLUTION Like the preceding problem, we translate and then solve.

Eight is to 15 as 20 is to what?	Given
$\dfrac{8}{15} = \dfrac{20}{x}$	Translate
$\dfrac{15}{8} = \dfrac{x}{20}$	Invert both sides to get x into numerator
$20 \cdot \dfrac{15}{8} = \cancel{20} \cdot \dfrac{x}{\cancel{20}}$	Multiply both sides by 20 to get x alone
$\dfrac{300}{8} = x$	Simplify
$37.5 = x$	Divide

Notice that we first had to invert both sides of the equations before we could solve for x. This is because x must be in the numerator.

EXAMPLE 51 Seven is to 2 as what is to 30?

SOLUTION Again, we translate and solve.

Seven is to 2, as what is to 30?

$$\frac{7}{2} = \frac{x}{30}$$

$$30 \cdot \frac{7}{2} = \cancel{30} \cdot \frac{x}{\cancel{30}}$$

$$\frac{210}{2} = x$$

$$105 = x$$

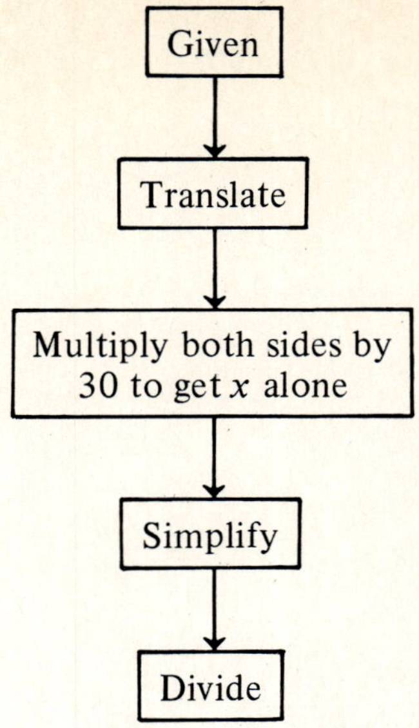

PROBLEM SET Solve the following proportion problems. For problems 1 to 8, leave the answers as fractions; for problems 9 to 15, leave as one-place decimals.

See Examples 49 to 51

1. Solve for x: $\frac{x}{8} = \frac{3}{4}$.
2. Solve for x: $\frac{x}{6} = \frac{7}{2}$.
3. Solve for x: $\frac{4}{x} = \frac{9}{13}$. (Be sure to invert first.)
4. Solve for x: $\frac{10}{32} = \frac{35}{x}$. (Be sure to invert first.)
5. What is to 9 as 8 is to 12?
6. Four is to 15 as what is to 20?
7. Nine is to what as 15 is to 5?
8. Fifteen is to 7 as 27 is to what?
9. What is to 29 as 13 is to 58?
10. Seventy-two is to what as 45 is to 10?
11. One hundred is to 40 as what is to 150?
12. Nine is to 16 as 200 is to what?
13. What is to 100 as 36 is to 90?
14. Fifty-three is to 80 as 500 is to what?
15. Twenty is to what as 30 is to 34?

ANSWERS **1.** 6 **2.** 21 **3.** $5\frac{7}{9}$ **4.** 112 **5.** 6 **6.** $5\frac{1}{3}$ **7.** 3 **8.** $12\frac{3}{5}$ **9.** 6.5 **10.** 16 **11.** 375 **12.** 355.6 **13.** 40 **14.** 754.7 **15.** 22.7

> To solve verbal proportion problems:
>
> 1. Identify the unknown, and label it x.
> 2. With x in the numerator, write the ratio comparing x to the other quantity (which goes in the denominator).
> 3. Make a proportion by setting the first ratio (with x) equal to a ratio of the given quantities. Be sure to put the same units in both numerators, and the other units in the denominators, such as $\frac{\text{dollars}}{\text{pound}} = \frac{\text{dollars}}{\text{pound}}$.
> 4. Solve for x by multiplying. (We can drop the units at this point.)

EXAMPLE 52 Bob drove 212 miles on 8.3 gallons of gasoline. How much gas does he need to go 850 miles?

SOLUTION We are given two quantities: *gallons* and *miles*. We compare them with ratios, and we assume that the ratios are equal to get our proportion.

Let x = gallons needed to go 850 miles	Identify unknown
$\frac{x \text{ gal}}{850 \text{ miles}} = \frac{8.3 \text{ gal}}{212 \text{ miles}}$	Set up proportion with x in numerator
$\cancel{850} \cdot \frac{x}{\cancel{850}} = \frac{850 \cdot (8.3)}{212}$	Multiply both sides by 850 to get x alone
$x = \frac{7055}{212}$	Simplify
$x = 33.3$ gallons	Divide

Notice that both ratios were the same comparison: gallons to miles.

EXAMPLE 53 If Renee can earn \$6200 in 5 months, how much can she earn in 1 year (12 months)?

SOLUTION Here we are comparing the quantities *dollars* and *months*.

Let x = dollars that Renee can earn in 12 months	Identify unknown
$\frac{x \text{ dollars}}{12 \text{ months}} = \frac{6200 \text{ dollars}}{5 \text{ months}}$	Set up proportion with x in numerator

$\cancel{12} \cdot \dfrac{x}{\cancel{12}} = \dfrac{12 \cdot (6200)}{5}$	Multiply both sides by 12 to get x alone
$x = \dfrac{74{,}400}{5}$	Simplify
$x = \$14{,}880$	Divide

Again notice that both ratios were the same: dollars to months.

EXAMPLE 54 Doug is the manager of a town-house group interested in wall siding. He learns that it will cost \$289,000 to do 129 units of another group. From this, about what will it cost to do Doug's 93 units?

SOLUTION We are comparing the quantities *dollars* and *units*.

Let x = cost to side 93 units	Identify unknown
$\dfrac{x \text{ dollars}}{93 \text{ units}} = \dfrac{289{,}000 \text{ dollars}}{129 \text{ units}}$	Set up proportion with x in numerator
$\cancel{93} \cdot \dfrac{x}{\cancel{93}} = \dfrac{93 \cdot (289{,}000)}{129}$	Multiply both sides by 93 to get x alone
$x = \dfrac{26{,}877{,}000}{129}$	Multiply
$x = \$208{,}348.84$	Divide

EXAMPLE 55 Fran, a social worker, spends 180 hours with 50 clients. If she has 900 hours to spend with clients, how many can she see?

SOLUTION This is a proportion problem with two quantities, *hours* and *clients*.

Let x = clients Fran can see in 900 hours	Identify unknown

$$\frac{x \text{ clients}}{900 \text{ hours}} = \frac{50 \text{ clients}}{180 \text{ hours}}$$

Set up proportion with x in numerator

$$\cancel{900} \cdot \frac{x}{\cancel{900}} = \frac{900 \cdot (50)}{180}$$

Multiply both sides by 900 to get x alone

$$x = \frac{45{,}000}{180}$$

Multiply

$$x = 250 \text{ clients}$$

Divide

EXAMPLE 56 One kilogram is 2.2 pounds. Howie weighs 160 pounds. How many kilograms is this?

SOLUTION Here we have the two quantities *kilograms* and *pounds*.

Let x = Howie's weight in kilograms

Identify unknown

$$\frac{x \text{ kg}}{160 \text{ lb}} = \frac{1 \text{ kg}}{2.2 \text{ lb}}$$

Set up proportion with x in numerator

$$\cancel{160} \cdot \frac{x}{\cancel{160}} = \frac{160 \cdot (1)}{2.2}$$

Multiply both sides by 160 to get x alone

$$x = 72.7 \text{ kg}$$

Divide

EXAMPLE 57 If 50,000 Americans out of 220,000,000 are killed every year in traffic accidents, how many are killed per 100,000 people?

SOLUTION Rates are frequently given per 100,000. Here the quantities are people *killed* and *total* people.

Let x = number killed per 100,000

Identify unknown

$$\frac{x \text{ killed}}{100{,}000 \text{ total}} = \frac{50{,}000 \text{ killed}}{220{,}000{,}000 \text{ total}}$$

Set up proportion with x in numerator

$$\cancel{100{,}000} \cdot \frac{x}{\cancel{100{,}000}} = \frac{100{,}000 \cdot (50{,}000)}{220{,}000{,}000}$$

Multiply both sides by 100,000 to get x alone

$$x = \frac{5{,}000{,}000{,}000}{220{,}000{,}000}$$

Multiply

$$x = 22.7$$

Divide

Thus about 23 out of every 100,000 Americans will be killed in traffic accidents in a year.

EXAMPLE 58 A building casts a 21.7-meter shadow. Next to it, a 1-meter stick casts a 0.55-meter shadow. How tall is the building?

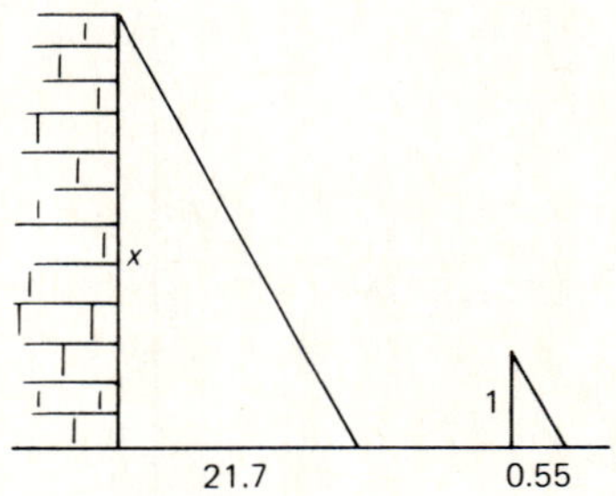

SOLUTION Since the triangles formed by the shadows have the same shape, we can assume that the sides are proportional: *height* to *shadow*. (This is a fact from geometry.)

Let x = height of building

Identify unknown

$$\frac{x \text{ (height)}}{21.7 \text{ (shadow)}} = \frac{1 \text{ (height)}}{0.55 \text{ (shadow)}}$$

Set up proportion with x in numerator

$$\cancel{21.7} \cdot \frac{x}{\cancel{21.7}} = \frac{21.7 \cdot (1)}{0.55}$$

Multiply both sides by 21.7 to get x alone

$$x = 39.45$$

Divide

YES	NO
$\dfrac{x\text{ (height)}}{21.7\text{ (shadow)}} = \dfrac{1\text{ (height)}}{0.55\text{ (shadow)}}$	~~$\dfrac{x\text{ (height)}}{21.7\text{ (shadow)}} \quad \dfrac{0.55\text{ (shadow)}}{1\text{ (height)}}$~~

The proportion must have the *same* units in the numerator and the *same* units in the denominator.

EXAMPLE 59 Clyde Arbuckle's will leaves his entire estate to his sons: Nigel, Clarence, and Kent. They are to split the \$482,000 estate into a 5 : 4 : 3 ratio (Nigel gets 5 parts, Clarence gets 4, and Kent gets 3). What is each son's share?

SOLUTION The numbers, 5, 4, and 3, each represent a *part* of the whole, while the \$482,000 is the *total* estate. So we add $5 + 4 + 3 = 12$ to get the total. Now we can say that Nigel's share is 5 parts in 12, Clarence's share is 4 parts in 12, and Kent's share is 3 parts in 12. We solve for each share separately.

Let x = Nigel's share	Identify unknown
$\dfrac{x\text{ (Nigel)}}{482{,}000\text{ (total)}} = \dfrac{5\text{ (Nigel)}}{12\text{ (total)}}$	Set up proportion with x in numerator
$\cancel{482{,}000} \cdot \dfrac{x}{\cancel{482{,}000}} = \dfrac{482{,}000 \cdot (5)}{12}$	Multiply both sides by 482,000 to get x alone
$x = \dfrac{2{,}410{,}000}{12}$	Multiply
$x = \$200{,}833.33$	Divide

We now compute Clarence's and Kent's share the same way, except that we call their unknown shares y and z.

Let y = Clarence's share	Identify unknown
$\dfrac{y\text{ (Clarence)}}{482{,}000\text{ (total)}} = \dfrac{4\text{ (Clarence)}}{12\text{ (total)}}$	Set up proportion with y in numerator
$\cancel{482{,}000} \cdot \dfrac{y}{\cancel{482{,}000}} = \dfrac{482{,}000 \cdot (4)}{12}$	Multiply both sides by 482,000 to get y alone

$$y = \frac{1{,}928{,}000}{12}$$ **Multiply**

$$y = \$160{,}666.67$$ **Divide**

Let z = Kent's share **Identify unknown**

$$\frac{z\text{ (Kent)}}{482{,}000\text{ (total)}} = \frac{3\text{ (Kent)}}{12\text{ (total)}}$$ **Set up proportion with z in numerator**

$$\cancel{482{,}000} \cdot \frac{z}{\cancel{482{,}000}} = \frac{482{,}000 \cdot (3)}{12}$$ **Multiply both sides by 482,000 to get z alone**

$$z = \frac{1{,}446{,}000}{12}$$ **Multiply**

$$z = \$120{,}500$$ **Divide**

We can check these three answers by adding them together.

Nigel	$200,833.33
Clarence	160,666.67
Kent	120,500.00
Total estate	$482,000.00

This is the original estate, so the answers check.

PROBLEM SET Solve the following proportion problems.

See Examples 52 to 59

1. Apples cost 87 cents for 3 pounds. How much will 10 pounds cost?
2. A bag of 200 nails costs $1.25. How many nails can you buy for $3.00?
3. Thirty miles per hour is equivalent to 44 feet per second. If a person is driving at 45 miles per hour, how fast is this in feet per second?
4. In her luxury car, Pam drove 200 miles on 18 gallons of gas. How far can she drive on 45 gallons of gas?
5. In his subcompact car, Tom drove 300 miles on 12 gallons of gas. How many gallons does he need to make a 2500-mile trip?
6. It costs Judy $195 to carpet 20 square yards of floor. What will it cost her to carpet 50 square yards of floor?
7. It costs $7.70 for enough wallpaper to cover 70 square feet. How much will it cost to wallpaper 500 square feet?

8. If a \$10,000 life insurance policy costs \$210 per year, how much does a \$35,000 policy cost?

9. If 1 inch is 2.54 centimeters, how many inches are in 100 centimeters?

10. If 1 quart is 0.95 liter, how many liters are in 20 quarts?

11. If 1 square mile is 640 acres, how many square miles are in 200 acres?

12. If 1 kilogram is 2.2 pounds, how many kilograms is a 185-pound man?

13. If 1 pound is 454 grams, how many grams are in a $\frac{1}{2}$ pound?

14. If 1 foot is 0.305 meter, how many meters is a 6-foot man?

15. How many meters is a 5-foot 6-inch woman?

16. If 2 cups of flour will make 15 pancakes, how many cups of flour are needed to make 300 pancakes?

17. In a group of 148, there are 40 joggers. At this rate, how large should a group be in order to get 100 joggers?

18. The ratio of men to women in a certain field is 5 to 4. JM Industries employs 630 people. How many women should they employ to have the same ratio as in the field?

19. A 2-meter stick casts a 1.1-meter shadow. How tall is a tree that casts a 37-meter shadow?

20. A meter stick casts a 1.7-meter shadow. How tall is a building that casts a 19-meter shadow?

21. A survey indicates that 9 out of 150 people are unemployed. Using this, how many are unemployed out of a labor force of 95,000,000 people?

22. Art and Bob form an investment partnership. Art puts up \$5000, and Bob puts up \$9000. Their investments are now worth \$42,000. What is Art's share? What is Bob's share?

23. Cindy, Donna, and Ellen open a store together. Cindy puts up \$4000, Donna puts up \$6500, and Ellen puts up \$9500. After some time, the store is worth \$55,000. What is Cindy's share? Donna's share? Ellen's share?

24. If there are 1,000,000 divorces a year out of a total of 220,000,000 Americans, what is the divorce rate per 100,000?

ANSWERS 1. 290 cents 2. 480 3. 66 4. 500 miles 5. 100
6. \$487.50 7. \$55 8. \$735 9. 39.37 10. 19 11. 0.3125
12. 84.1 13. 227 14. 1.83 15. 1.68 16. 40 17. 370
18. 280 19. 67.3 meters 20. 11.2 meters 21. 5,700,000
22. Art: \$15,000; Bob: \$27,000
23. Cindy: \$11,000; Donna: \$17,875; Ellen: \$26,125 24. 455 per 100,000

Miscellaneous Exercises 4.4

Write the following statements as ratios.

See Examples 43 and 44

1. Debbie's car went 312 miles on 12.6 gallons of gas.
2. One kilogram is 2.2 pounds.

Translate the following statements into equations.

See Example 45 **3.** What is to 8 as 3 is to 19?

4. Four is to what as 5 is to 23?

Find x in the following equations.

See Examples 46 to 48

5. $\frac{x}{4} = 5$ **6.** $\frac{x}{12} = 3$ **7.** $\frac{x}{0.7} = 8$ **8.** $\frac{x}{10} = \frac{12}{15}$

Translate and solve the following statements.

See Examples 49 to 51

9. x is to 14 as 6 is to 21.

10. What is to 7 as 3 is to 2?

11. Twelve is to what as 5 is to 3?

12. Eight is to 15 as what is to 9?

13. Nine is to 10 as 24 is to what?

See Examples 52 to 59 The following applications are typical proportion problems. Solve them using the methods shown in Examples 52 to 59.

Health applications

14. Ten milliliters of a solution contain 5,000,000 units of penicillin. How many units of penicillin will 0.5 milliliter contain?

15. Fifty capsules contain 0.375 grain of an active medication. How many capsules are needed to provide 0.1 grain of medication?

16. In medicine, it is common to write 1 : 20 to mean a $\frac{1}{20}$ ratio. How much pure glucose must be used to produce 100 milliliters of a 1 : 20 glucose solution? The glucose is $\frac{1}{20}$ of the solution.

Chemistry application

17. Mercury (Hg) has an atomic weight of about 200. Oxygen (O) has an atomic weight of about 16. How much oxygen is needed to produce 1000 grams of mercury oxide (HgO). (*Hint:* Find the atomic weight of HgO, and then compare ratios of oxygen to total weight.)

Technical applications

18. The resistance in 250 feet of wire is 7 ohms. Find the resistance in 700 feet of wire.

19. Drafters use proportions when they make scale drawings. Suppose that an architectural sketch uses the scale 0.25 inch represents 1 foot. If a wall is actually 7.5 feet tall, what would its size be in the scale drawing?

Business applications

20. A cost accountant decides that the departments in an office should share the cost of the utilities in proportion to their floor space. One month, the utilities were \$4000. Find the share that each of the following departments should pay.

Department	Floor Space (square feet)
W	2200
X	1700
Y	3300
Z	2800

21. A partnership buys an apartment building on June 1. They agree to pay the real estate taxes for the year's last 7 months. If one year's taxes are $7200, what is the partnership's share of the tax?

Consumer applications

22. Frank drove 300 miles on 18 gallons of gasoline. How many gallons will he need to drive 1000 miles?

23. Helen drove 450 miles on 25 gallons of gasoline. How far can she drive on 60 gallons?

24. With 1 quart of paint, Nancy was able to paint $1\frac{1}{2}$ walls. How much paint will she need to paint 4 walls?

25. Susan knows from experience that 2 pounds of hamburger feed 5 people. How much hamburger will she need to feed a party of 28 people?

Metric applications

In Chapter 5, we discuss the metric system of measurement. Here, we give a few problems that use proportions to convert between the English system (our current system) and the metric system (the coming system).

26. One kilogram is 2.2 pounds. How many kilograms is a 125-pound woman?

27. One quart is 0.95 liter. How many liters are in 1 gallon (4 quarts)?

28. One mile is 1.61 kilometers. Two European towns are 300 kilometers apart. How many miles is this?

29. One ounce is 28.3 grams. How many grams are in a 12-ounce can of beer?

Now complete Post-Test 1 for Section 4.4. on page T63.

CHAPTER 4 SUMMARY

Important Words and Phrases

amount (4.3)
base (4.3)
percent (4.1)
percentage (4.3)
proportion (4.4)
ratio (4.4)
translating (4.3)

Important Ideas

(4.1) Writing a percent as a reduced fraction.
Writing a percent as a decimal.

(4.2) Writing a fraction as a percent.
Writing a decimal as a percent.

(4.3) Translating English into mathematics.
Solving percent problems by translation.
Solving verbal percentage problems.

(4.4) Solving proportion problems by translation.
Solving verbal proportion problems.

Choose the correct (or closest) answer.

1. Write 79% as a decimal.
(a) 7900 (b) 79 (c) 7.9 (d) 0.079 (e) 0.79

2. Write 0.05% as a decimal.
(a) 5.0 (b) 0.5 (c) 0.0005 (d) 0.005 (e) 0.05

3. Write 18% as a fraction.
(a) $1\frac{8}{100}$ (b) $18\frac{18}{100}$ (c) $1\frac{8}{10}$ (d) $\frac{9}{50}$ (e) $\frac{18}{10}$

4. Write 375% as a mixed number.
(a) $37\frac{1}{2}$ (b) $3\frac{3}{4}$ (c) $\frac{3}{8}$ (d) $37\frac{5}{100}$ (e) $375\frac{75}{100}$

5. Write 0.183 as a percent.
(a) 18.3% (b) 1.83% (c) 183%
(d) 0.183% (e) 0.00183%

6. Write 2.1 as a percent.
(a) 21% (b) 210% (c) 2.1% (d) 0.21% (e) 0.021%

7. Write $\frac{4}{25}$ as a percent.
(a) 160% (b) 40% (c) 4.25% (d) 16% (e) 4%

8. Write $\frac{3}{13}$ as a percent.
(a) 433.3% (b) 4.3% (c) 3.13% (d) 23.1% (e) 3%

9. Translate the following into a mathematical equation:

Eighteen is 2% of what number?

(a) $18 \cdot n = 2$ (b) $18 \cdot n = 0.02$ (c) $18 = 0.02 \cdot n$
(d) $18 = 2 \cdot n$ (e) $n = 0.02 \cdot 18$

10. 780 is 15% of what number?
(a) 52 (b) 5200 (c) 117 (d) 11,700 (e) 1923

11. Ralph Smith wins an election, 58 to 53. What percent of the vote did he get?
(a) 52.3% (b) 58% (c) 47.7% (d) 109% (e) 54.1%

12. Barbara buys a \$300 camera. She gets a 15% discount. How much does she pay for the camera?
(a) \$255 (b) \$45 (c) \$200 (d) \$285 (e) \$280

13. What number is to 10 as 14 is to 40?
(a) 56 (b) 3.5 (c) 28.6 (d) 4.2 (e) 1.4

14. Bill drives 215 miles on 11.1 gallons of gasoline. How far can he drive on 40 gallons of gasoline?

(a) 59.7 **(b)** 774.8 **(c)** 95460 **(d)** 597 **(e)** 954.6

15. One kilogram is 2.2 pounds. How many kilograms is a 242-pound football player?

(a) 532.4 **(b)** 121 **(c)** 90.9 **(d)** 110 **(e)** 142

Now complete Post-Test 1 for Chapter 4 on page T65.

5

MEASUREMENT AND METRIC GEOMETRY

Before you begin this chapter, complete the Chapter 5 Pre-Test on page T69.

5.1 MEASUREMENT AND THE METRIC SYSTEM

SECTION OBJECTIVES

After completing this unit, you will be able to:

1. Understand the English and the metric systems of measurement.
2. Convert a measurement to other units in the same system or the other system.

TYPICAL PROBLEMS

You will be able to work problems like the ones below.

(a) A record album is approximately how many centimeters across?
A. 1 B. 5 C. 30 D. 100

(b) A five-year-old-child weighs approximately how many kilograms?
A. 2 B. 20 C. 200 D. 2000

(c) Ten feet is how many meters?

(d) Fifty kilograms is how many pounds?

(e) One liter is how many cups?

ANSWERS **(a)** 30 cm **(b)** 20 kg **(c)** 3.05 m **(d)** 110 lb **(e)** 4.2 cups

We use numbers every day to make **measurements**. When we use a number to describe a quantity, such as length, weight, liquid capacity, temperature, time, and so on, we are **measuring**. Some measurements are a matter of opinion, such as scoring a person of the opposite sex a "10" or giving the odds of a team winning a game.

The measurements that we discuss are very exact. For instance, the meter was once 1/10,000,000 the distance from the North Pole to the equator. Now the meter is defined to be 1,650,763.73 wavelengths of red-orange light of excited krypton of mass number 86. Although this is very precise, it is of little use in measuring the walls of a kitchen. Therefore, for practical purposes, we consider the following to be true:

(a) *Length* is the measurement we get from an ordinary ruler.

(b) *Weight* is the measurement we get from a bathroom or grocery scale.

(c) *Liquid capacity* is the measurement we get from a measuring cup.

(d) *Time* is the measurement that we get from a clock or watch.

(e) *Temperature* is the measurement that we get from a thermometer.

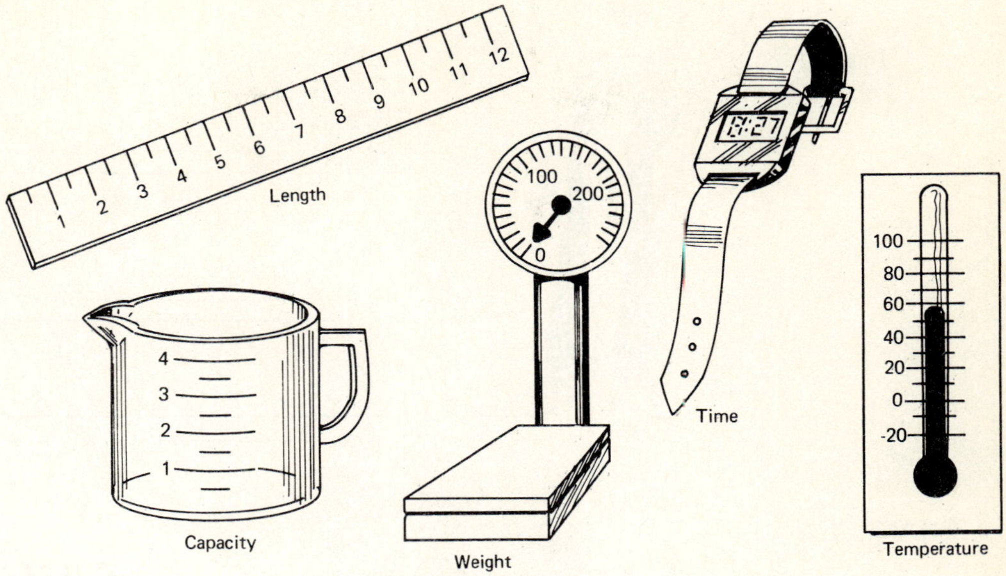

Typical Measuring Devices

Before we talk about the metric system of measurement, we should first discuss our current system, the **English system**. Table 5.1 summarizes the major units and their relations.

Table 5.1

English System of Measurements

Length	1 ft = 12 in.	in. = inch
	1 yd = 3 ft	ft = foot (feet)
	1 mi = 5280 ft	yd = yard
	1 mi = 1760 yd	mi = mile
Weight	1 lb = 16 oz	oz = ounce
	1 ton = 2000 lb	lb = pound
Liquid Capacity	1 pt = 2 cups	pt = pint
	1 qt = 2 pt	qt = quart
	1 gal = 4 qt	gal = gallon
Time	1 min = 60 sec	sec = second
	1 hr = 60 min	min = minute
	1 day = 24 hr	hr = hour
	1 yr = 365 day	yr = year

EXAMPLE 1

Typical Lengths

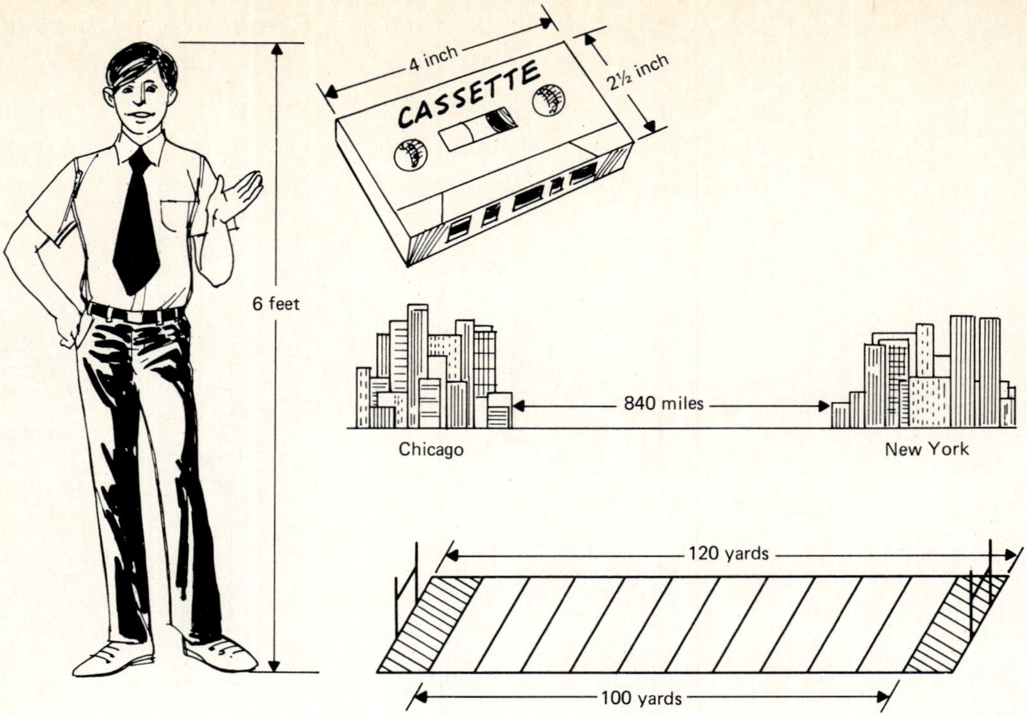

EXAMPLE 2

Typical Weights

EXAMPLE 3

Typical Liquid Capacities

PROBLEM SET With a ruler, measure the following lines in inches.

See Examples 1 to 3

1. 2.

3. 4.

After each item, indicate the approximate measurement in the units given.

5.	Buffalo (pounds)	**(a)** 18	**(b)** 180	**(c)** 1800	**(d)** 18,000
6.	Family car (feet)	**(a)** 2	**(b)** 7	**(c)** 14	**(d)** 28
7.	Bathtub (gallons)	**(a)** 25	**(b)** 150	**(c)** 750	**(d)** 3000
8.	Record album (inches across)	**(a)** $\frac{1}{2}$	**(b)** 3	**(c)** 12	**(d)** 30

ANSWERS 1. $1\frac{1}{2}$ in. 2. 2 in. 3. $\frac{3}{4}$ in. 4. 3 in. 5. 1800 lb 6. 14 ft 7. 150 gal 8. 12 in.

Frequently, we need to convert from one unit to another. For instance, we might wish to express 48 feet in yards. To do these conversions, we use Table 5.1 and the *multiplying-by-1 trick*. This means multiplying the given unit by a special fraction (called a **unit fraction**) that is equal to 1. As examples, the following are unit fractions (= 1):

$$\frac{12 \text{ inches}}{1 \text{ foot}}, \quad \frac{1 \text{ pound}}{16 \text{ ounces}}, \quad \frac{4 \text{ quarts}}{1 \text{ gallon}}$$

since each numerator equals its denominator. We choose the unit fraction in such a way as to cancel the given units and leave the desired units.

given units	×	unit fraction	=	desired units

To convert units:

1. Put the given units on the left side of the equal sign; put the desired units on the right side.
2. Choose a conversion fact(s) from Table 5.1 to connect the given and the desired units.
3. Arrange the units as a fraction to cancel the given units and leave the desired units.
4. Fill in the numbers and simplify.

EXAMPLE 4 The Sears Tower in Chicago is 1454 feet tall. How many miles is this?

SOLUTION Here we are given a measurement in feet, and we want it in miles. We see in Table 5.1 that 1 mile = 5280 feet. We use this fact to make our unit fraction.

$$\text{ft} \times \boxed{\;?\;} = \text{mi}$$

Put given unit on left; put desired unit on right

$$\cancel{\text{ft}} \times \frac{\text{mi}}{\cancel{\text{ft}}} = \text{mi}$$

Set up units to cancel properly

$$1454\,\cancel{\text{ft}} \times \boxed{\frac{1 \text{ mi}}{5280\,\cancel{\text{ft}}}} = \frac{1454}{5280}\text{ mi}$$

Fill in numbers; multiply

$$= 0.275 \text{ mi}$$

Divide

Thus, the Sears Tower is about 0.275 mile tall.

EXAMPLE 5 A certain car weighs 2 tons. How many ounces is this?

SOLUTION Here we have to make two conversions: from tons to pounds, and from pounds to ounces. We use the facts that 1 ton = 2000 pounds and 1 pound = 16 ounces to make the two needed unit fractions.

$$\text{tons} \times \boxed{\;?\;} = \text{oz}$$

Put given unit on left; put desired unit on right

$$\cancel{\text{tons}} \times \frac{\cancel{\text{lb}}}{\cancel{\text{tons}}} \times \frac{\text{oz}}{\cancel{\text{lb}}} = \text{oz}$$

Set up units to cancel properly

$$2\,\cancel{\text{ton}} \times \boxed{\frac{2000\,\cancel{\text{lb}}}{1\,\cancel{\text{ton}}}} \times \boxed{\frac{16 \text{ oz}}{1\,\cancel{\text{lb}}}} = 64{,}000 \text{ oz}$$

Fill in numbers; multiply

EXAMPLE 6 How many cups are in a gallon of milk?

SOLUTION As in Example 5, we will need more than one unit fraction. We convert gallons $\longrightarrow$ quarts $\longrightarrow$ pints $\longrightarrow$ cups.

$$\text{gal} \times \boxed{?} = \text{cup}$$

Put given unit on left; put desired unit on right

$$\cancel{\text{gal}} \times \frac{\cancel{\text{qt}}}{\cancel{\text{gal}}} \times \frac{\cancel{\text{pt}}}{\cancel{\text{qt}}} \times \frac{\text{cup}}{\cancel{\text{pt}}} = \text{cup}$$

Set up units to cancel properly

$$1\ \cancel{\text{gal}} \times \frac{4\ \cancel{\text{qt}}}{1\ \cancel{\text{gal}}} \times \frac{2\ \cancel{\text{pt}}}{1\ \cancel{\text{qt}}} \times \frac{2\ \text{cup}}{1\ \cancel{\text{pt}}} = 16\ \text{cups}$$

Fill in numbers; multiply

EXAMPLE 7 How many seconds are in one year?

SOLUTION We assume that this is a 365-day year and use all the time facts in Table 5.1.

$$\text{yr} \times \boxed{?} = \text{sec}$$

Put given unit on left; put desired unit on right

$$\cancel{\text{yr}} \times \frac{\cancel{\text{day}}}{\cancel{\text{yr}}} \times \frac{\cancel{\text{hr}}}{\cancel{\text{day}}} \times \frac{\cancel{\text{min}}}{\cancel{\text{hr}}} \times \frac{\text{sec}}{\cancel{\text{min}}} = \text{sec}$$

Set up units to cancel properly

$$1\ \cancel{\text{yr}} \times \frac{365\ \cancel{\text{day}}}{1\ \cancel{\text{yr}}} \times \frac{24\ \cancel{\text{hr}}}{1\ \cancel{\text{day}}}$$

$$\times \frac{60\ \cancel{\text{min}}}{1\ \cancel{\text{hr}}} \times \frac{60\ \text{sec}}{1\ \cancel{\text{min}}} = 31{,}536{,}000\ \text{sec}$$

Fill in numbers; multiply

PROBLEM SET Complete the following tables. (You need not do a box with an X in it.)

See Examples 4 to 7

	inches	feet	yards	miles
1.		6		X
2.	X		880	
3.	X			2.5
4.	96			X

	ounces	pounds	tons
5.	24		X
6.		1500	
7.			2.7
8.		$2\frac{1}{2}$	X

	cups	pints	quarts	gallons
9.	4			
10.				15,000
11.			24	
12.		40		

	seconds	minutes	hours	days	years
13.	120			X	X
14.		5		X	X
15.	X		4		X
16.	X	X		30	
17.	X	X	X		$\frac{1}{2}$

ANSWERS 1. 72 in.; 2 yd 2. 2640 ft; $\frac{1}{2}$ mi 3. 13,200 ft; 4400 yd

4. 8 ft; $2\frac{2}{3}$ yd 5. 1.5 lb 6. 24,000 oz; $\frac{3}{4}$ ton

7. 86,400 oz; 5400 lb 8. 40 oz 9. 2 pt; 1 qt; $\frac{1}{4}$ gal

10. 240,000 cups; 120,000 pt; 60,000 qt 11. 96 cups; 48 pt; 6 gal

12. 80 cups; 20 qt; 5 gal 13. 2 min; $\frac{1}{30}$ hr 14. 300 sec; $\frac{1}{12}$ hr

15. 240 min; $\frac{1}{6}$ day 16. 720 hr; 0.082 yr 17. 182.5 days

The **English system** is our current system. It is the system that we Americans are used to and comfortable with. However, it can also be a very awkward system since it often involves multiplying and dividing by numbers such as 2, 3, 4, 9, 12, 16, 27, 36, 144, and 1728. These are not pleasant numbers to deal with. (The English system came about without much plan: the inch was three barleycorns laid end to end; a mile was a thousand double-steps of a Roman soldier; and so on.)

Enter the **metric system**. Carefully planned, the metric system is based totally on 10's, 100's, 1000's, and so on. Any conversion within the metric system becomes a simple shift of the decimal point. There are special prefixes for every power and fractional power of 10. However, we list here only the major prefixes.

Common metric prefixes:

$$\text{kilo} = 1000$$

$$\text{mega} = 1{,}000{,}000$$

$$\text{centi} = \frac{1}{100}$$

$$\text{milli} = \frac{1}{1000}$$

$$\text{micro} = \frac{1}{1{,}000{,}000}$$

Table 5.2 summarizes the basic relations within the metric system as well as the approximate conversions with the English system.

Table 5.2

Common Units Conversions

Length	1 m = 1000 mm	1 in. = 2.54 cm	mm = millimeter $\left(\frac{1}{1000}\text{ meter}\right)$
	1 m = 100 cm	1 ft = 0.3 m	cm = centimeter $\left(\frac{1}{100}\text{ meter}\right)$
	1 km = 1000 m	1 mi = 1.6 km	m = meter
	1 cm = 10 mm	1 m = 1.1 yd	km = kilometer (1000 meters)
Weight	1 g = 1000 mg	1 oz = 28 g	mg = milligram $\left(\frac{1}{1000}\text{ gram}\right)$
	1 kg = 1000 g	1 kg = 2.2 lb	g = gram
	1 MT = 1000 kg		kg = kilogram (1000 grams)
			MT = metric ton
			= megagram (1,000,000 grams)
Liquid Capacity	1 L = 1000 ml	1 L = 1.06 qt	ml = milliliter $\left(\frac{1}{1000}\text{ liter}\right)$
		1 gal = 3.8 L	L = liter
Temperature	$F = \frac{9}{5}C + 32$		F = degrees Fahrenheit
	$C = \frac{5}{9}(F - 32)$		C = degrees Celsius

The United States is the only major nation in the world not to use the metric system exclusively. American industry is changing over to the metric system in order to compete in world trade markets. As America goes metric, people are encouraged to *think metric*.

EXAMPLE 8

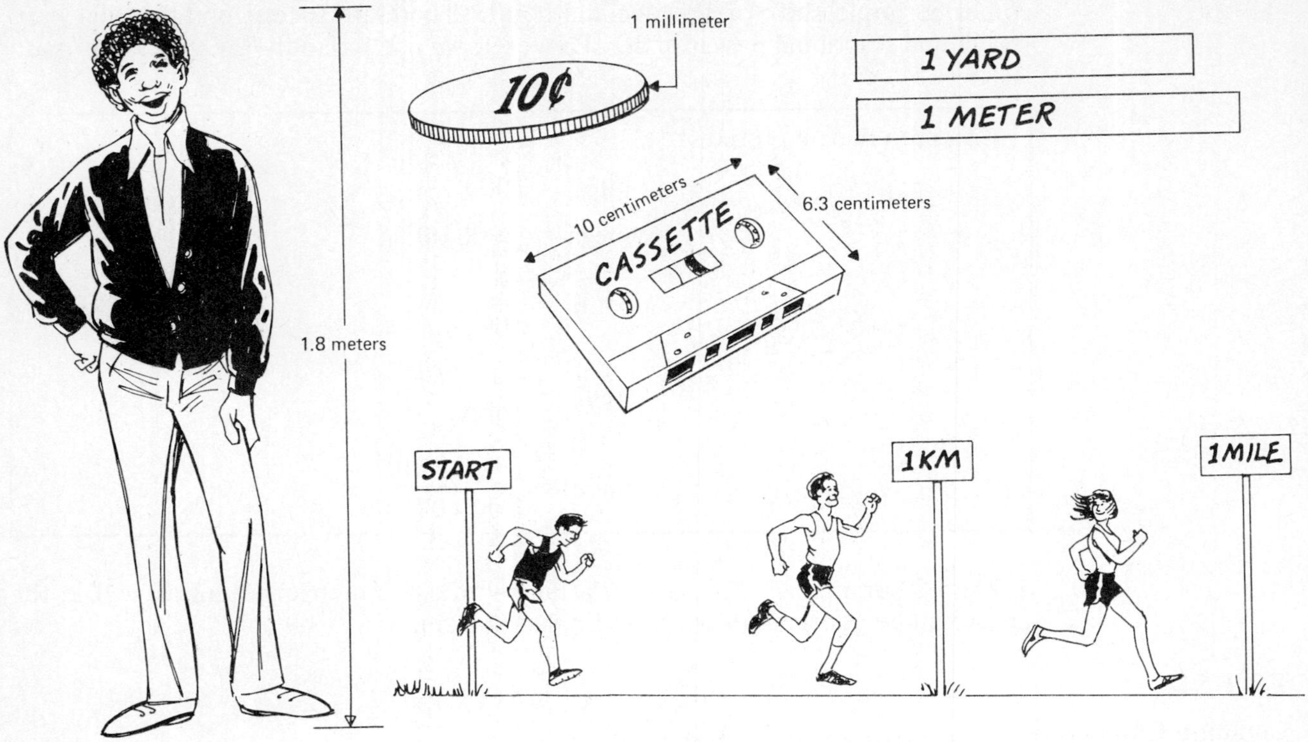

Typical Metric Lengths

EXAMPLE 9

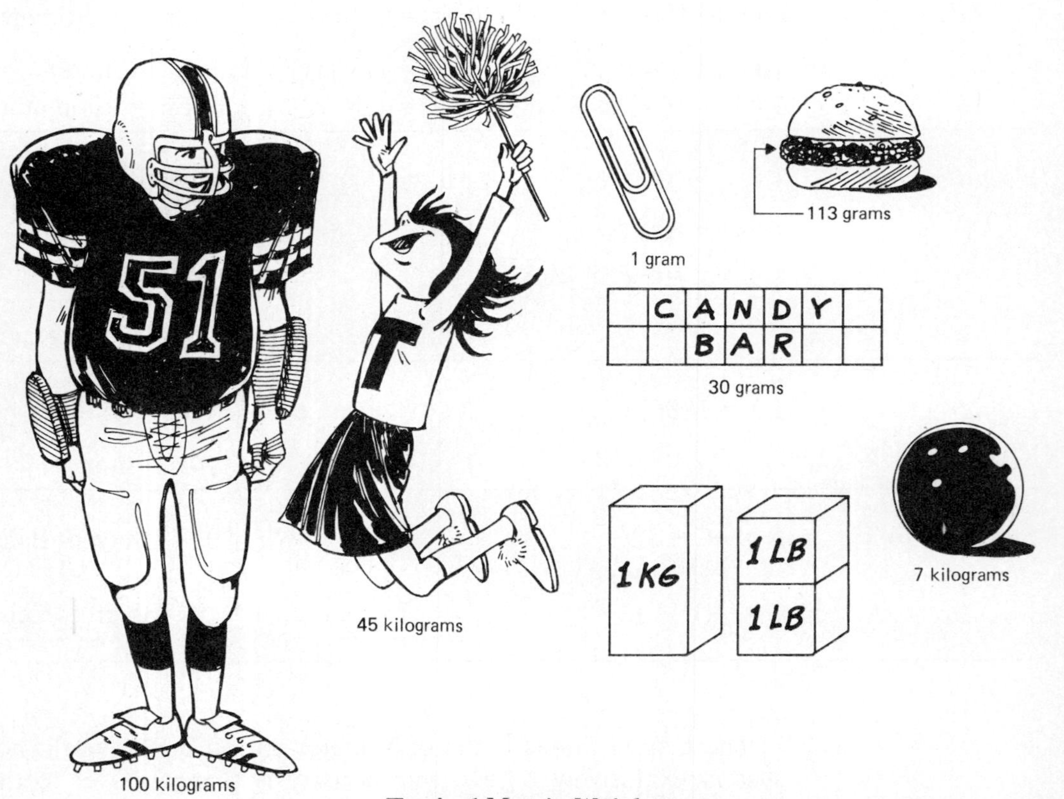

Typical Metric Weights

EXAMPLE 10

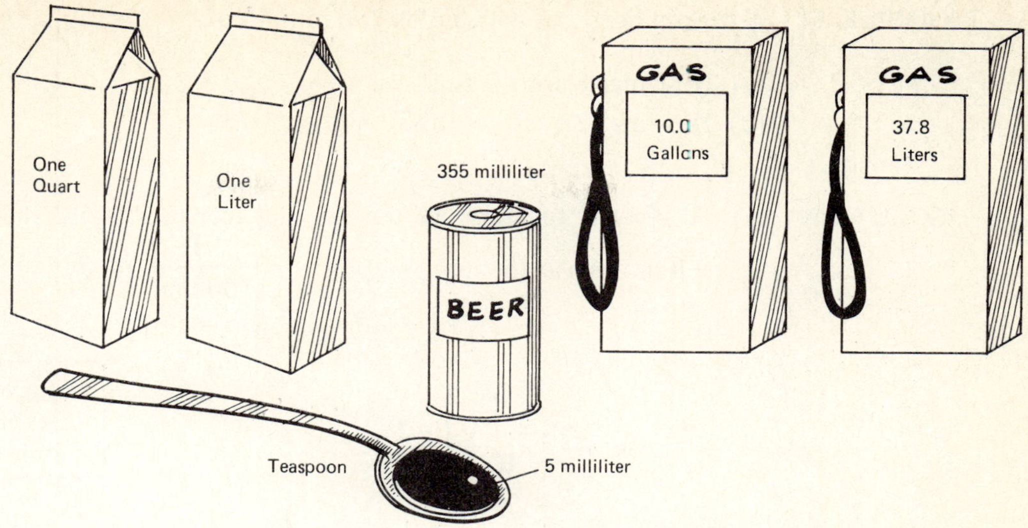

Typical Metric Liquid Capacities

PROBLEM SET With a ruler, measure the following lines in centimeters.

See Examples 8 to 10

1. 2.

3. 4.

After each item, choose the approximate measurement in the indicated units.

5.	Family car (meters)	(a) 1	(b) 4	(c) 10	(d) 40
6.	Distance from Chicago to St. Louis (km)	(a) 4.2	(b) 42	(c) 420	(d) 4200
7.	Ball-point pen (cm)	(a) 0.2	(b) 1	(c) 3	(d) 15
8.	Average adult man (kg)	(a) 8	(b) 40	(c) 80	(d) 400
9.	Dime (grams)	(a) 0.2	(b) 2	(c) 20	(d) 100
10.	Newborn baby (kg)	(a) 1	(b) 3.5	(c) 10	(d) 35
11.	Gallon of wine (L)	(a) 3.8	(b) 38	(c) 0.38	(d) 380
12.	Cup of coffee (ml)	(a) 2.4	(b) 24	(c) 100	(d) 240
13.	15-gallon gasoline fill-up (L)	(a) 5.7	(b) 32	(c) 57	(d) 320

ANSWERS 1. 4 cm 2. 5.5 cm 3. 2.5 cm 4. 7 cm 5. 4 m
6. 420 km 7. 15 cm 8. 80 kg 9. 2 g 10. 3.5 kg
11. 3.8 L 12. 240 ml 13. 57 L

Although it is preferable to think and measure in the metric system, it is still wise to know how to convert between our present English system and the coming metric system. We use the same *multiplying-by-1 trick* in which we construct the proper unit fraction(s) (equal to 1) from the facts in Tables 5.1 and 5.2.

EXAMPLE 11 A certain bottle of wine holds 750 milliliters.

(a) How many liters is this?
(b) How many quarts is this?

SOLUTION (a) To convert to liters, we use the fact from Table 5.2 that 1 L = 1000 ml. We will use this as the unit fraction: $1 = \dfrac{1\text{ L}}{1000\text{ ml}}$

$$\text{ml} \times \boxed{?} = \text{L}$$

Put given unit on left; put desired unit on right

$$\cancel{\text{ml}} \times \frac{\text{L}}{\cancel{\text{ml}}} = \text{L}$$

Set up units to cancel properly

$$750\ \cancel{\text{ml}} \times \frac{1\text{ L}}{1000\ \cancel{\text{ml}}} = \frac{750}{1000}\text{ L}$$

Fill in numbers; multiply

$$= 0.75\text{ L}$$

Divide

Notice that this was the same as moving the decimal point three places to the left: 750. = 0.750.

(b) Now let us convert 0.75 liter to quarts by using the fact that 1 liter = 1.06 quarts (see Table 5.2).

$$\text{L} \times \boxed{?} = \text{qt}$$

Put given unit on left; put desired unit on right

$$\cancel{\text{L}} \times \frac{\text{qt}}{\cancel{\text{L}}} = \text{qt}$$

Set up units to cancel properly

$$0.75\ \cancel{\text{L}} \times \frac{1.06\text{ qt}}{1\ \cancel{\text{L}}} = 0.795\text{ qt}$$

Fill in numbers; multiply

EXAMPLE 12 The speed limit on U.S. highways is 55 miles per hour. What is this in kilometers per hour?

SOLUTION We use the fact (from Table 5.2) that 1 mile = 1.6 km. (We do not have to convert time; they are both in hours.)

$$\frac{\text{mi}}{\text{hr}} \times \boxed{\;?\;} = \frac{\text{km}}{\text{hr}}$$

$$\frac{\cancel{\text{mi}}}{\text{hr}} \times \frac{\text{km}}{\cancel{\text{mi}}} = \frac{\text{km}}{\text{hr}}$$

$$55\,\frac{\cancel{\text{mi}}}{\text{hr}} \times \boxed{\frac{1.6 \text{ km}}{1\,\cancel{\text{mi}}}} = 88\,\frac{\text{km}}{\text{hr}}$$

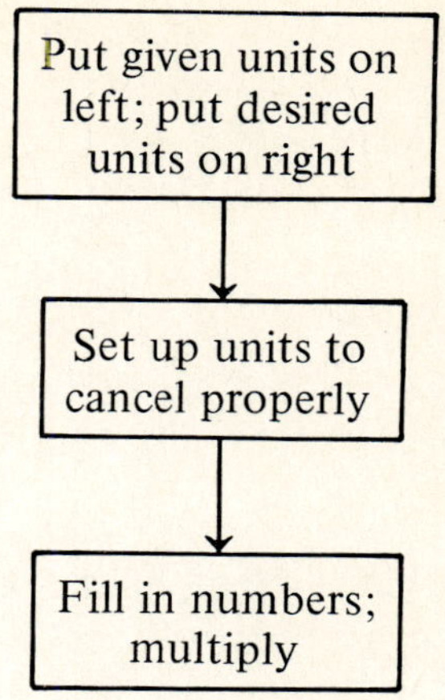

Thus, the speed limit on American roads is 88 kilometers per hour.

EXAMPLE 13 If a coal miner digs 16 tons of coal from the earth, how many kilograms is this?

SOLUTION We first convert tons to pounds and then pounds to kilograms.

$$\text{tons} \times \boxed{\;?\;} = \text{kg}$$

$$\cancel{\text{tons}} \times \frac{\cancel{\text{lb}}}{\cancel{\text{tons}}} \times \frac{\text{kg}}{\cancel{\text{lb}}} = \text{kg}$$

$$16\,\cancel{\text{tons}} \times \boxed{\frac{2000\,\cancel{\text{lb}}}{1\,\cancel{\text{ton}}}} \times \boxed{\frac{1 \text{ kg}}{2.2\,\cancel{\text{lb}}}} = \frac{32{,}000}{2.2}\text{ kg}$$

$$= 14{,}545.5 \text{ kg}$$

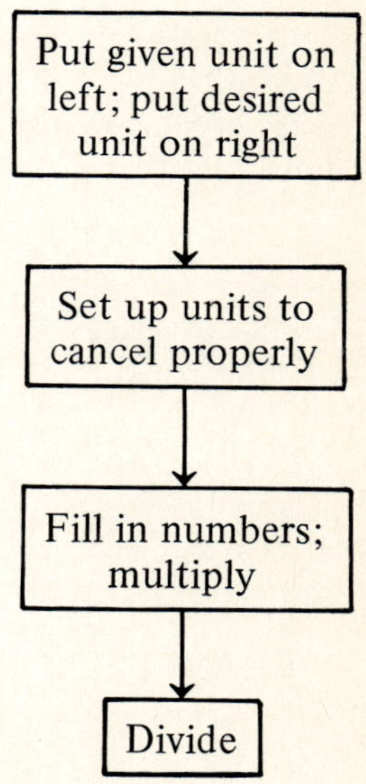

Even as America goes metric, we presume that folk singers will continue to sing: "You load sixteen tons and what do you get . . . ."

EXAMPLE 14 Sue buys 1.5 kilograms of coffee.

(a) How many grams is this?

(b) How many pounds is this?

SOLUTION (a) To find the number of grams, we use the fact that 1 kilogram = 1000 grams.

$$\text{kg} \times \boxed{?} = \text{g}$$

Put given unit on left; put desired unit on right

↓

$$\cancel{\text{kg}} \times \frac{\text{g}}{\cancel{\text{kg}}} = \text{g}$$

Set up units to cancel properly

↓

$$1.5\ \cancel{\text{kg}} \times \boxed{\frac{1000\text{ g}}{1\ \cancel{\text{kg}}}} = 1500\text{ g}$$

Fill in numbers; multiply

Notice that this conversion within the metric system is merely a shift of the decimal point, in this case three places to the right: 1.500 = 1500.

(b) To write 1.5 kilograms in pounds, we use the fact that 1 kilogram = 2.2 pounds.

$$\text{kg} \times \boxed{?} = \text{lb}$$

Put given unit on left; put desired unit on right

↓

$$\cancel{\text{kg}} \times \frac{\text{lb}}{\cancel{\text{kg}}} = \text{lb}$$

Set up units to cancel properly

↓

$$1.5\ \cancel{\text{kg}} \times \boxed{\frac{2.2\text{ lb}}{1\ \cancel{\text{kg}}}} = 3.3\text{ lb}$$

Fill in numbers; multiply

EXAMPLE 15 Carol buys a pair of 160-centimeter skis.

(a) How many inches is this?
(b) How many feet is this?

SOLUTION (a) To write this in inches, we use the fact that 1 inch = 2.54 centimeters.

$$\text{cm} \times \boxed{?} = \text{in.}$$

Put given unit on left; put desired unit on right

↓

$$\cancel{\text{cm}} \times \frac{\text{in.}}{\cancel{\text{cm}}} = \text{in.}$$

Set up units to cancel properly

$$160\ \cancel{\text{cm}} \times \frac{1 \text{ in.}}{2.54\ \cancel{\text{cm}}} = \frac{160}{2.54} \text{ in.}$$

Fill in numbers; multiply

$$= 63 \text{ in.}$$

Divide

(b) To convert this to feet, we use the fact that 1 foot = 12 inches.

$$\text{in.} \times \boxed{?} = \text{ft}$$

Put given unit on left; put desired unit on right

$$\cancel{\text{in.}} \times \frac{\text{ft}}{\cancel{\text{in.}}} = \text{ft}$$

Set up units to cancel properly

$$63\ \cancel{\text{in.}} \times \frac{1 \text{ ft}}{12\ \cancel{\text{in.}}} = \frac{63}{12} \text{ ft}$$

Fill in numbers; multiply

$$= 5\frac{1}{4} \text{ ft}$$

Write as a mixed number

EXAMPLE 16 Jeni's car gets about 20 miles per gallon. How much is this in kilometers per liter?

SOLUTION We have to convert both miles to kilometers, and gallons to liters.

$$\frac{\text{mi}}{\text{gal}} \times \boxed{?} = \frac{\text{km}}{\text{L}}$$

Put given units on left; put desired units on right

$$\frac{\cancel{\text{mi}}}{\cancel{\text{gal}}} \times \frac{\text{km}}{\cancel{\text{mi}}} \times \frac{\cancel{\text{gal}}}{\text{L}} = \frac{\text{km}}{\text{L}}$$

Set up units to cancel properly

$$20\ \frac{\cancel{\text{mi}}}{\cancel{\text{gal}}} \times \frac{1.6 \text{ km}}{1\ \cancel{\text{mi}}} \times \frac{1\ \cancel{\text{gal}}}{3.8 \text{ L}} = \frac{32}{3.8}\ \frac{\text{km}}{\text{L}}$$

Fill in numbers; multiply

$$= 8.4 \frac{\text{km}}{\text{L}}$$ Divide

Thus an automobile that gets 20 miles per gallon gets about 8.4 kilometers per liter.

PROBLEM SET Complete the following tables. (You need not do any box with an X in it.)

See Examples 11 to 16 and Tables 5.1 and 5.2

	mm	cm	m	km
1.		30		X
2.	35			X
3.	X			2.5
4.			6	

	ml	L
5.	600	
6.		2.3

	mg	g	kg
7.	500		X
8.		40	
9.	X		2

10. A car weighs 3000 pounds. How many kilograms is this? How many metric tons is this?
11. Kathy weighs 130 pounds. How many kilograms is this?
12. An Olympic event is the 1500-meter run. How many yards is this? How many miles is this?
13. A certain swimming pool holds 60,000 gallons. How many liters is this?
14. Two towns are 100 miles apart. How many kilometers is this?
15. A board is 150 centimeters long. How many meters is this? How many inches is this?
16. A bottle holds 0.3 liter. How many milliliters is this? How many cups is this?
17. A can of cat food holds 6 ounces. How many grams is this?
18. Roger's subcompact car gets 30 miles per gallon. How many kilometers per liter is this?

ANSWERS 1. 300 mm; 0.3 m 2. 3.5 cm; 0.035 m 3. 250,000 cm; 2500 m
4. 6000 mm; 600 cm; 0.006 km 5. 0.6 L 6. 2300 ml
7. 0.5 g 8. 40,000 mg; 0.040 kg 9. 2000 g
10. 1363.6 kg; 1.36 metric tons 11. 59 kg 12. 1650 yd; 0.94 mi
13. 228,000 L 14. 160 km 15. 1.5 m; 59 in.
16. 300 ml; 1.27 cups 17. 168 g 18. 12.6 km/L

Miscellaneous Exercises 5.1

Choose the closest approximation for each of the following in the indicated units.

See Examples 1 to 3

1. Portable TV set (pounds) (a) 1 (b) 7 (c) 20 (d) 100
2. Table height (inches) (a) 10 (b) 30 (c) 50 (d) 100
3. Pitcher of beer (quarts) (a) $\frac{1}{4}$ (b) 2 (c) 10 (d) 40

See Examples 8 to 10

4. Frisbee (grams) (a) 5 (b) 20 (c) 100 (d) 400
5. Football field (meters) (a) 90 (b) 300 (c) 30 (d) 900
6. Pitcher of beer (liters) (a) 2 (b) 20 (c) 150 (d) 400
7. Petite woman (kilograms) (a) 600 (b) 20 (c) 100 (d) 45
8. Cigarette (centimeters) (a) 2 (b) 10 (c) 50 (d) 100
9. Tablespoon (milliliters) (a) 15 (b) 50 (c) 150 (d) 500

Complete the following tables.

See Examples 4, 12, 15, and 16

	cm	m	in.	ft	yd
10.	100				
11.				10	
12.			36		

	km	mi
13.	30	
14.		200

See Examples 5, 13, and 14

	mg	g	kg	oz	lb
15.		56			
16.				4	
17.			3		

See Examples 6, 11, and 16

	ml	L	qt	gal
18.			2	
19.		5		

Health applications

20. In the apothecaries' system, the smallest unit of weight is the *grain*. Health science professionals use the following approximation:

$$1 \text{ gram} = 15 \text{ grains}$$

(a) An aspirin tablet contains 5 grains of medication. How many grams is this? How many milligrams is this?
(b) A tablet contains 500 milligrams of vitamin C. How many grams is this? How many grains is this?

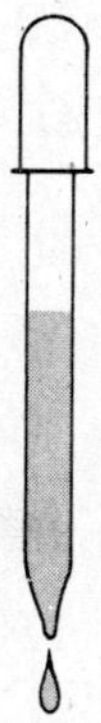
1 minim

21. In the apothecaries' system, the *minim* is equal to 1 drop of liquid. It is related to other units as follows:

$$1 \text{ teaspoon} = 60 \text{ minims}$$

$$1 \text{ milliliter} = 15 \text{ minims}$$

(a) If 100 milliliters remain in a bottle of IV (intravenous) fluid, how many minims are there?
(b) A patient is to receive 30 minims of medicine per day. How many teaspoons is this?

Technical applications

22. A job calls for $\frac{1}{2}$-inch-head rivets. How large are the heads in centimeters?

23. A certain sheet of No. 8 gauge copper weighs 30 kilograms. How many pounds is this?

24. A certain cylindrical drum holds 30 gallons of oil. How many liters is this?

Business applications

25. A chain of lumberyards sells a certain shelf board for 35 cents per foot. How much is this in cents per meter? Set this up as follows:

$$\frac{\text{cents}}{\text{ft}} \times \boxed{\;?\;} = \frac{\text{cents}}{\text{m}}$$

26. A company receives a foreign order for 5000 megagrams (metric tons). How many tons is this?

Life science application

27. For the following animals, the brain weight (in grams) and the body weight (in kilograms) are given. Convert these measurements into ounces and pounds as indicated.

Animal	Brain Weight (grams)	Brain Weight (ounces)	Body Weight (kilograms)	Body Weight (pounds)
Human	1300		60	
Elephant	6000		7000	
Gorilla	600		250	
Deer	210		60	
Beaver	30		6	

Consumer applications

28. Barbara buys ten 2-liter bottles of soda pop. How many liters is this? How many quarts is this? How many cups is this?

29. Mike's car gets 25 miles per gallon. How much is this in kilometers per liter?

30. Sharon buys 150 centimeters of material. How many yards is this?

31. Len weighs 100 kilograms and is 1.65 meters tall. Is he overweight, underweight, or average?

32. Peanut butter provides 13 grams of protein per 50 grams of peanut butter $\left(\frac{13 \text{ g P}}{50 \text{ g}}\right)$. How many grams of protein per pound is this? $\left(\textit{Hint:}\right.$ Set this up as follows: $\left.\frac{\text{g P}}{\text{g}} \times \boxed{?} = \frac{\text{g P}}{\text{lb}}.\right)$

Cliché rewriting

Our folklore is full of expressions that involve the units of the English system of measure. For instance,

"I'd walk a million miles for one of your smiles"

might someday be changed to its metric equivalent,

"I'd walk 1,600,000 kilometers for one of your smiles"

(Not very poetic, is it?) Take the following expressions and clichés and change all the English units to the indicated metric equivalents.

33. Give him an *inch* (x-centimeters), and he'll take a *mile* (y-kilometers).

34. An *ounce* (x-grams) of prevention is worth a *pound* (y-kilograms) of cure.

35. Seats on the *fifty-yard* (x-meters) line.

36. The cowboy wore a *ten-gallon* (x-liters) hat.

37. It hit me like a *ton* (x-kilograms) of bricks.

38. A miss is as good as a *mile* (x-kilometers).

39. A *pint* (x-liters) is a *pound* (y-kilograms), the whole world 'round.

Now complete Post-Test 1 for Section 5.1 on page T71.

5.2 ANGLES AND LENGTHS

SECTION OBJECTIVES

After finishing this unit, you will be able to:

1. Understand and measure angles.
2. Compute the perimeter of a polygon.
3. Compute the circumference of a circle (or semicircle).
4. Compute the perimeter of a combination polygon/circle figure.

TYPICAL PROBLEMS You will be able to work problems like the ones below.

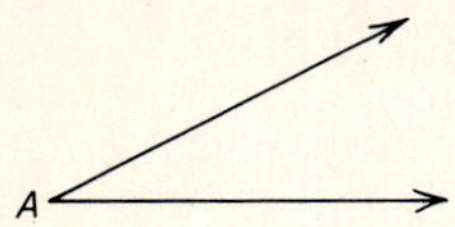

(a) What size angle is $\frac{1}{4}$ of a complete revolution?

(b) What is the measure of the angle A at the left?

Find the perimeters of the figures.

(c)

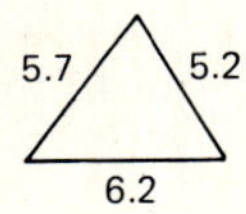

(d)

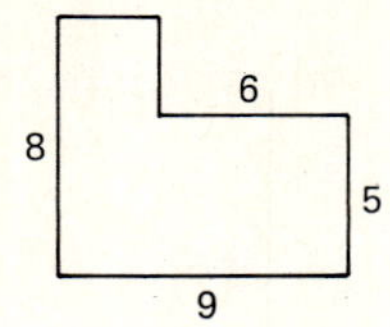

(e)

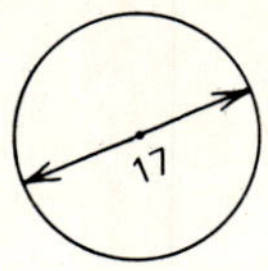

(f)

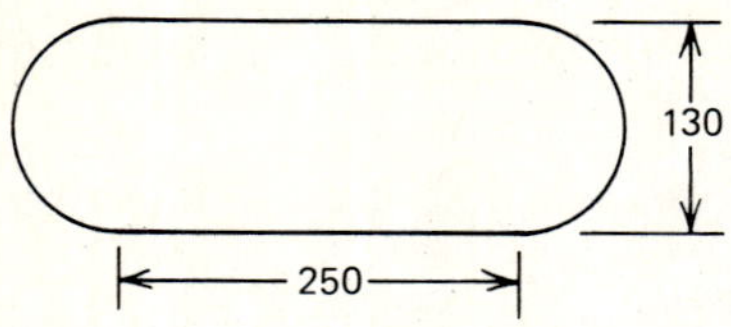

ANSWERS **(a)** 90° **(b)** 27° **(c)** 17.1 **(d)** 34 **(e)** 53.38 **(f)** 908.2

In Section 5.1, we studied the idea of measurement and the two common systems of measurement. In this section, we look at length in more detail, but first we study angles. An **angle** consists of a point (called the **vertex**) and two rays (called the **legs** or **sides**) going out from the points.

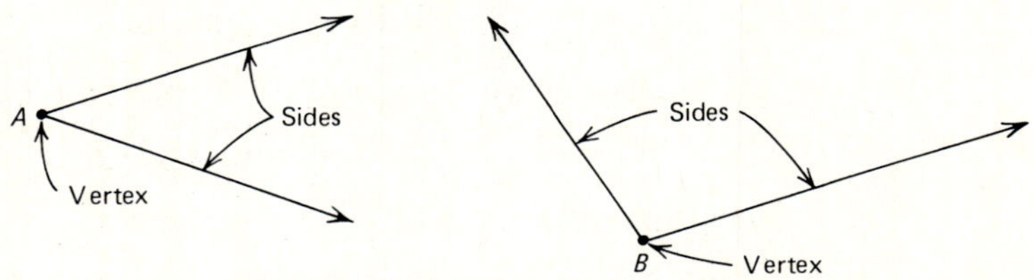

Typical Angles

We commonly label an angle by its vertex. For instance, the angles above are labeled $\angle A$ and $\angle B$.

We measure an angle by how much we have to rotate to go from one side to the other. We measure angles with **degrees**, and one complete revolution (or circle) is 360 degrees, written 360°.

1 complete revolution = 360 degrees = 360°

EXAMPLE 17 The following angles are simple fractions of a complete revolution (or circle). Notice that all the measures are based on 360°, the measure of one complete revolution.

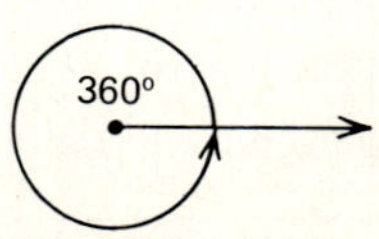

1 revolution

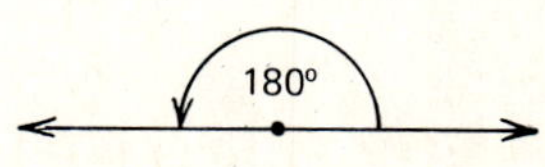

$\frac{1}{2}$ revolution

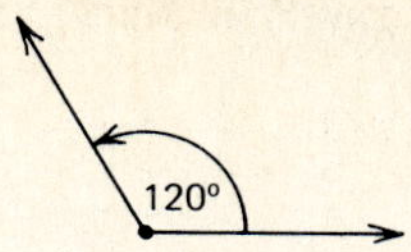

$\frac{1}{3}$ revolution

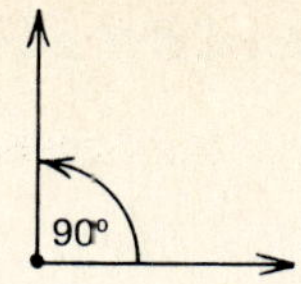

$\frac{1}{4}$ revolution

(This angle is also called a **right angle.**)

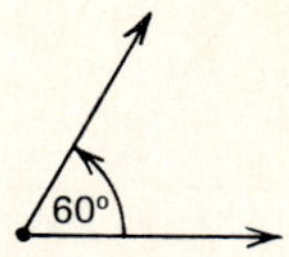

$\frac{1}{6}$ revolution

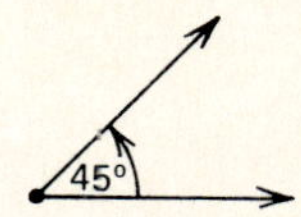

$\frac{1}{8}$ revolution

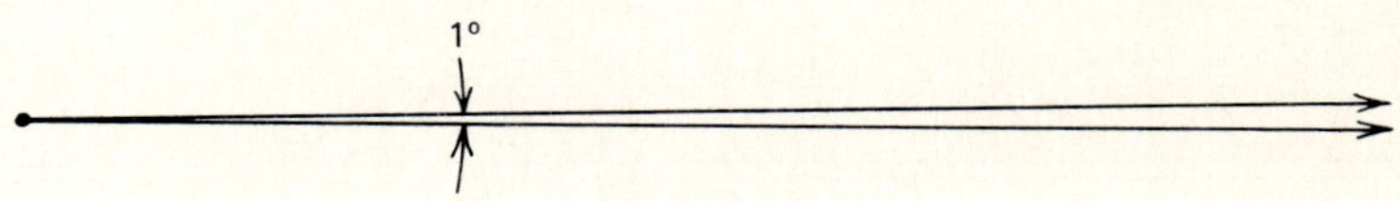

$\frac{1}{360}$ revolution

To measure any angle, we use a **protractor** like the one shown below.

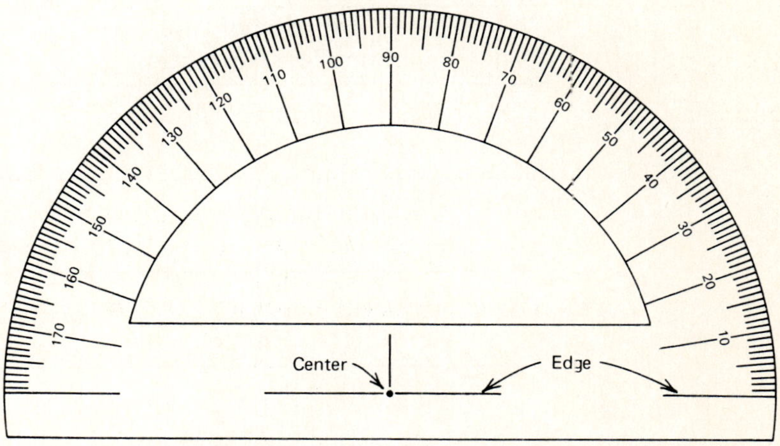

To measure an angle:

1. Put the center of the protractor on the vertex of the angle.
2. Put one edge of the protractor on one side of the angle. (The sides of an angle can be extended indefinitely, if necessary.)
3. Take the reading where the other side of the angle crosses the protractor.

EXAMPLE 18 $\angle A = 42°$

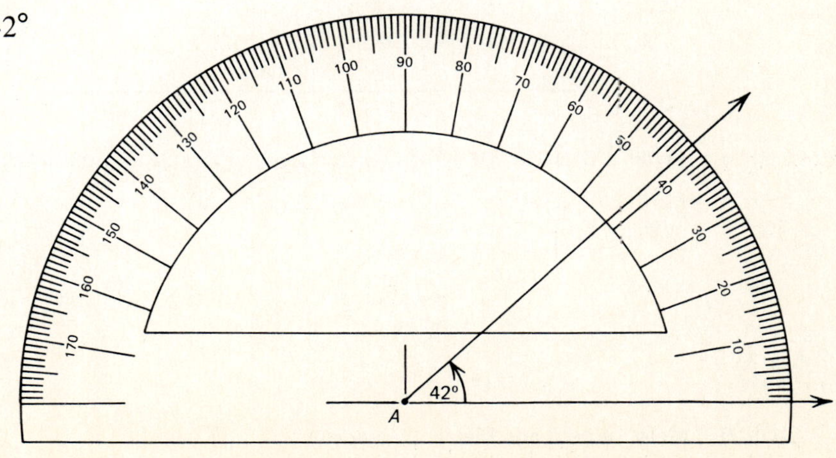

EXAMPLE 19 $\angle B = 127°$

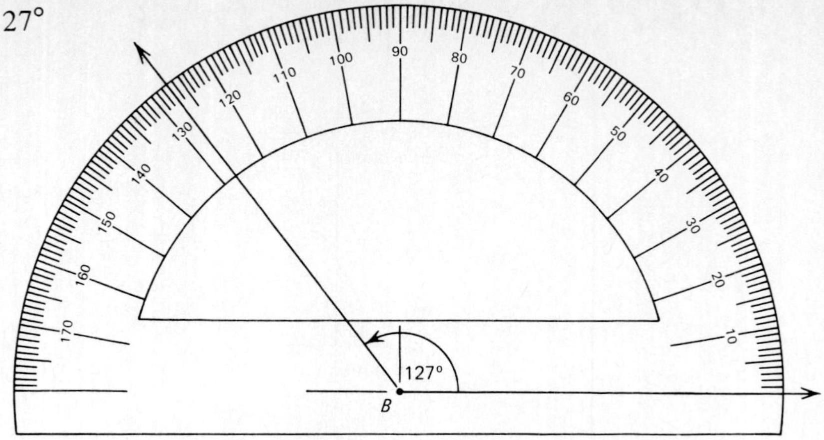

PROBLEM SET Complete the following table.

		Fraction of One Complete Revolution	Angle Measure
See Examples 17 to 19	**1.**	$\frac{1}{2}$	
	2.	$\frac{1}{3}$	
	3.	$\frac{3}{4}$	
	4.	$\frac{2}{3}$	
	5.	$\frac{1}{5}$	
	6.		30°
	7.		36°
	8.		40°

Use a protractor to measure the following angles. (Extend the sides, if necessary.)

9.

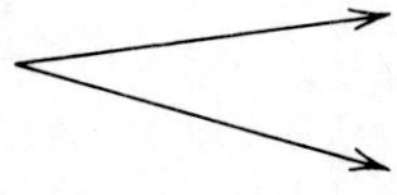

10.

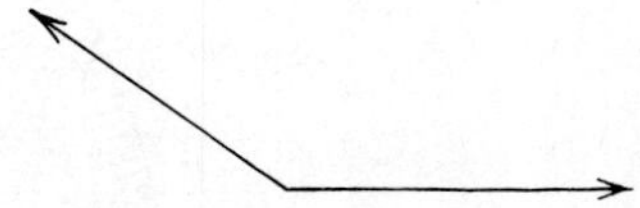

11. 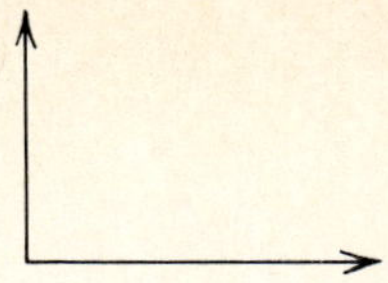

12. 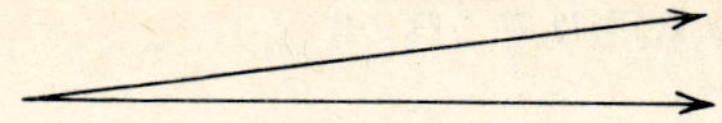

13. 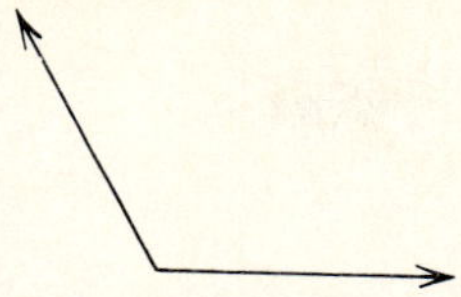

14. 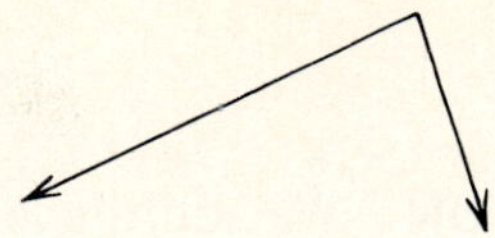

ANSWERS 1. 180° 2. 120° 3. 270° 4. 240° 5. 72° 6. $\frac{1}{12}$
7. $\frac{1}{10}$ 8. $\frac{1}{9}$ 9. 23° 10. 145° 11. 90° 12. 8°
13. 120° 14. 82°

PROBLEM SET Use a protractor to construct the following angles.

See Examples 18 and 19

1. 15° 2. 68° 3. 100° 4. 159°

ANSWERS

1. 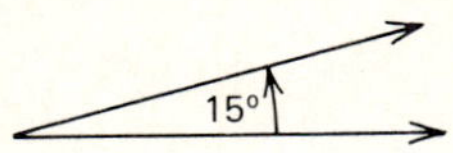

2. 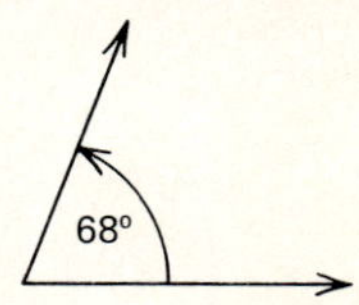

3. 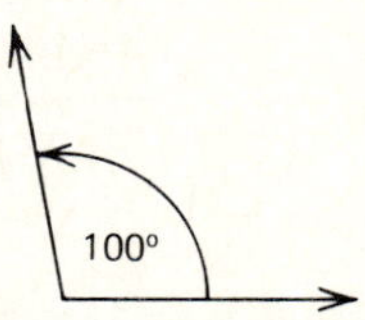

4. 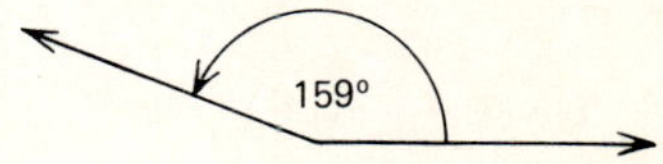

Now let us look at **length**. A closed figure composed of line segments attached together is called a **polygon**.

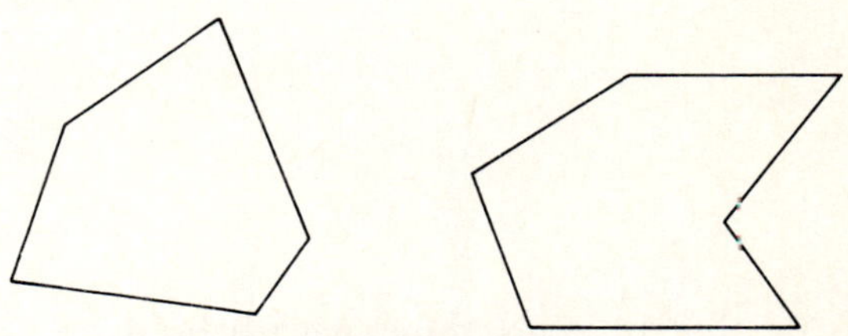

Typical Polygons

The **perimeter** of a polygon is the total distance around the figure. (Think of an imaginary bug walking once around the figure.) We find the perimeter by adding the lengths of all the sides.

> To determine the perimeter of a polygon:
>
> 1. Determine the length of each of the sides.
> 2. Add up all the lengths of the sides.

EXAMPLE 20 Find the perimeter of the triangle shown.

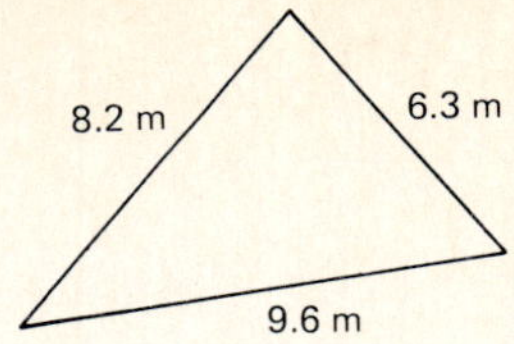

SOLUTION We add the lengths together.

$$\begin{array}{r} 8.2 \text{ m} \\ 6.3 \text{ m} \\ +9.6 \text{ m} \\ \hline \text{perimeter} = 24.1 \text{ m} \end{array}$$

Perimeter = sum of lengths

EXAMPLE 21 Find the perimeter of the polygon shown.

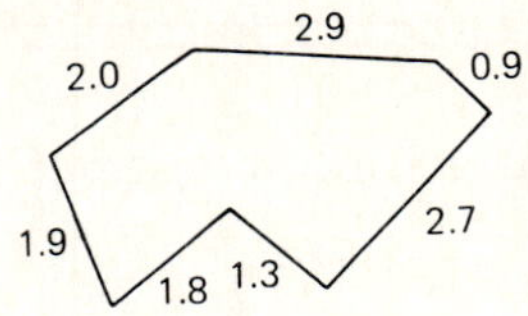

SOLUTION We add the lengths together.

$$\begin{array}{r} 2.0 \\ 2.9 \\ 0.9 \\ 2.7 \\ 1.3 \\ 1.8 \\ +1.9 \\ \hline \text{perimeter} = 13.5 \end{array}$$

Perimeter = sum of lengths

EXAMPLE 22 Find the perimeter of the room shown.

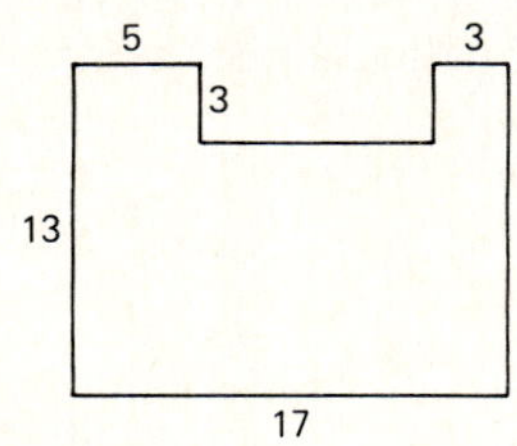

SOLUTION To find the perimeter, we fill in all the missing lengths. How did we get 9 in the middle? The length across is 17. The two lengths at the top are $5 + 3 = 8$. So that leaves $17 - 8 = 9$. So

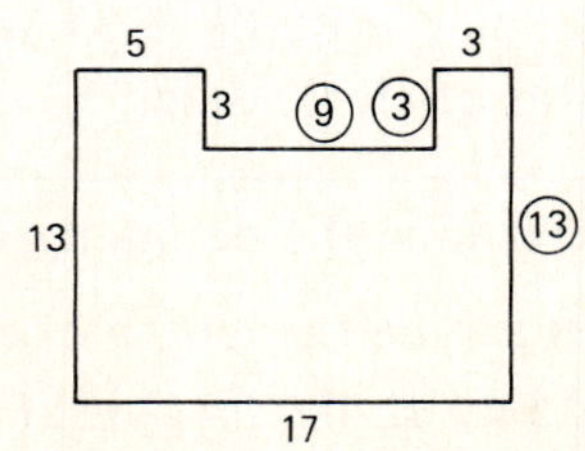

perimeter = 5 + 3 + 9 + 3 + 3 + 13 + 17 + 13 = 66

Perimeter = sum of lengths

PROBLEM SET Find the perimeters of the figures below.

See Examples 20 to 22

1.

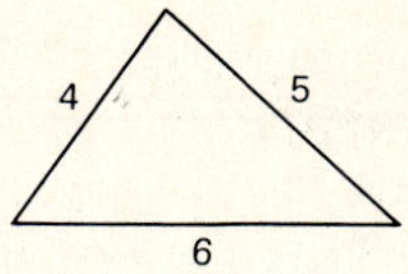

2.

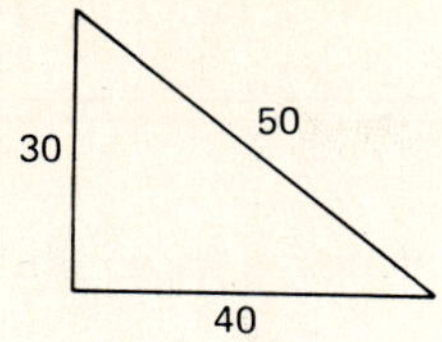

3.

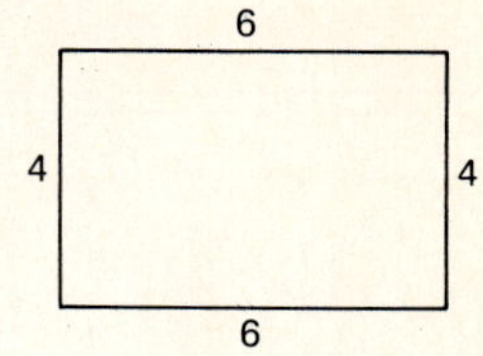

4.

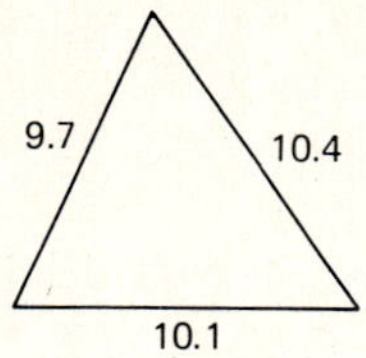

5.

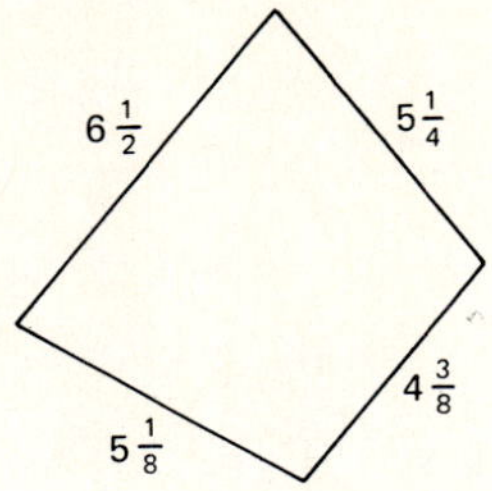

6.

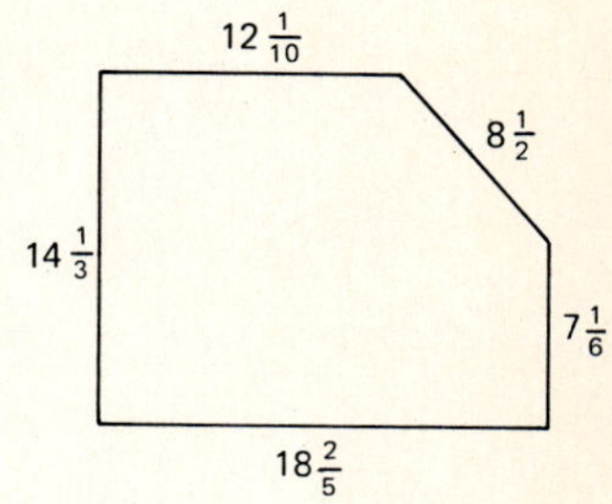

7.

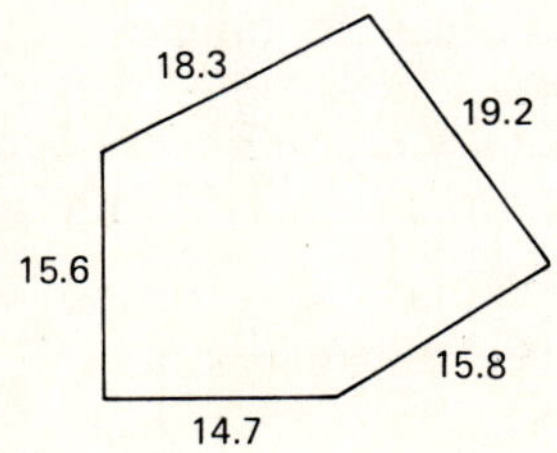

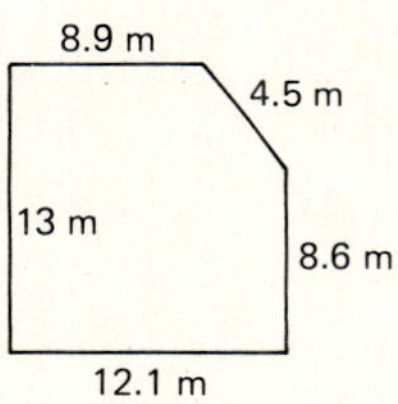

8. Larry is fencing in a lot like the one shown at the left. How much fence will he need?

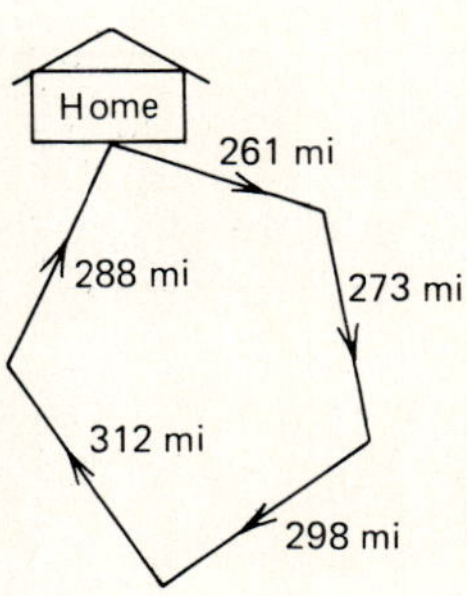

9. At the left is a sketch of the path Donna took on her last business trip. What is the perimeter?

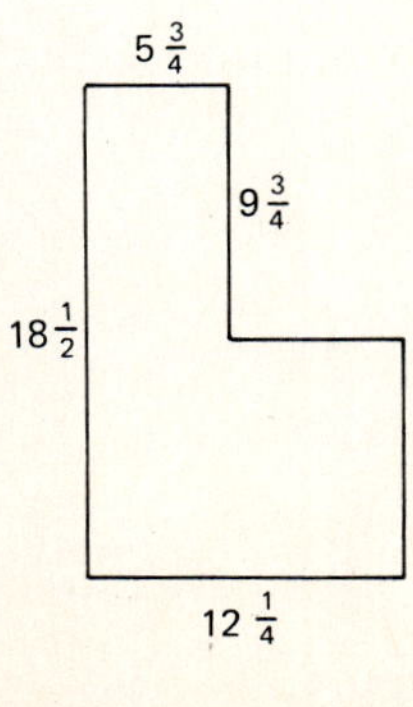

10. At the left is a sketch of the Freedmans' basement. To buy floor molding, they must find the perimeter. Find this perimeter.

ANSWERS **1.** 15 **2.** 120 **3.** 20 **4.** 30.2 **5.** $21\frac{1}{4}$ **6.** $60\frac{1}{2}$ **7.** 83.6 **8.** 47.1 **9.** 1432 **10.** $61\frac{1}{2}$

A special polygon is the **rectangle**. This is a four-sided polygon with four right angles.

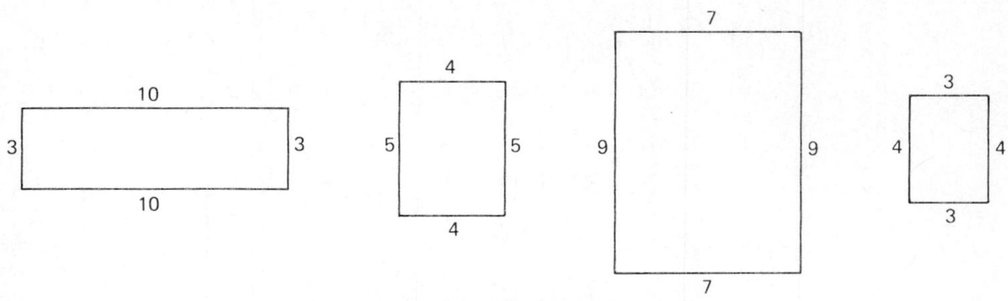

Typical Rectangles

Notice that *the opposite sides in each rectangle are equal*. This makes computing the perimeter easier. We call the dimensions of a rectangle the **length (L)** and the **width (W)**.

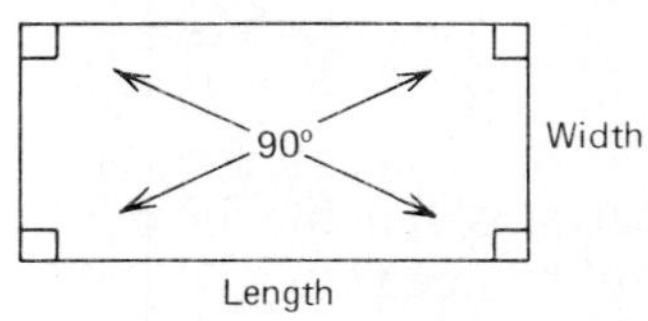

The perimeter of a rectangle is *two lengths and two widths*. As a formula, this is as follows.

$$P_{\text{rectangle}} = 2 \cdot L + 2 \cdot W$$

where L is the length and W is the width.

EXAMPLE 23 The perimeters of the following rectangles are computed by adding twice the length and twice the width.

(a)

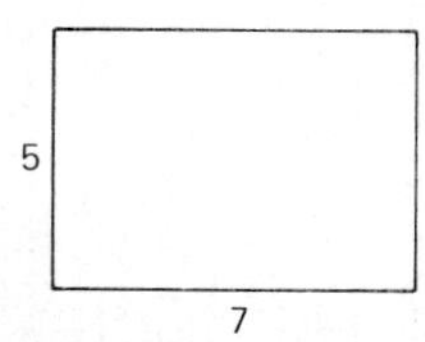

$$\begin{aligned} P &= 2 \times 5 + 2 \times 7 \\ &= 10 + 14 \\ &= 24 \end{aligned}$$

(b)

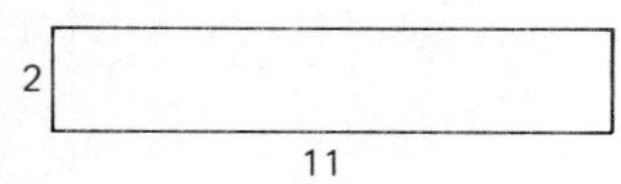

$$\begin{aligned} P &= 2 \times 2 + 2 \times 11 \\ &= 4 + 22 \\ &= 26 \end{aligned}$$

(c)

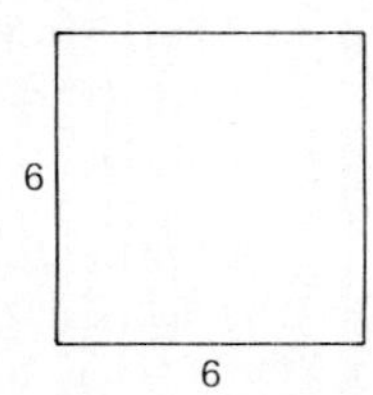

$$\begin{aligned} P &= 2 \times 6 + 2 \times 6 \\ &= 12 + 12 \\ &= 24 \end{aligned}$$

Example 23(c) is an example of a special rectangle: all four sides are equal. This is called a **square**. The perimeter is the sum of the four equal sides, or *four times the length of a side*.

$$P_{\text{square}} = 4 \cdot s$$

where s is the side of a square.

EXAMPLE 24 The perimeters of the following squares are computed as four times the length of a side.

(a) 

$P = 4 \times 5 = 20$

(b) 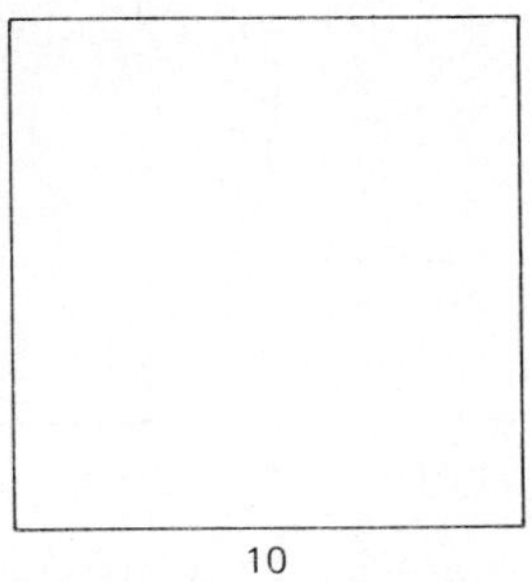

$P = 4 \times 10 = 40$

(c) 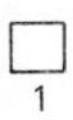

$P = 4 \times 1 = 4$

PROBLEM SET Compute the perimeters of the following rectangles and squares.

See Examples 23 and 24

1. 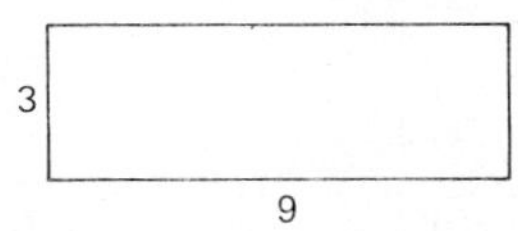

2. 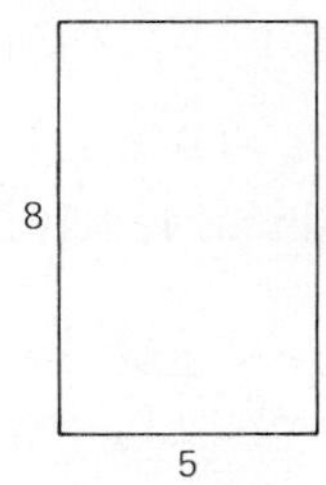

3. A rectangle: 5 by 6.
4. A rectangle: 2 by 17.
5. A square: side = 3.
6. A square: side = 7.

ANSWERS 1. 24 2. 26 3. 22 4. 38 5. 12 6. 28

Below is a **circle** with **center** at 0. The **diameter** D is the distance across the entire circle through the center. The **radius** r is the distance from the center to the circle.

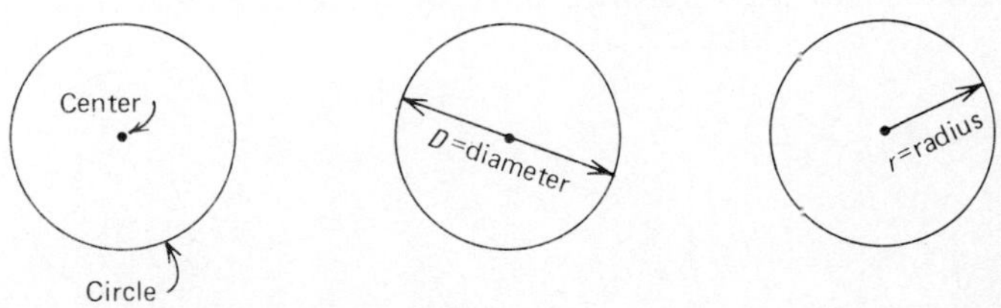

Notice that *the diameter is always twice the radius*. (Or, *the radius is half the diameter*.)

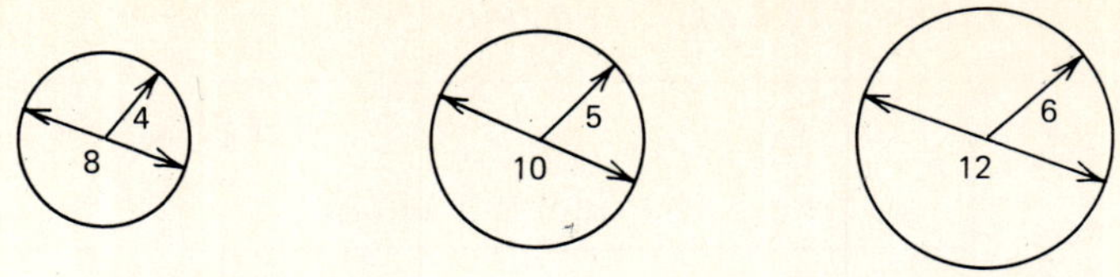

The distance around a circle is called its **circumference** C. It turns out that the ratio of the circumference of a circle to its diameter is the same for all circles. We call this ratio π **(pi)**, which is *approximately* 3.14.

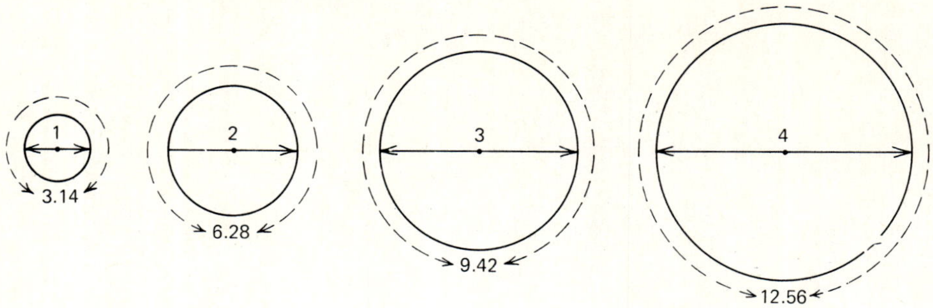

$$\frac{C}{D} = \frac{3.14}{1} = \frac{6.28}{2} = \frac{9.42}{3} = \frac{12.56}{4} = 3.14\ (= \pi)$$

$$C = \pi \cdot D \qquad \text{or} \qquad C = 2 \cdot \pi \cdot r$$

where D = diameter, r = radius, and π = 3.14 (approximately).

EXAMPLE 25 Find the circumference of the circle shown.

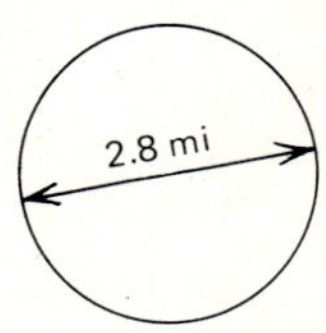

SOLUTION Since we have the diameter, we use the formula $C = \pi \cdot D$.

$$C = \pi \cdot D$$
$$= (3.14) \cdot (2.8) \longrightarrow$$
$$= 8.792 \longleftarrow$$

```
  3.14
X  2.8
 2 512
 6 28
 8.792
```

So the circumference = 8.792 miles.

EXAMPLE 26 Find the circumference of the circle shown.

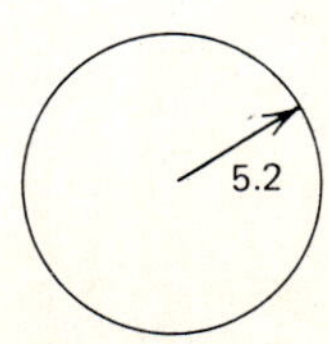

SOLUTION Since we have the radius, we use the formula $C = 2 \cdot \pi \cdot r$.

$$\begin{aligned} C &= 2 \cdot \pi \cdot r \\ &= 2 \cdot (3.14) \cdot (5.2) \longrightarrow \\ &= 32.656 \longleftarrow \end{aligned} \qquad \boxed{\begin{array}{r} 3.14 \\ \times\ 5.2 \\ \hline 628 \\ 15\ 70 \\ \hline 16.328 \\ \\ 16.328 \\ \times\ 2 \\ \hline 32.656 \end{array}}$$

So the circumference is 32.656. (Also, we could have first found the diameter as 10.4 and then used $C = \pi \cdot D$.)

EXAMPLE 27 Find the circumference of the semicircle shown.

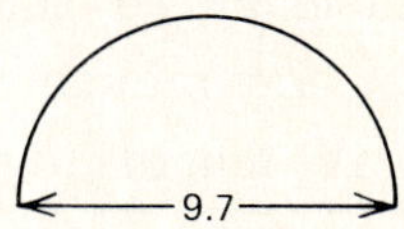

SOLUTION This is half a circle, so we take half the circumference of a whole circle. Since we have the diameter, we use $C = \pi \cdot D$.

$$\begin{aligned} \frac{1}{2}C &= \frac{1}{2} \cdot \pi \cdot D \\ &= \frac{(3.14) \cdot (9.7)}{2} \longrightarrow \\ &= 15.229 \longleftarrow \end{aligned} \qquad \boxed{\begin{array}{r} 3.14 \\ \times\ 9.7 \\ \hline 2\ 198 \\ 28\ 26 \\ \hline 30.458 \\ \\ 15.229 \\ 2\,\overline{)30.458} \end{array}}$$

> Circumference of a semicircle is $\frac{1}{2}$ circumference of whole circle

So the circumference of the semicircle is 15.229.

PROBLEM SET Find the circumferences of the circles and semicircles.

See Examples 25 to 27

1.

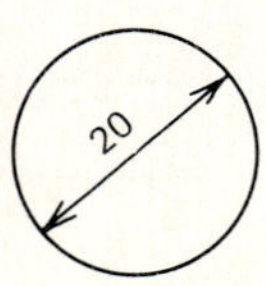

2.

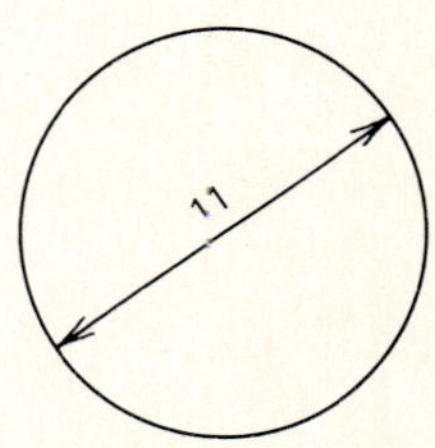

3. A circle: diameter $= 7\frac{1}{2}$.

4. A circle: radius = 8.

5. A circle: radius = 4.9.

6. A circle: radius $= 2\frac{1}{4}$.

7. 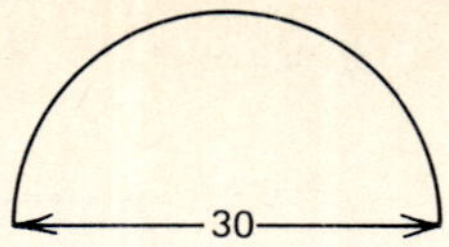

8. A semicircle: diameter = 42.

9. A semicircle: radius = 9.6.

10. Jackie is planting a circular garden of diameter 14.5. How much fence will she need to enclose it?

ANSWERS 1. 62.8 2. 34.54 3. 23.55 4. 50.24 5. 30.772 6. 14.13 7. 47.1 8. 65.94 9. 30.144 10. 45.53

We can often find the perimeter of a complicated figure by breaking it into smaller, simpler parts.

> To find the perimeter of a combination of line segments and circles:
> 1. Break the figure into its line-segment and circle parts.
> 2. Find the lengths of the line segments.
> 3. Find the circumference of the circles (or parts of circles).
> 4. Add all the lengths together.

EXAMPLE 28 Find the perimeter of the figure shown.

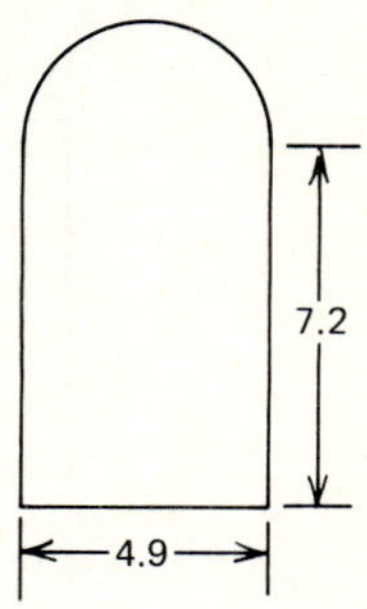

SOLUTION We cut the figure into three line segments and a semicircle. We need to find the circumference of the semicircle. Notice that the diameter of the semicircle is the same as the base, 4.9.

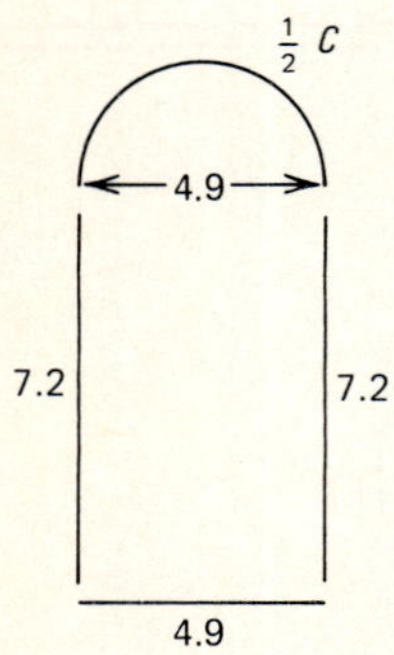

$$\frac{1}{2}C = \frac{1}{2}\cdot\pi\cdot D$$

$$= \frac{(3.14)(4.9)}{2}$$

$$= 7.693$$

3.14
× 4.9
2 826
12 56
15.386
7.693
2 ⟌ 15.386

Circumference of semicircle is $\frac{1}{2}$ circumference of whole circle

Perimeter = 26.993

7.693
7.200
7.200
+4.900
26.993

Perimeter = sum of lengths

EXAMPLE 29 The outdoor track at Joque State College is shown. What is one lap (the perimeter)?

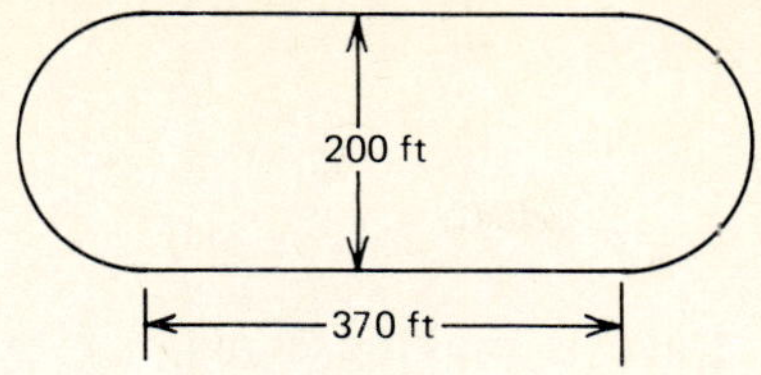

SOLUTION We cut the track into four parts: two straightaways, and two semicircle turns. Let us find the circumference of a turn.

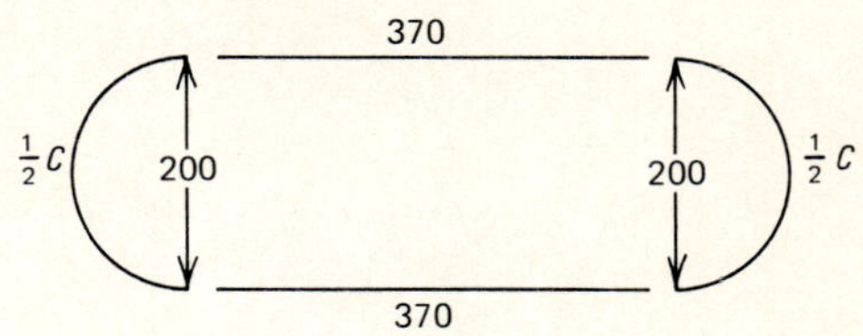

$$\frac{1}{2}C = \frac{1}{2} \cdot \pi \cdot D$$

$$= \frac{(3.14) \cdot (200)}{2} \rightarrow \quad \begin{array}{r} 3.14 \\ \times\ 200 \\ \hline 628.00 \end{array} \qquad 2\overline{)628} = 314$$

$$= 314$$

Circumference of a semicircle is $\frac{1}{2}$ circumference of a whole circle

$$\text{perimeter} = 1368 \quad \begin{array}{r} 370 \\ 370 \\ 314 \\ +314 \\ \hline 1368 \end{array}$$

Perimeter = sum of lengths

PROBLEM SET Find the perimeters of the following figures.

See Examples 28 and 29

1.

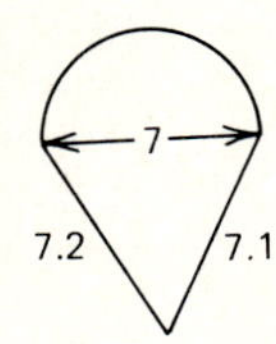

2.

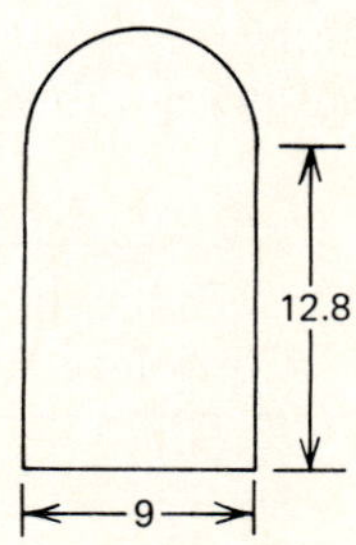

3.

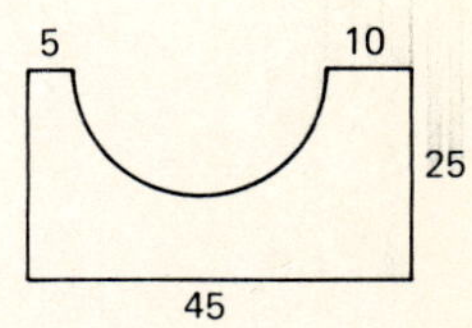

4.

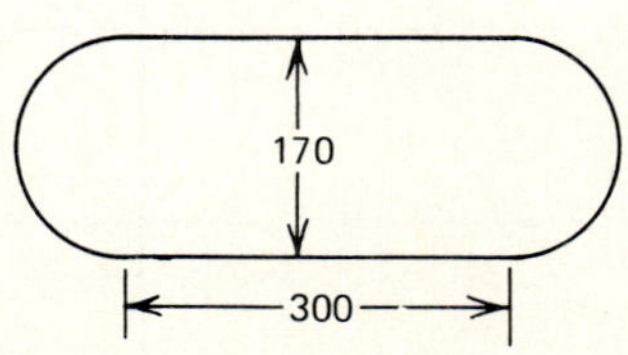

5.

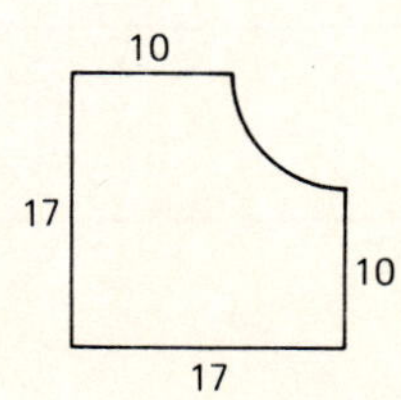

6.

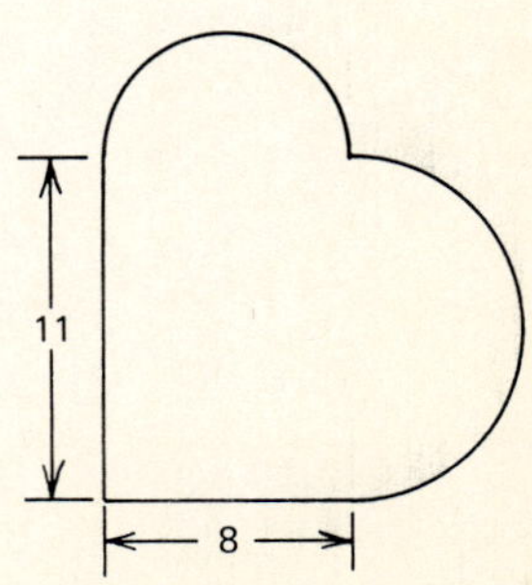

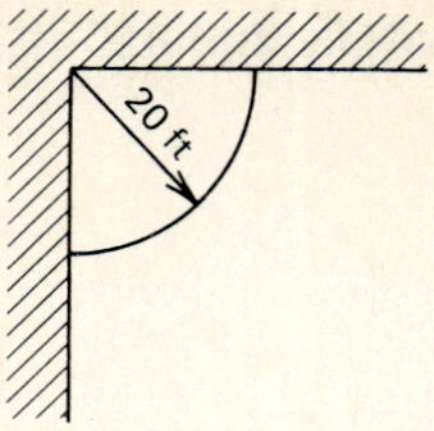

7. Jane is planting a quarter-circular garden 20 feet in radius where two walls meet. How much fence does she need?

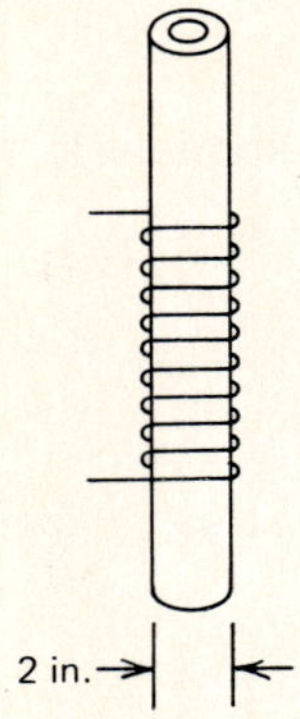

8. Dave is making a magnet by wrapping wire around a 2-in.-diameter metal bar 10 times. How much wire is needed just to do the wrapping around the bar?

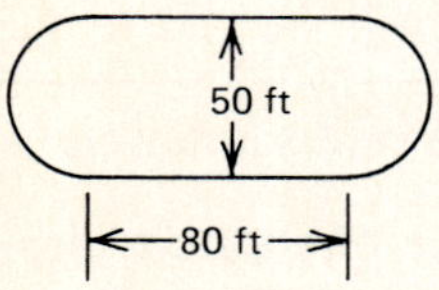

9. At the left is a sketch of a jogging track at the Raccoonville YMCA. What is the length of one lap around the track?

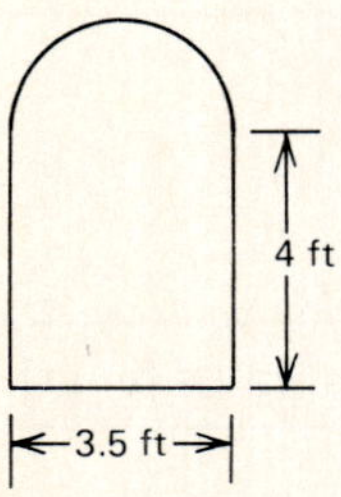

10. Dave and Linda are caulking a window like the one shown at the left. Find its perimeter.

ANSWERS **1.** 25.29 **2.** 48.73 **3.** 157.1 **4.** 1133.8 **5.** 64.99 **6.** 48.83 **7.** 31.4 **8.** 62.8 **9.** 317 **10.** 16.995

Miscellaneous Exercises 5.2

Complete the following table.

	Fraction of One Complete Revolution	Angle Measure
See Example 17 1.	$\frac{1}{2}$	
2.	$\frac{1}{10}$	
3.		60°
4.		90°

Measure the following angles with a protractor.

See Examples 18 and 19

5. 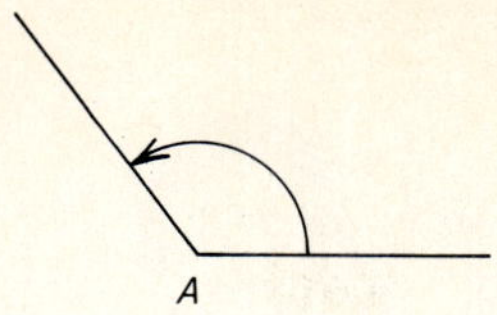

6. 

7. 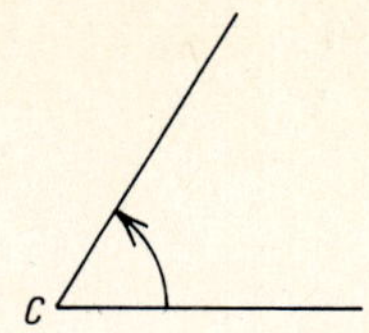

Approximate the following angles *without* a protractor.

See Examples 18 and 19

8. 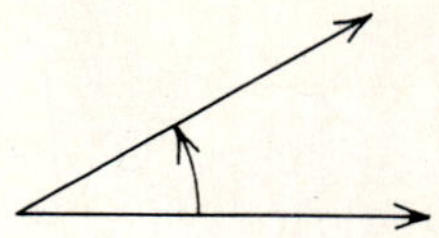

(a) 5° **(b)** 30° **(c)** 85° **(d)** 155°

9. 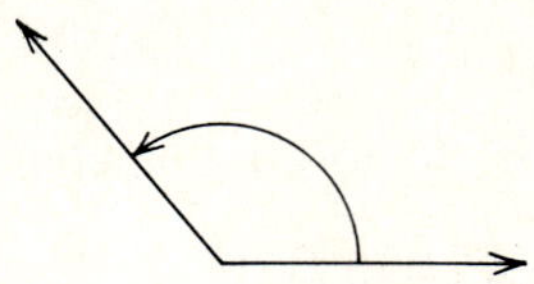

(a) 10° **(b)** 75° **(c)** 130° **(d)** 175°

Compute the following perimeters.

See Examples 20 to 22

10. 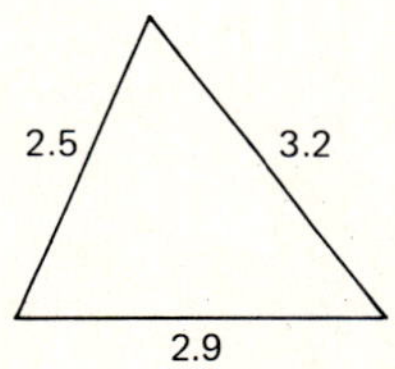

11. 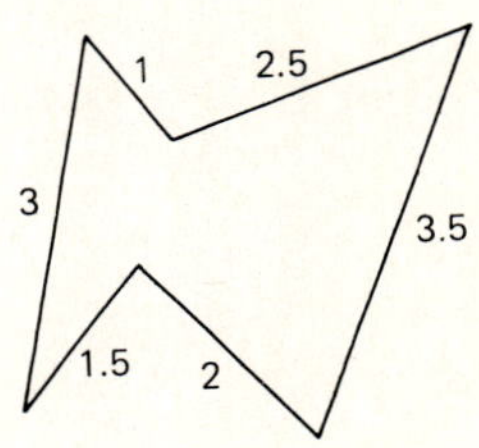

12. 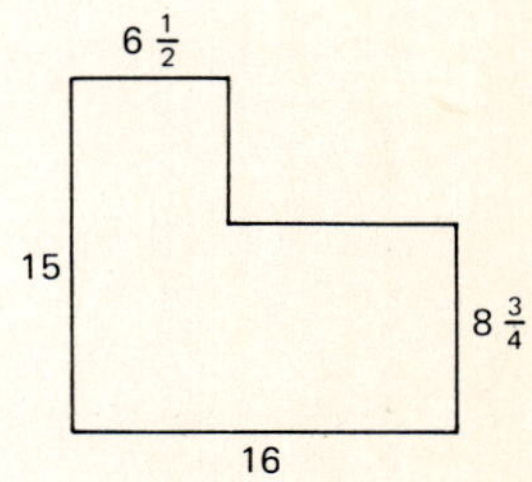

See Examples 23 and 24

13. A rectangle: 10 by 4.

14. A square: side = 3.

Compute the following circumferences.

See Examples 25 to 27

15. A circle: diameter = 7.

16. A circle: radius = 4.

17. A semicircle: radius = 3.

Compute the following perimeters.

See Examples 28 and 29

18.

19.

Health applications

20. Arm slings are set at various angles. For the sketches below, determine the approximate angle made by the forearm and the upper arm.

(a) (b) (c)

21. A man's leg has a diameter of about 5 inches. (Assume that the leg is circular.)
   (a) What is the circumference of the leg?
   (b) If a nurse has 200 inches of bandage, how many times can it be wrapped around the patient's leg?

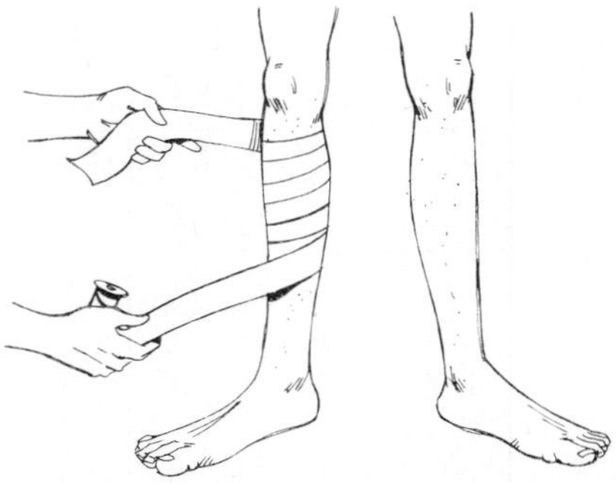

Business applications

22. A manufacturing plant in Chicago must ship a load of goods by truck to New York, Dallas, Denver, Los Angeles, and back to Chicago. Use the distances (approximate) on the map below to find the order of delivery that will get to all the cities and back to Chicago in the least number of miles. (Use trial and error: try the different routes.)

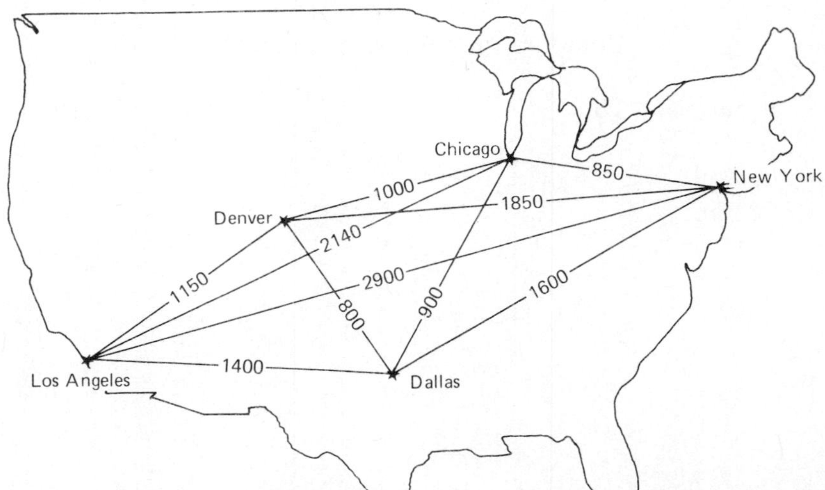

23. Businesses frequently use pie (or circle) diagrams to display information. The pie is sliced in proportion to each of the portions. Draw a pie diagram for the information presented in the table. Use the proportion

$$\frac{\text{desired angle}}{360^\circ} = \frac{\text{portion}}{\text{total}}$$

Section	Employees
Sales	225
Manufacturing	300
Research	200
Administration	275
Total	1000

Technical applications

24. Below is the sketch of a pulley. Find the length of the belt that would be used to go around the pulley.

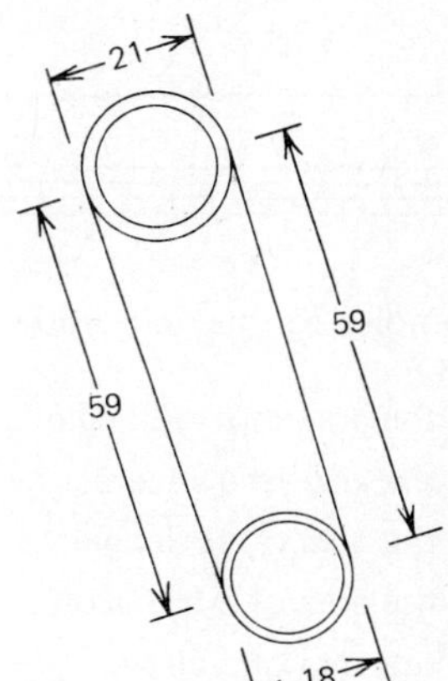

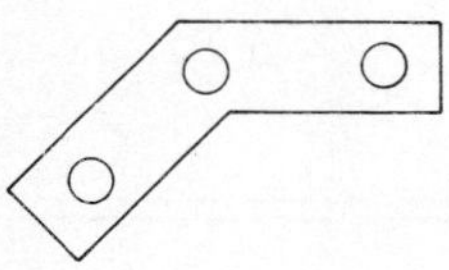

25. At the left is the sketch of an offset bracket. Find the angle that it makes.

Earth science application

26. The diameter of the earth is about 8000 miles. Find the circumference of the earth (assume that the equator is a circle).

Chemistry application

27. Below is a sketch of the molecular structure of water (H_2O). Find the angle between the hydrogen bonds.

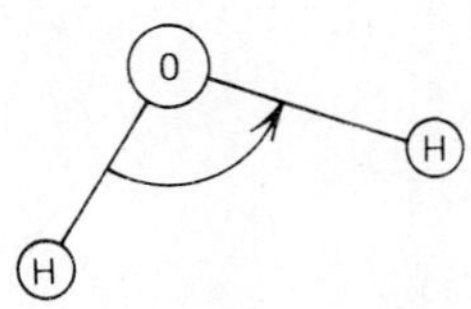

Consumer applications

28. The Smiths' family room is sketched below. Compute the perimeter, the amount of floor molding that they need to go around the walls.

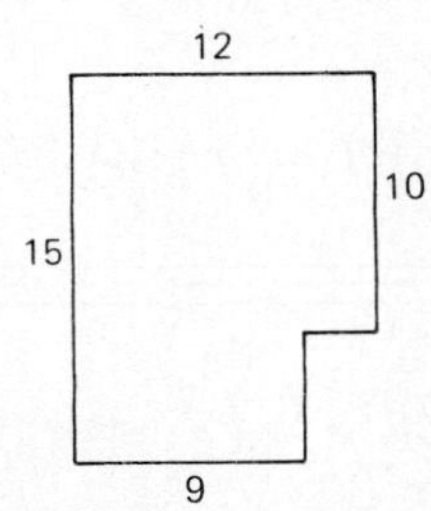

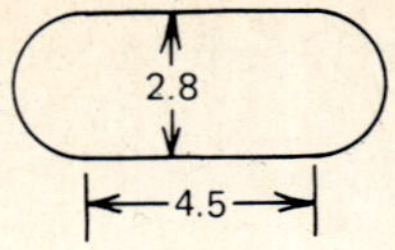

29. Tim is planting a semicircular garden, with radius 10, against his house. How much fence does he need to enclose the garden?

30. Jane and Bill buy a tabletop as sketched at the left. They need to buy trim to put around the outside of the tabletop. Compute the perimeter.

Geometry application

31. For each of the following triangles, compute the perimeter and the total measure of all the angles.

(a)

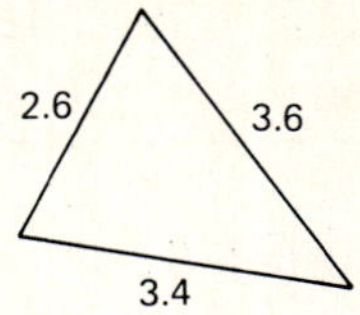

(b)

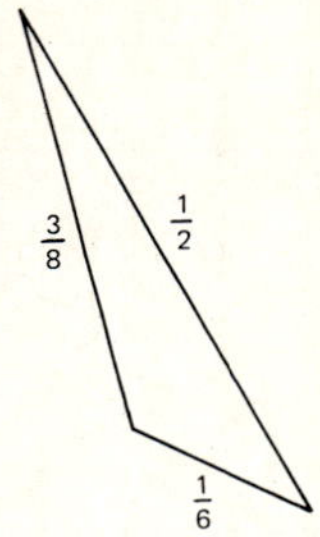

(c)

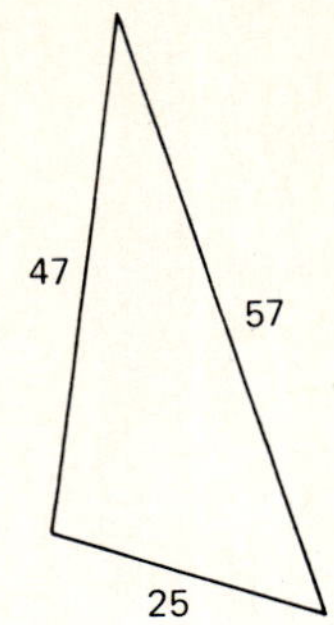

Now complete Post-Test 1 for Section 5.2 on page T73.

5.3 AREA

SECTION OBJECTIVES

After completing this unit, you will be able to:

1. Find the area of a rectangle.
2. Find the area of a square.
3. Find the area of a triangle.
4. Find the area of a trapezoid.
5. Find the area of a circle.
6. Find the area of combinations of figures.

TYPICAL PROBLEMS

You will be able to work problems like the ones below. Compute the areas of the figures.

(a)

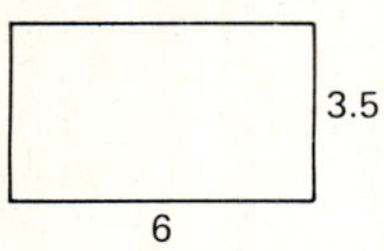

(b)

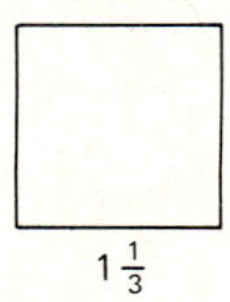

(c)

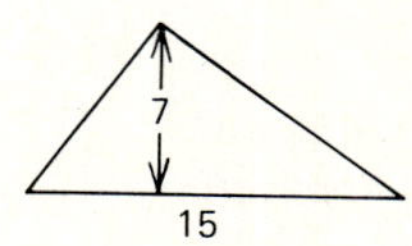

(d)

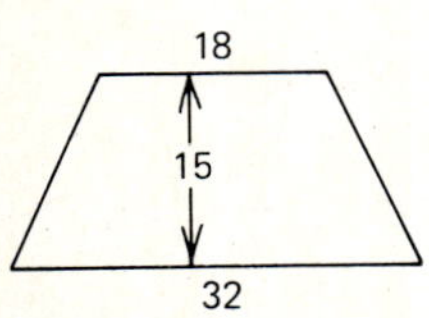

(e)

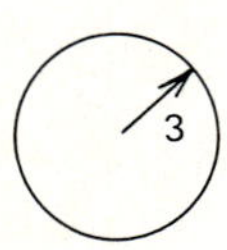

(f)

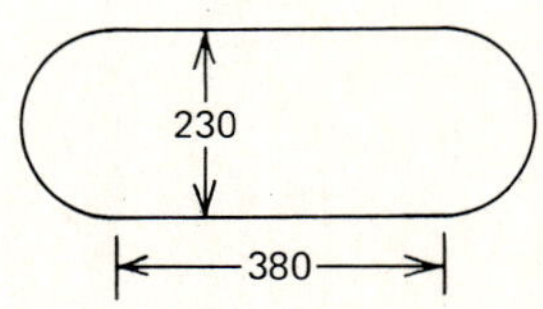

ANSWERS **(a)** 21 **(b)** $1\frac{7}{9}$ **(c)** 52.5 **(d)** 375 **(e)** 28.26 **(f)** 128,926.5

Area is a measure of surface. It is measured in **square units**: square inches, square feet, square meters, and so on. Each of these is a square with a side of one inch, one foot, one meter, and so on. Below are two such units: one square inch (abbreviated 1 sq in. or 1 in.^{2}) and one square centimeter (abbreviated 1 sq cm or 1 cm^2).

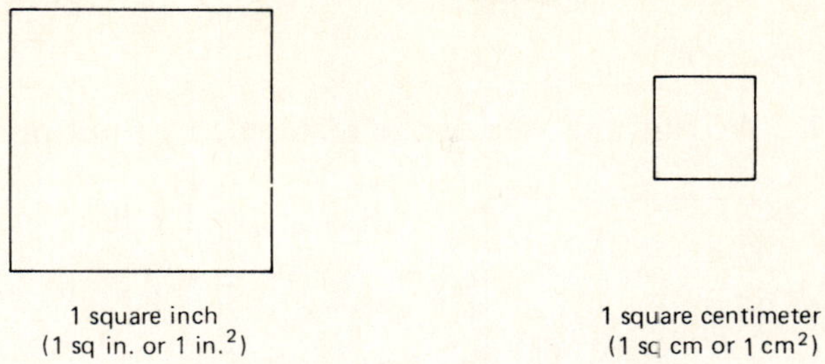

Two Square Units

Most floor tiles come one foot by one foot (or one square foot). When we find the area of a floor in square feet, we are finding how many one-square-foot tiles cover the floor. This is a typical application of area: determining how many squares it takes to cover a surface.

In Section 5.2, we talked about the **rectangle**, which has four sides (the opposite sides are equal) and all the angles are 90°. The two dimensions are called **length** and **width**.

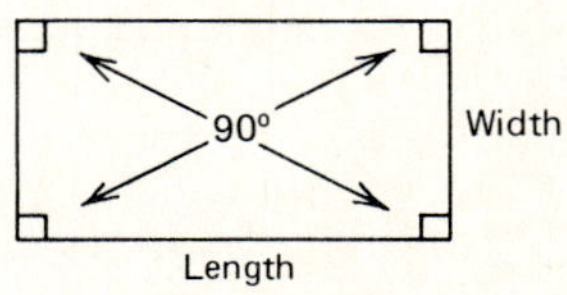

The Rectangle

To find the area of a rectangle, we wish to find how many square units cover it. For instance, consider the rectangles sketched below.

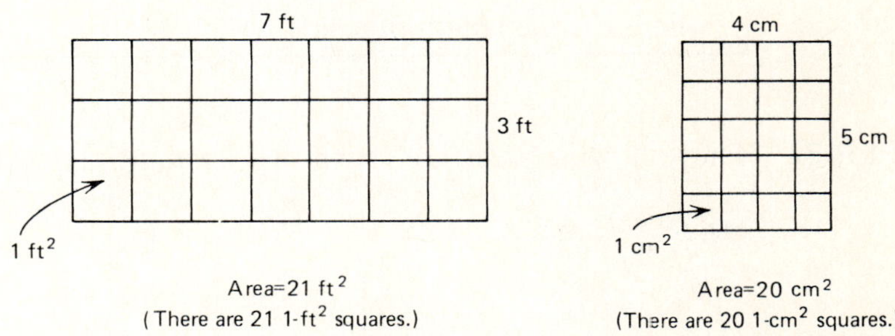

Notice that the area is simply the product of the *length times the width*. This is easy to see when the dimensions are whole numbers, but it is also true for any length and width.

$$A_{\text{rectangle}} = L \cdot W$$

where L is the length and W is the width of the rectangle.

EXAMPLE 30 We find the area of the rectangle by multiplying length times width.

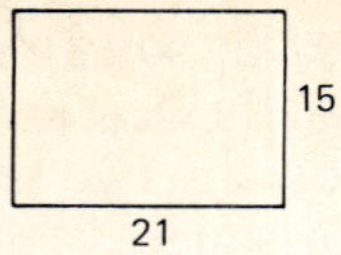

$$A_{\text{rectangle}} = L \cdot W = (21) \cdot (15) = 315$$

$$\begin{array}{r} 21 \\ \times\ 15 \\ \hline 105 \\ 21 \\ \hline 315 \end{array}$$

EXAMPLE 31 We find the area of the rectangle by multiplying length times width.

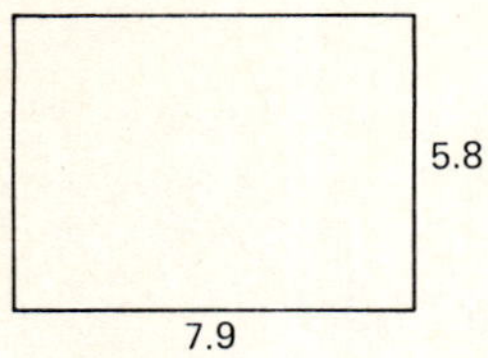

$$A_{\text{rectangle}} = L \cdot W = (7.9) \cdot (5.8) = 45.82$$

$$\begin{array}{r} 7.9 \\ \times\ 5.8 \\ \hline 6\ 32 \\ 39\ 5 \\ \hline 45.82 \end{array}$$

As we discussed in Section 5.2, the **square** is a special rectangle with all its sides equal.

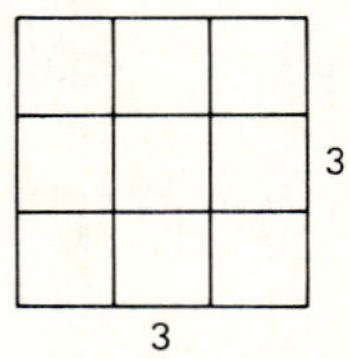

Area=3x3=9

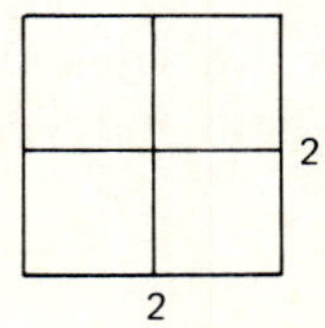

Area=2x2=4

If each side has length s, then the area is the *length of a side times itself*, or $s \cdot s$. We have another way to write this: $s \cdot s = s^2$ (read "s squared").

$$A_{\text{square}} = s^2 = s \cdot s$$

where s is the length of a side.

EXAMPLE 32 The following areas are computed by multiplying the length of the side by itself.

(a) $A = 7^2 = 7 \times 7 = 49$

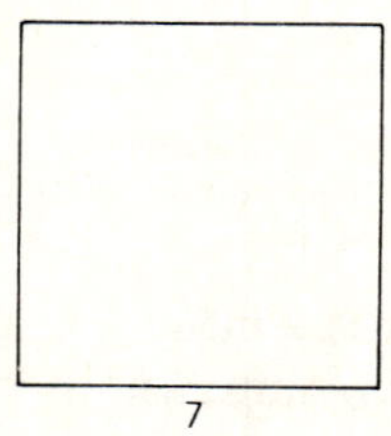

(b) $A = 4^2 = 4 \times 4 = 16$

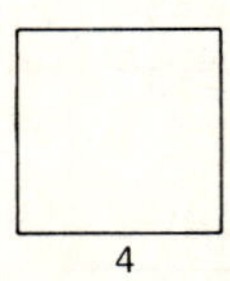

(c) $A = 5.2^2 = (5.2) \times (5.2) = 27.04$

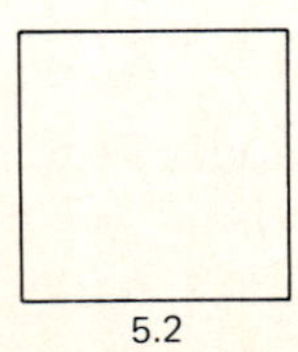

PROBLEM SET Find the areas of the following figures.

See Examples 30 to 32

1.

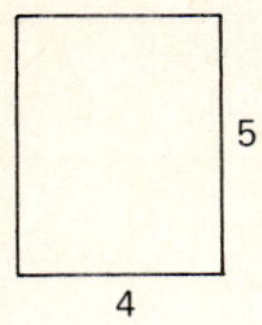

2.

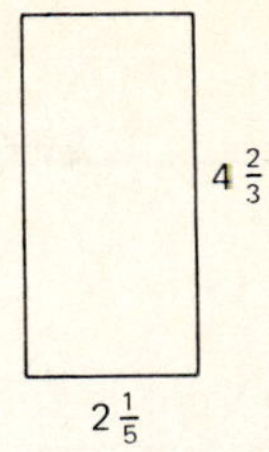

3. A rectangle: 16.2 by 14.7.

4. A square: side = 10.

5. A square: side = 1.6.

6. A square: side = $1\frac{1}{2}$.

7. A rectangular lawn that the Wileys are sodding is 24 feet by 35 feet. Find the area.

8. Judy is painting a wall 13.5 feet by 7.9 feet. Find its area.

9. The Wilsons are carpeting a rectangular floor $4\frac{1}{3}$ meters by $3\frac{2}{3}$ meters. Find its area.

10. Joe is painting over the square sign shown at the left. Find its area.

ANSWERS 1. 20 2. $10\frac{4}{15}$ 3. 238.14 4. 100 5. 2.56 6. $2\frac{1}{4}$
7. 840 ft^2 8. 106.65 ft^2 9. $15\frac{8}{9}$ m^2 10. 17.64 ft^2

A **triangle** is a polygon with three sides. Triangles come in all shapes and sizes.

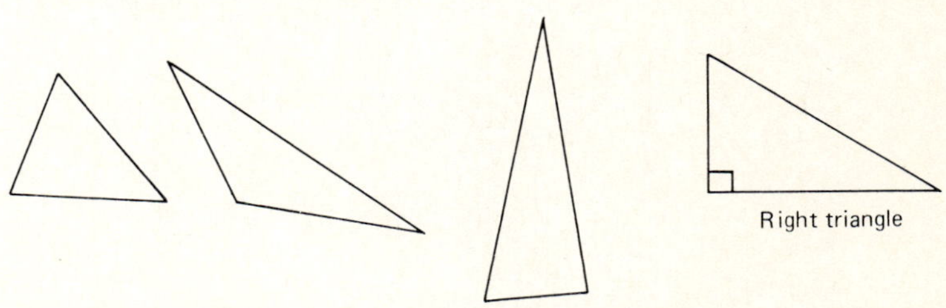

Typical Triangles

The triangle on the far right is a special triangle. It is called a **right triangle** since it contains a right angle (90°). The dimensions are called the **base** (*b*) and the **height** (*h*). The area of a right triangle is easy to compute when we view it as one-half of a rectangle with length b and width h.

Since the area of the entire rectangle is base times height, the area of the triangle is *one-half the base times the height.* In fact, this is true for any triangle: The area is one-half the base times the height.

$$A_{\text{triangle}} = \frac{1}{2} \cdot b \cdot h$$

where b is the base and h is the height.

EXAMPLE 33 We compute the areas of the following triangles by using the formula

$$A = \frac{1}{2} \cdot b \cdot h$$

(a) $A = \frac{1}{2} \cdot (4) \cdot (6) = 12$

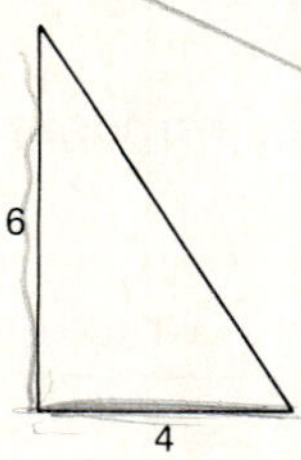

(b) $A = \frac{1}{2} \cdot (10) \cdot (8) = 40$

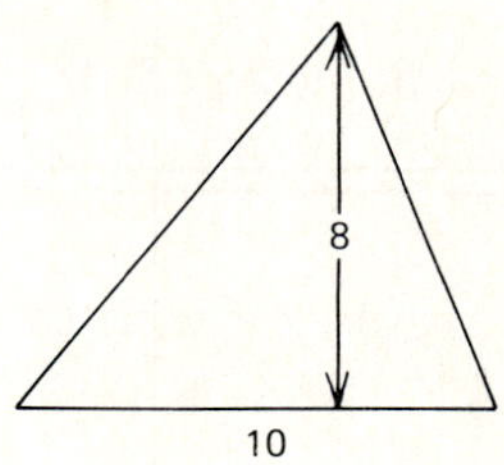

(c) $A = \frac{1}{2} \cdot (2) \cdot (2) = 2$

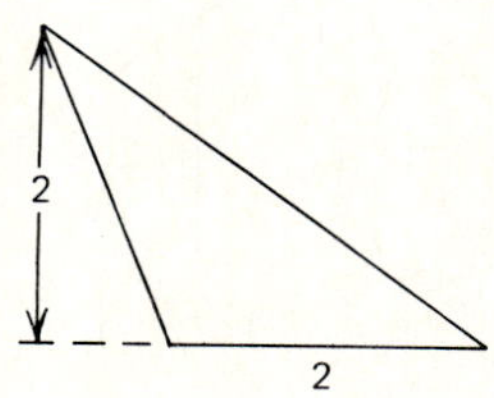

EXAMPLE 34 Find the area of the triangle shown.

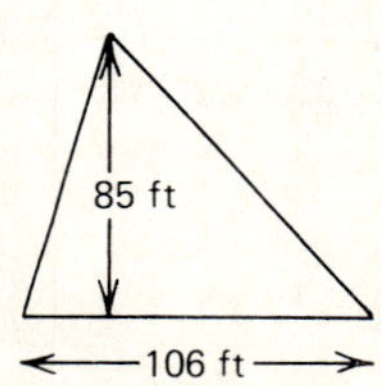

SOLUTION We use the formula $A = \frac{1}{2} \cdot b \cdot h$.

$$A_{\text{triangle}} = \frac{1}{2} \cdot b \cdot h$$

$$= \frac{(106) \cdot (85)}{2}$$

$$= 4505$$

```
   106
 ×  85
   530
  848
  9010

   4505
2 ) 9010
```

EXAMPLE 35 Find the area of the triangle shown.

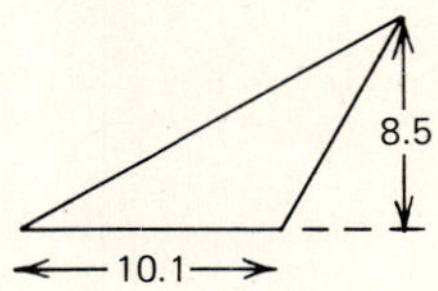

SOLUTION We use the formula $A = \frac{1}{2} \cdot b \cdot h$.

$$A_{\text{triangle}} = \frac{1}{2} \cdot b \cdot h$$

$$= \frac{(10.1) \cdot (8.5)}{2}$$

$$= 42.925$$

```
   10.1
 ×  8.5
   5 05
  80 8
  85.85

   42.925
2 ) 85.850
```

PROBLEM SET Compute the area of each of the following triangles.

See Examples 33 to 35

1.

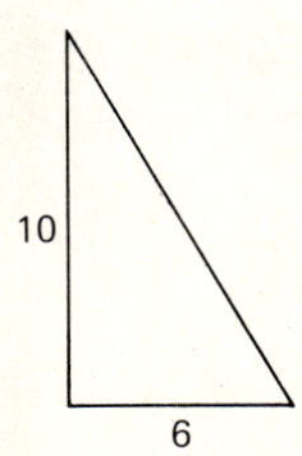

2.

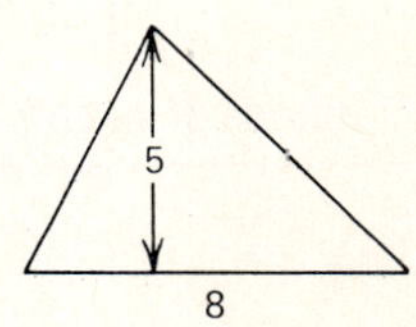

3. A triangle: base = 4, height = 6.
4. A triangle: base = 83, height = 75.
5. A triangle: base = $2\frac{1}{4}$, height = $2\frac{1}{2}$.
6. A triangle: base = 6.2, height = 6.5.
7. Gene is planting a triangular garden like the one shown at the left. Find its area.

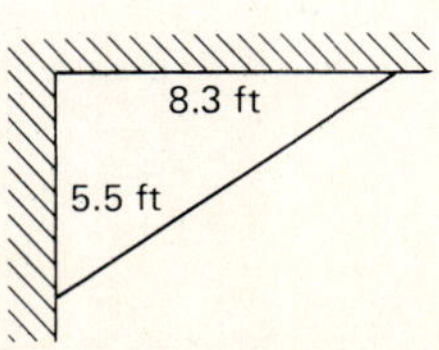

ANSWERS 1. 30 2. 20 3. 12 4. 3112.5 5. $2\frac{13}{16}$ 6. 20.15 7. 22.825

The last polygon that we discuss is the **trapezoid.** The trapezoid has four sides, two of which are parallel (they will never meet even if extended).

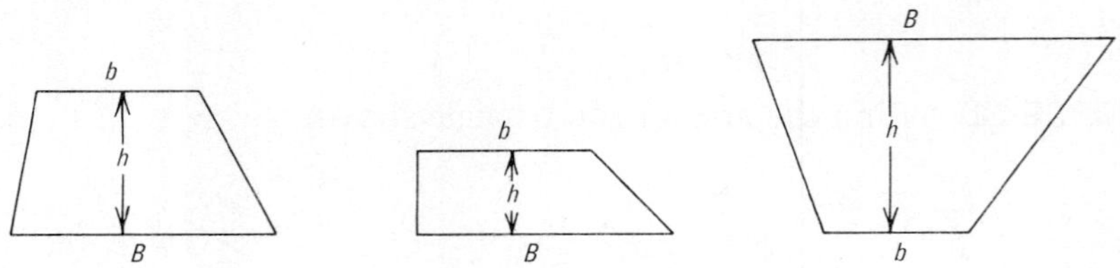

Typical Trapezoids

Notice that each trapezoid has two parallel bases: a **big base** (B) and a **little base** (b). The distance between these bases is called the **height** (h). It is possible to view a trapezoid as one-half of a rectangle with length $(B + b)$ and width h.

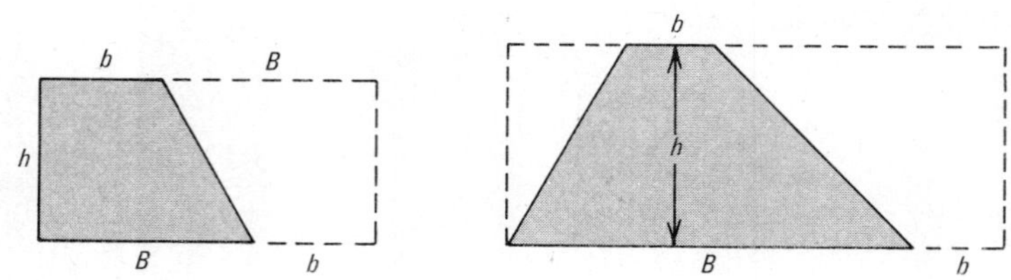

Thus the area of a trapezoid can be computed as *one-half the product of* $(B + b)$ *and h.*

$$A_{\text{trapezoid}} = \frac{1}{2} \cdot (B + b) \cdot h$$

where B is the big base, b is the little base, and h is the height.

EXAMPLE 36 Find the area of the trapezoid shown.

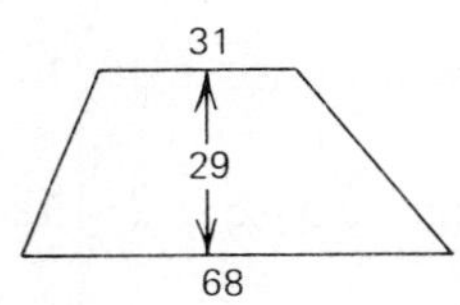

SOLUTION We use the formula $A = \frac{1}{2} \cdot (B + b) \cdot h$.

$$A_{\text{trapezoid}} = \frac{1}{2} \cdot (B + b) \cdot h$$

$$= \frac{(68 + 31) \cdot (29)}{2} \longrightarrow$$

$$= 1435.5 \longleftarrow$$

$68 + 31 = 99$
99×29: 891, 198, 2871
$2871.0 \div 2 = 1435.5$

EXAMPLE 37 Find the area of the trapezoid shown.

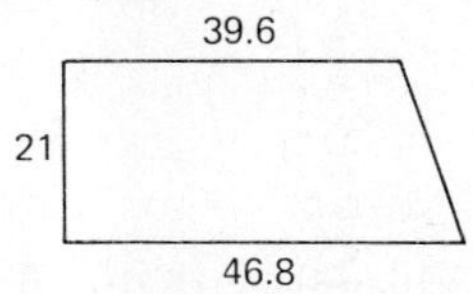

SOLUTION We use the formula $A = \frac{1}{2} \cdot (B + b) \cdot h$.

$$A_{\text{trapezoid}} = \frac{1}{2} \cdot (B + b) \cdot h$$

$$= \frac{(46.8 + 39.6) \cdot (21)}{2} \longrightarrow$$

$$= 907.2 \longleftarrow$$

$46.8 + 39.6 = 86.4$
86.4×21: 86 4, 1728, 1814.4
$1814.4 \div 2 = 907.2$

PROBLEM SET Compute the area of each of the following trapezoids.

See Examples 36 and 37

1.

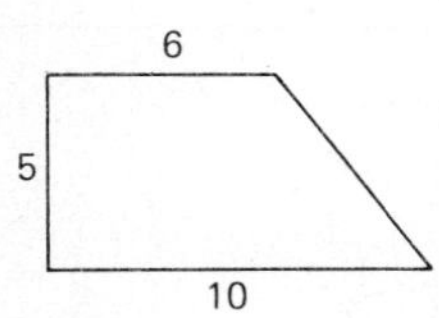

2.

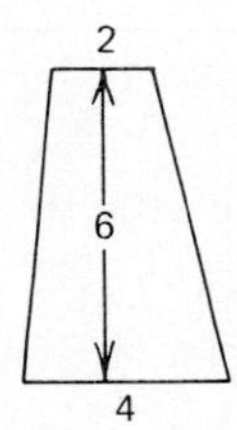

3. A trapezoid: big base = 20, little base = 10, and height = 4.
4. A trapezoid: big base = 65, little base = 42, and height = 37.

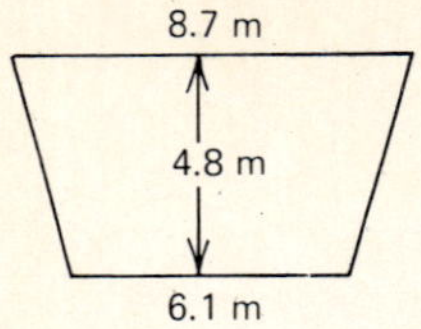

5. A trapezoid: big base = $3\frac{1}{2}$, little base = $1\frac{1}{4}$, and height = $4\frac{1}{2}$.
6. A trapezoid: big base = 18.2, little base = 14, and height = 8.
7. A theater group is painting a stage shaped like the sketch at the left. Find its area.

ANSWERS 1. 40 2. 18 3. 60 4. 1979.5 5. $10\frac{11}{16}$ 6. 128.8 7. 35.52 m^2

In Section 5.2, we discussed the **circle**, with **radius** r and **diameter** D.

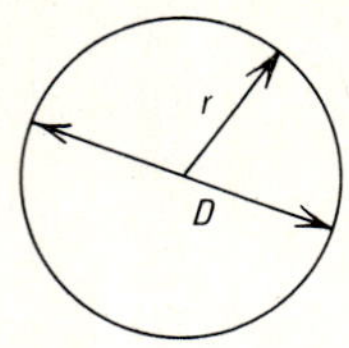

Recall that the ratio of the circumference C to the diameter D is always the same: $\frac{C}{D} = \pi$ (pi) = 3.14 (approximately). The formula for the circumference is

$$C = \pi \cdot D = 2 \cdot \pi \cdot r$$

To study the area of a circle, let us cut it into small slices and rearrange them.

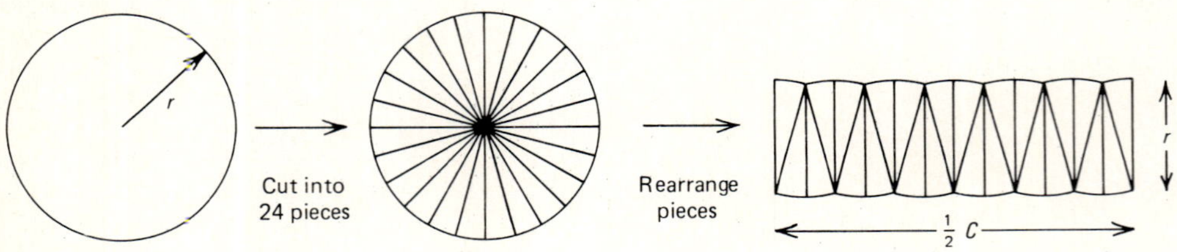

By cutting into small slices and rearranging, the circle can be make to look almost like a rectangle with length $\frac{1}{2}C$ $(= \pi \cdot r)$ and width r. This suggests that the formula for the area of a circle is $\pi \cdot r$ times r, or $\pi \cdot r^2$ (read "pi r squared").

$$A_{\text{circle}} = \pi \cdot r^2 = 3.14 \cdot r \cdot r$$

where r is the radius.

EXAMPLE 38 Find the area of the circle shown.

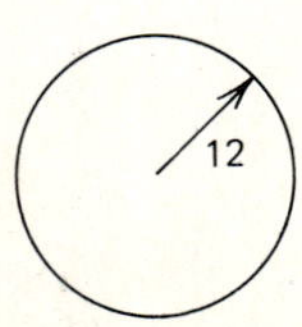

SOLUTION We use the circle formula: We square the radius and multiply by $\pi = 3.14$ (approximately).

$$A_{\text{circle}} = \pi \cdot r^2 = 3.14 \cdot r \cdot r$$
$$= (3.14) \cdot (12) \cdot (12) \longrightarrow$$
$$= 452.16 \longleftarrow$$

```
      12
    × 12
    ----
      24
     12
    ----
     144

     144
  ×  3.14
  -------
    5 76
   14 4
  432
  -------
  452.16
```

EXAMPLE 39 Find the area of the circle shown.

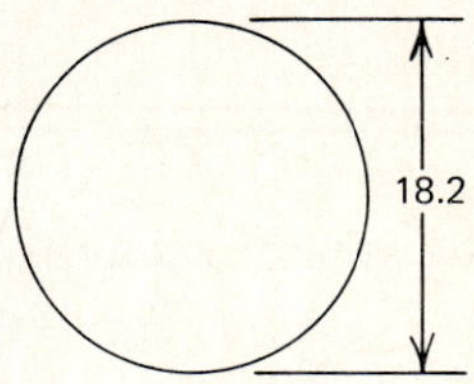

SOLUTION We need the radius in order to use the area formula. So we divide the diameter by 2: $r = \dfrac{18.2}{2} = 9.1$.

$$A_{\text{circle}} = \pi \cdot r^2 = 3.14 \cdot r \cdot r$$
$$= (3.14) \cdot (9.1) \cdot (9.1) \longrightarrow$$
$$= 260.0234 \longleftarrow$$

```
       9.1
     × 9.1
     -----
        91
      81 9
     -----
     82.81

     82.81
  ×   3.14
  --------
    3 3124
    8 281
  248 43
  --------
  260.0234
```

YES	NO
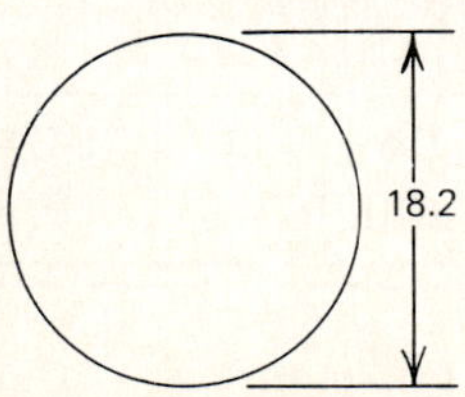 $D = 18.2$ $r = 9.1$ $A = (3.14) \cdot (9.1) \cdot (9.1)$	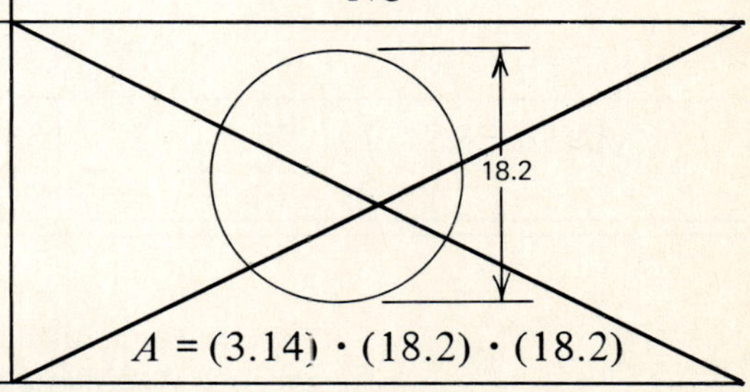 $A = (3.14) \cdot (18.2) \cdot (18.2)$

We must first find the *radius* to use the area formula.

EXAMPLE 40 Find the area of the semicircle shown.

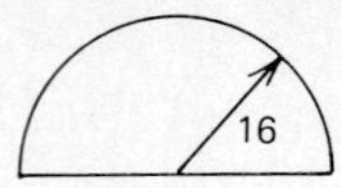

SOLUTION We first find the area of a whole circle and then divide by 2.

$$A_{\text{semicircle}} = \frac{1}{2} \cdot \pi \cdot r^2 = \frac{3.14 \cdot r \cdot r}{2}$$

$$= \frac{(3.14) \cdot (16) \cdot (16)}{2} \longrightarrow$$

$$= 401.92 \longleftarrow$$

```
   16           256
 X 16       X  3.14
   96         10 24
  16          25 6
  256        768
             803.84

             401.92
          2 )803.84
```

PROBLEM SET Find the area of each of the following circles and semicircles.

See Examples 38 to 40

1. 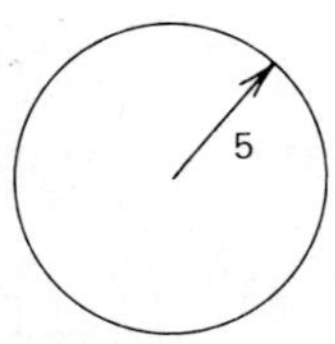

2. 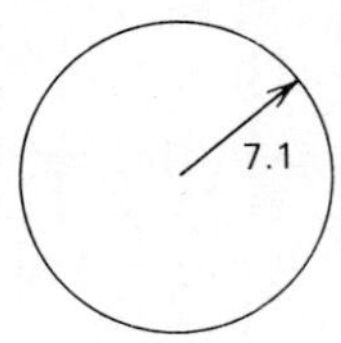

3. A circle: diameter = 12.
4. A circle: diameter = 16.6.
5. A semicircle: radius = 9.
6. A semicircle: diameter = 300.
7. An amusement park is installing a 50-meter-diameter merry-go-round. Find the area that this takes.
8. A local radio station broadcasts in an 8-mile radius. How many square miles does this cover?
9. The Morisons are installing a 20-foot-diameter semicircular patio by their house. Find the area.

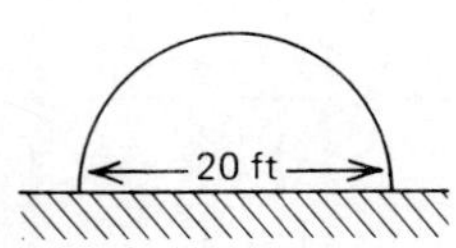

ANSWERS 1. 78.5 2. 158.2874 3. 113.04 4. 216.3146 5. 127.17 6. 35,325 7. 1962.5 m^2 8. 200.96 mi^2 9. 157 ft^2

If we have combinations of the figures described previously, we break the whole figure into easier pieces.

> To find the area of a combination of figures:
>
> 1. Break the figure into easier pieces: triangles, rectangles, circles, semicircles, and so on.
> 2. Find the area of each separate piece.
> 3. Add all the areas together and subtract missing portions.

EXAMPLE 41 Find the area of the figure shown.

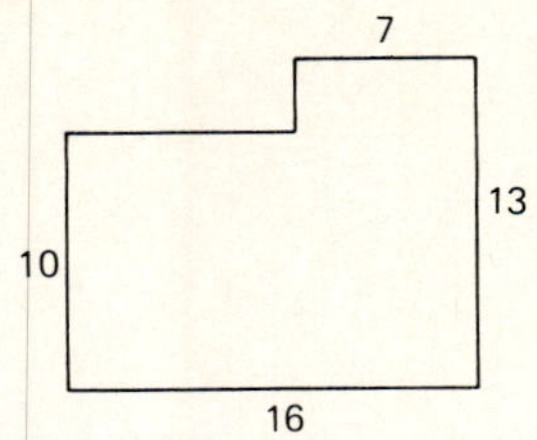

SOLUTION We cut the figure into two rectangles and fill in the missing dimension. As shown, we have cut this into rectangles I and II.

We find the area of each and then add up the areas to get a total.

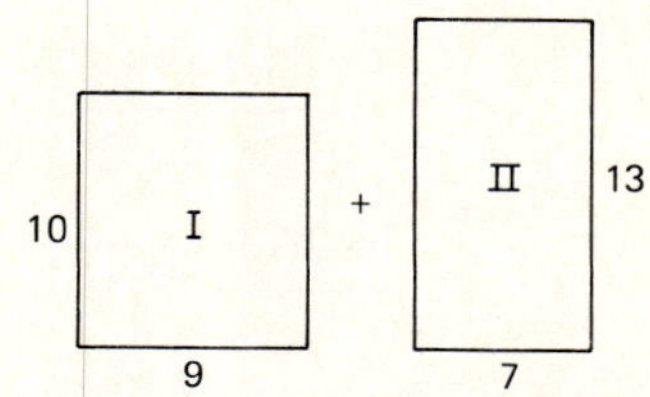

$$A_{\text{I}} = L \cdot W \qquad A_{\text{II}} = L \cdot W$$
$$= (10) \cdot (9) \qquad = (13) \cdot (7)$$
$$= 90 \qquad = 91$$

$$A_{\text{total}} = A_{\text{I}} + A_{\text{II}}$$
$$= 90 + 91$$
$$= 181$$

EXAMPLE 42 Find the area of the figure shown.

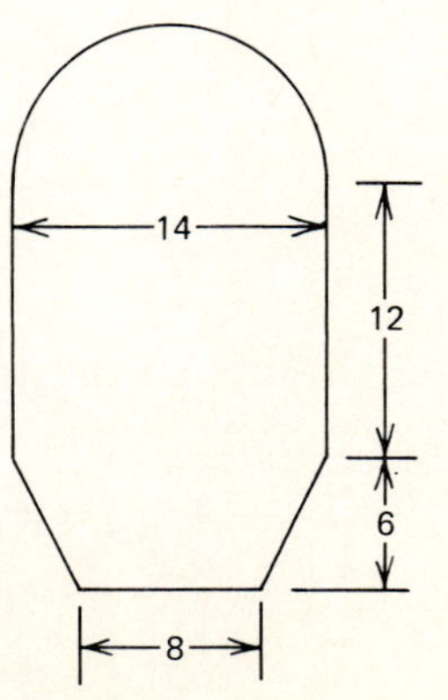

SOLUTION We first cut the figure up into three easier figures: a *semicircle* (I), a *rectangle* (II), and a *trapezoid* (III). We then find the areas of all three of these.

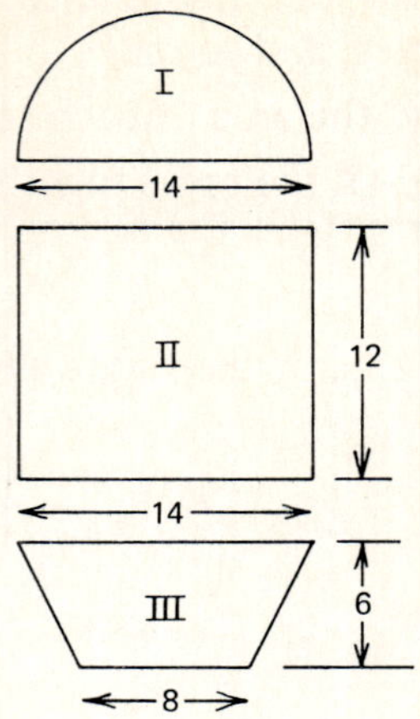

I. *Semicircle:* The diameter is 14, so the radius $r = 7$. We find the area of a whole circle and divide by 2.

$$A_{\text{semicircle}} = \frac{1}{2} \cdot \pi \cdot r^2$$
$$= \frac{(3.14) \cdot (7) \cdot (7)}{2}$$
$$= 76.93$$

7 × 7 = 49

49 × 3.14: 1 96, 4 9, 147 → 153.86

153.86 ÷ 2 = 76.93

II. *Rectangle:* The area of the rectangle is length times width.

$$A_{\text{rectangle}} = L \cdot W$$
$$= (14) \cdot (12)$$
$$= 168$$

14 × 12: 28, 14 → 168

III. *Trapezoid:* We use the trapezoid formula.

$$A_{\text{trapezoid}} = \frac{1}{2} \cdot (B + b) \cdot h$$
$$= \frac{(14 + 8) \cdot (6)}{2}$$
$$= 66$$

14 + 8 = 22

22 × 6 = 132

132 ÷ 2 = 66

Now we add all three of the areas.

$$A_{\text{total}} = A_{\text{I}} + A_{\text{II}} + A_{\text{III}}$$
$$= 76.93 + 168 + 66$$
$$= 310.93$$

76.93 + 168.00 + 66.00 = 310.93

EXAMPLE 43 Find the area of the unshaded region shown.

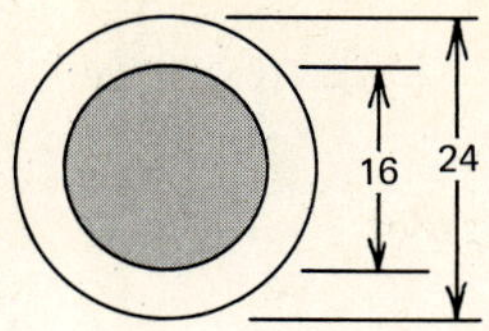

SOLUTION The best way to do this is to think of a big circle (radius = 12) with a small circle (radius = 8) cut out of it. We find both of the areas and then subtract the small area from the big area.

I. *Small circle:* The radius of the small circle is $\frac{16}{2} = 8$.

$$\begin{aligned} A_{\text{I}} &= \pi \cdot r^2 \\ &= (3.14)(8)(8) \longrightarrow \\ &= 200.96 \longleftarrow \end{aligned}$$

$$\begin{array}{r} 8 \\ \times\ 8 \\ \hline 64 \end{array} \qquad \begin{array}{r} 64 \\ \times\ 3.14 \\ \hline 2\,56 \\ 6\,4\ \\ 192\ \ \\ \hline 200.96 \end{array}$$

II. *Big circle:* The radius of the big circle is $\frac{24}{2} = 12$.

$$\begin{aligned} A_{\text{II}} &= \pi \cdot r^2 \\ &= (3.14)(12)(12) \longrightarrow \\ &= 452.16 \longleftarrow \end{aligned}$$

$$\begin{array}{r} 12 \\ \times\ 12 \\ \hline 24 \\ 12\ \\ \hline 144 \end{array} \qquad \begin{array}{r} 144 \\ \times\ 3.14 \\ \hline 5\,76 \\ 14\,4\ \\ 432\ \ \\ \hline 452.16 \end{array}$$

Now we subtract the two areas.

$$\begin{aligned} A_{\text{ring}} &= A_{\text{II}} - A_{\text{I}} \\ &= 452.16 - 200.96 \longrightarrow \\ &= 251.20 \longleftarrow \end{aligned}$$

$$\begin{array}{r} 452.16 \\ -\ 200.96 \\ \hline 251.20 \end{array}$$

PROBLEM SET Find the areas of the following figures. Shaded figures are cut out.

See Examples 41 to 43

1.

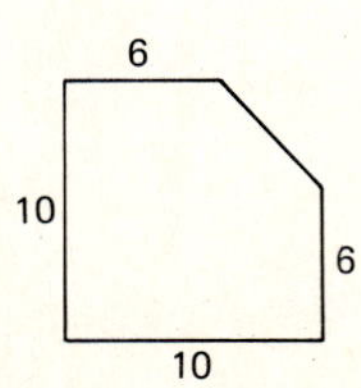

2.

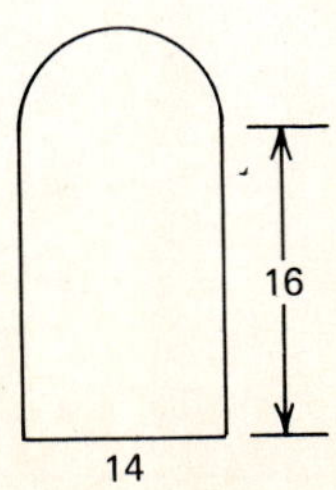

3.

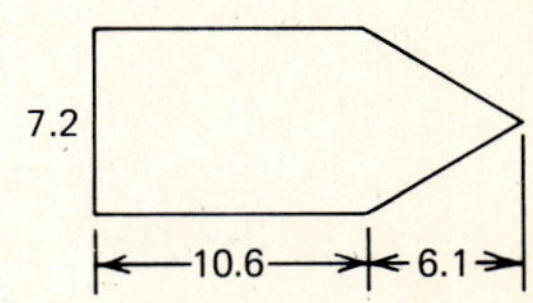

4.

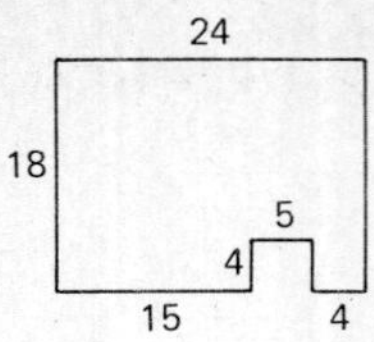

5.

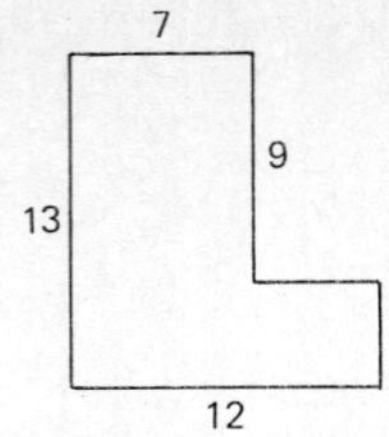

6.

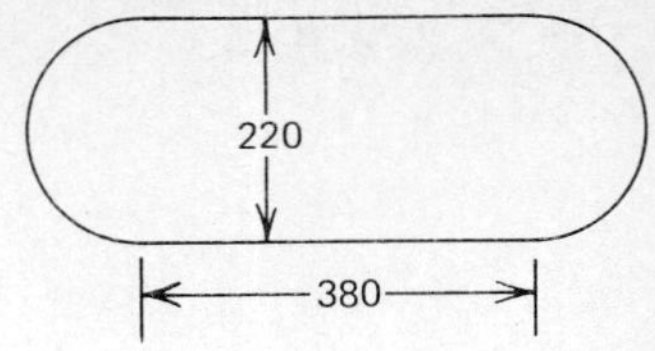

7.

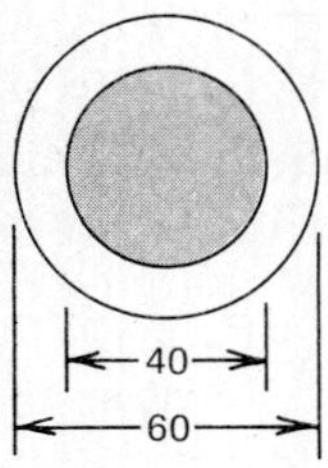

8.

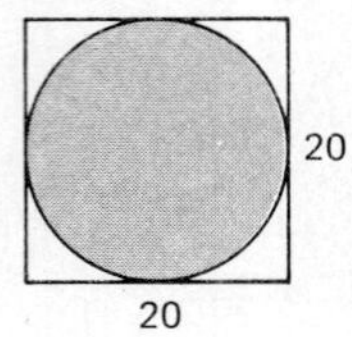

9.

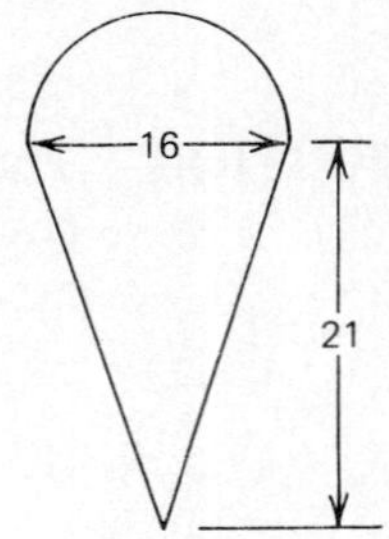

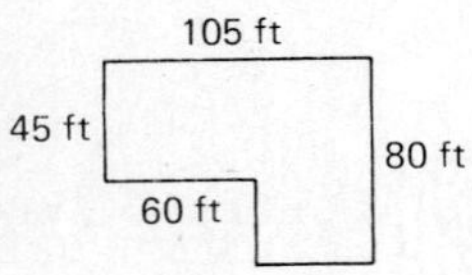

10. Jane and Bob's yard is sketched at the left. In order to buy enough grass seed, they must know the area. Find the area.

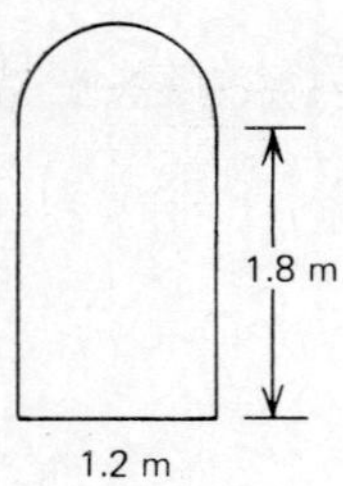

11. Michele has a window in her apartment like the one sketched at the left. In order to estimate her air-conditioning needs, she must know the area. (The more area, the more heat from sunlight.) What is the area?

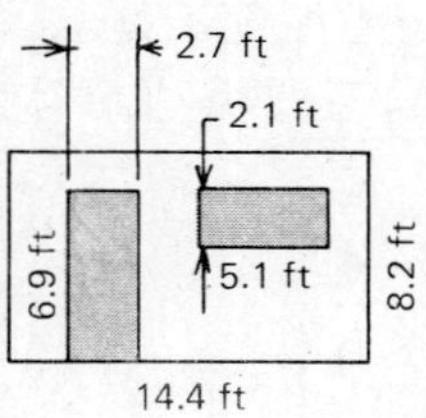

12. Nancy is going to paint a wall like the one drawn at the left. She does *not* have to paint the door or window. Find the area to be painted.

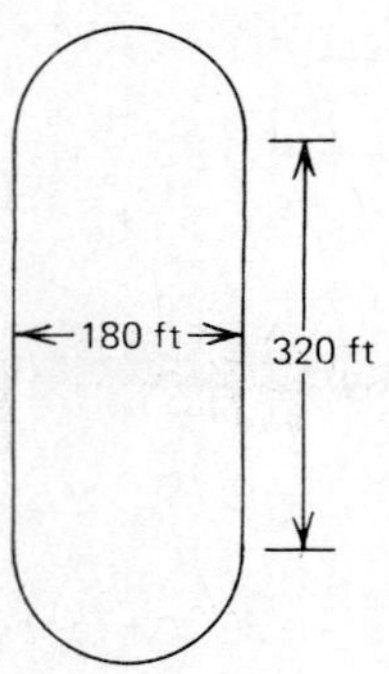

13. A sketch of a football field is shown at the left. Find the area.

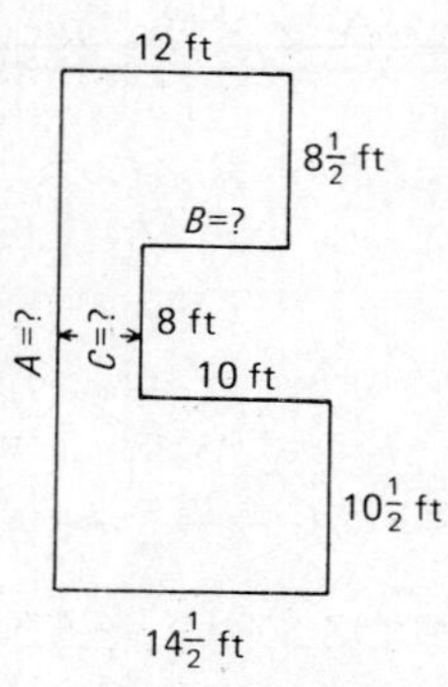

14. Pam and Rudy are going to carpet the two rooms and a hall as sketched at the left. Find the missing dimensions; then find the area.

ANSWERS 1. 92 2. 300.93 3. 98.28 4. 412 5. 111 6. 121,594 7. 1570 8. 86 9. 268.48 10. 6300 11. 2.7252 12. 88.74 13. 83,034 14. $A = 27$; $B = 7\frac{1}{2}$; $C = 4\frac{1}{2}$; area $= 290\frac{1}{4}$

Miscellaneous Exercises 5.3

Compute the following areas.

See Examples 30 and 31

1. A rectangle: 7 by 3.
2. A rectangle: 4.5 by 0.8.

See Example 32

3. A square: side = 5.
4. A square: side = 1.9.

See Examples 33 to 35

5. A triangle: base = 7, height = 4.
6. A triangle: base = $3\frac{1}{2}$, height = 2.
7. A triangle: base = 4.8, height = 0.8.
8. A triangle: base = $2\frac{3}{4}$, height = $2\frac{3}{4}$.

See Examples 36 and 37

9. A trapezoid: big base = 5, little base = 3, and height = 2.
10. A trapezoid: big base = 42, little base = 16, and height = 24.

See Examples 38 and 39

11. A circle: radius = 8.
12. A circle: diameter = 24.

See Example 40

13. A semicircle: diameter = 1.4.
14. A semicircle: radius = 200.

See Examples 41 and 42

15. 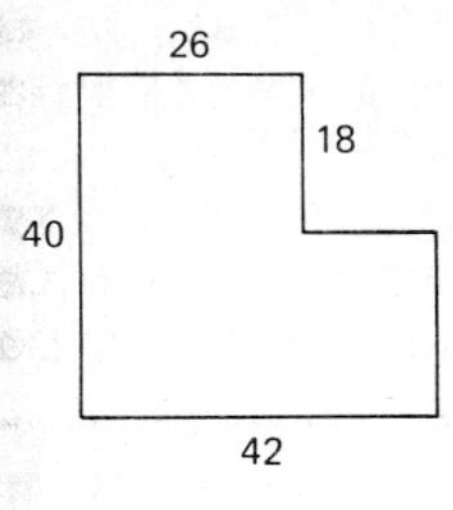

16. 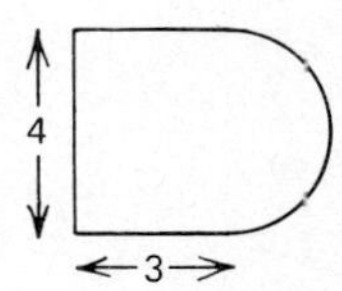

See Example 43

17. 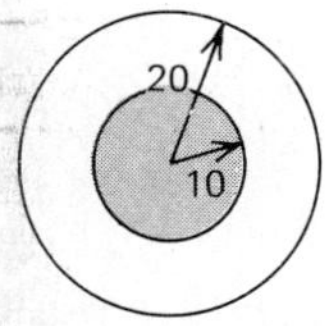

18. 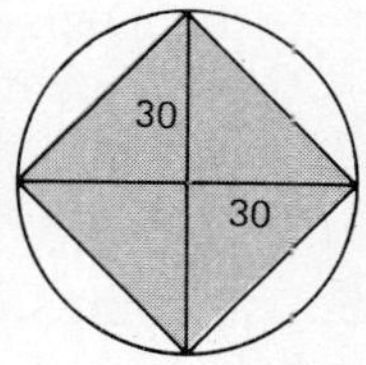

Health applications

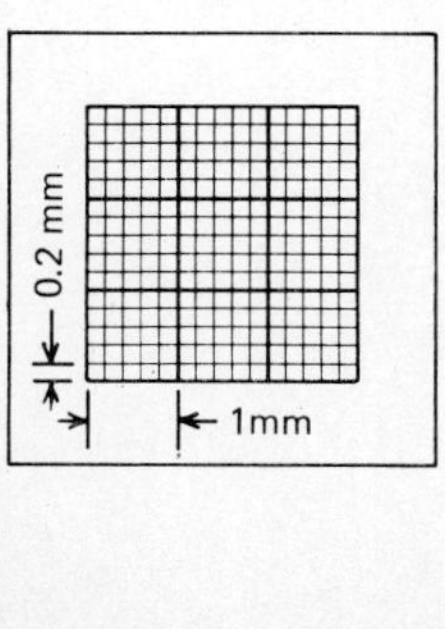

19. At the left is a sketch of a microscopic grid that might be used for counting blood cells.
    (a) What is the area of the whole grid?
    (b) What is the area of the smallest square?
20. A young boy has a rash that is 10 centimeters in diameter. What is the area of this rash?

Technical applications

21. A metal can (as sketched) is composed of two circles (the top and bottom) and a rectangle (bent into the side of the can). Notice that the length of the rectangle is the circumference of the circle.
(a) Find the area of the top and the bottom.
(b) Find the area of the side.
(c) Find the total area.

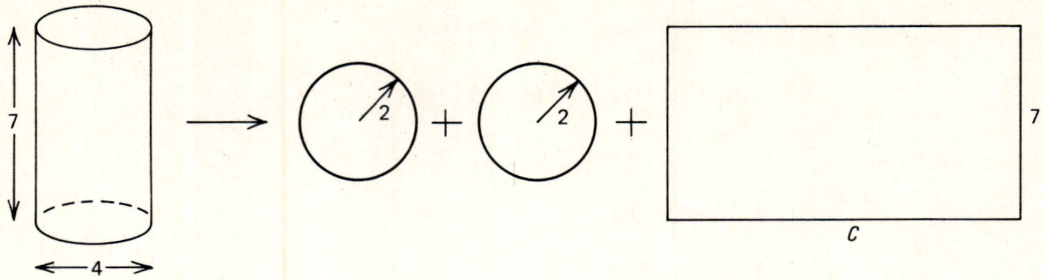

22. The wastebasket sketched can be formed from the metal pattern as shown. Find the area of the metal used.

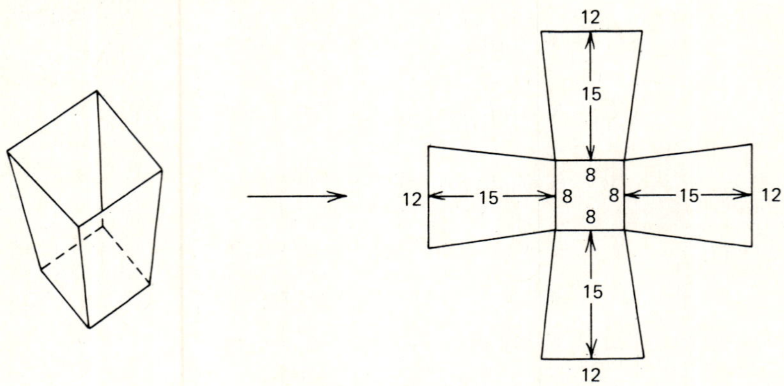

Business applications

23. A firm is considering advertising on a radio station that broadcasts with a 30-mile radius. What is the area of the broadcast?

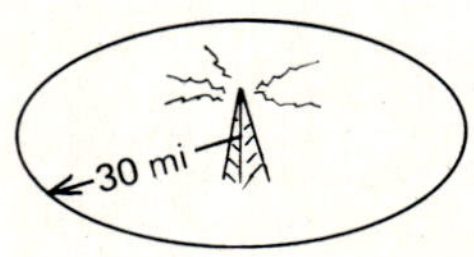

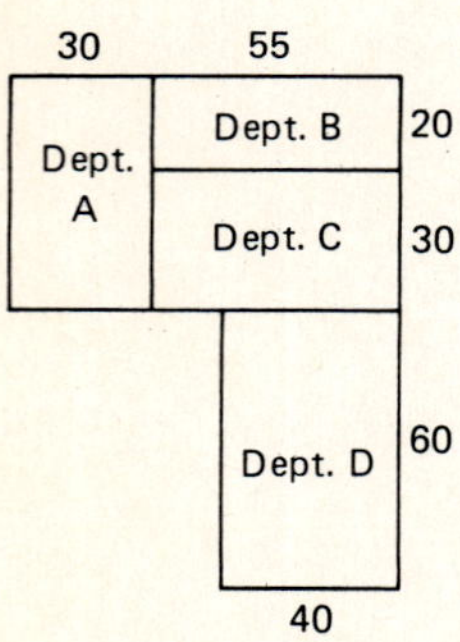

24. The floor space sketched at the left is divided among four departments as shown. Find the area for:
(a) Department *A*
(b) Department *B*
(c) Department *C*
(d) Department *D*
(e) All four departments

25. A company is planning a roller rink like the one sketched below. Find the area.

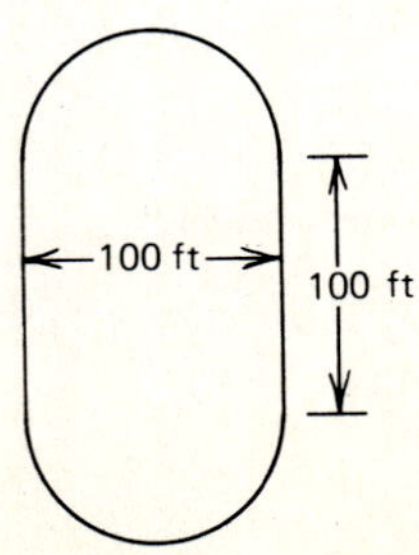

Environmental application

26. Oil spilling from a tanker forms a huge circle on the ocean with radius $1\frac{1}{2}$ miles. How many square miles does this spill cover? $\left(\text{Use } \pi = \frac{22}{7}.\right)$

Photographic applications

27. The size of the opening in a camera lens is measured by the *f/stop*, as shown below (*f*/2, *f*/2.8, *f*/4, and so on). Notice that the *f*/stops go up in number as the radii of the openings go down. Compute the areas of the openings.

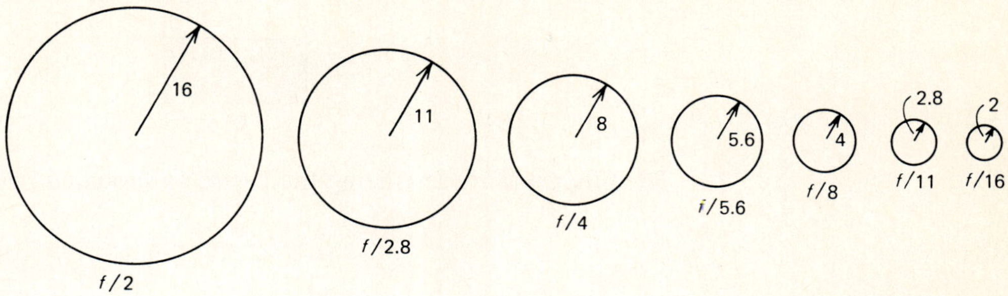

28. Judy buys developing paper, which comes in sheets 14 inches by 20 inches. She wants to cut each sheet into $3\frac{1}{2}$- by 5-inch pieces. How many can she get from each sheet?

Consumer applications

29. The Torres are carpeting a room like the one sketched below.
    (a) Find the area.
    (b) If 1 square yard is 9 square feet, how many square yards is this?
    (c) If the carpeting costs $12 per square yard, how much will this cost?

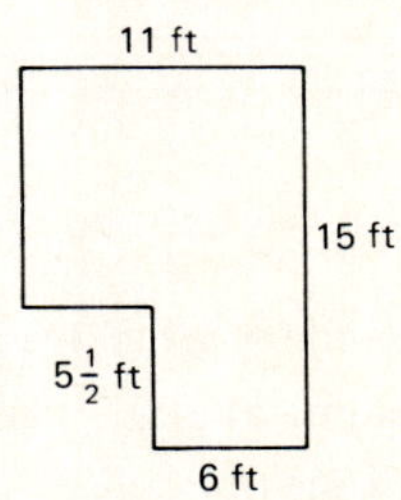

30. The Santos are painting a wall in their house. (They are not painting the door.) Find the area.

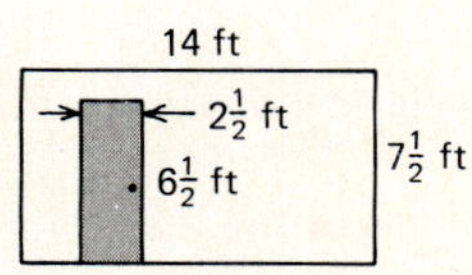

31. Jane and Bill are refinishing an old tabletop like the one shown below. (The units are in feet.) In order to find how much varnish to buy, they must find the area. Compute the area.

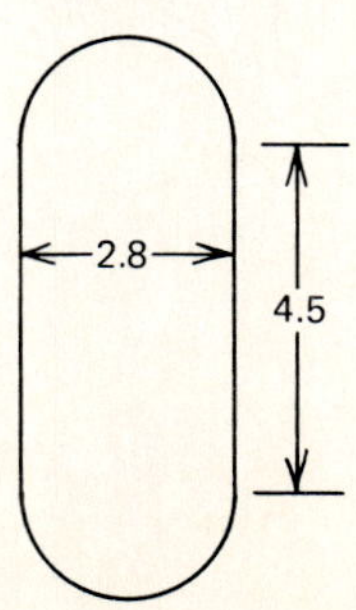

32. The Petersons are tiling their bathroom (sketched below with the units in feet). Find the area (exclude the toilet, sink, and tub).

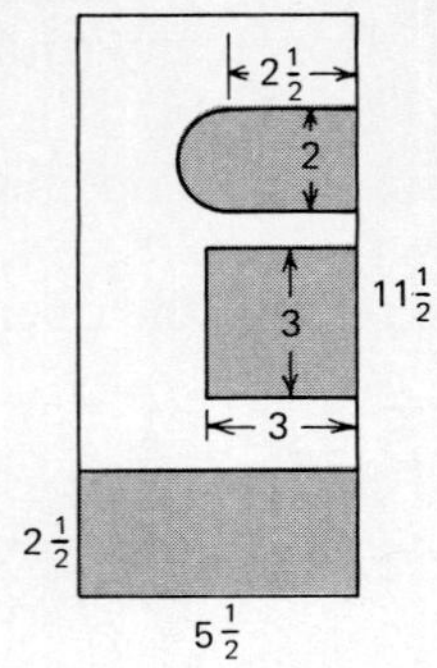

33. The Slivinskis are fertilizing their yard as sketched below. Find the area.

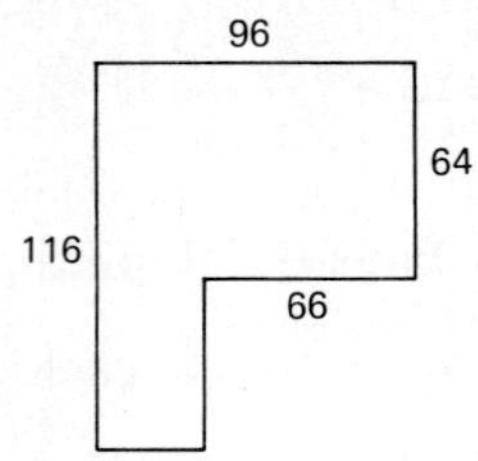

Now complete Post-Test 1 for Section 5.3 on page T75.

5.4 VOLUME

SECTION OBJECTIVES After completing this unit, you will be able to:

1. Find the volume of a rectangular box.
2. Find the volume of a cylinder.
3. Find the volume of a sphere.
4. Find the volume of a combination of the figures above.

TYPICAL PROBLEMS You will be able to work problems like the ones below.

Find the volumes of the following figures.

(a)

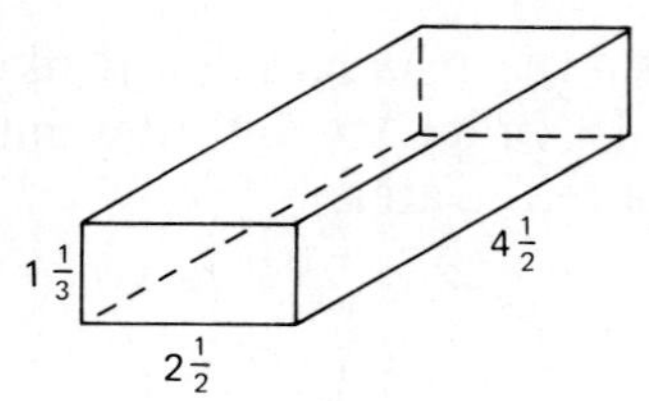

(b)

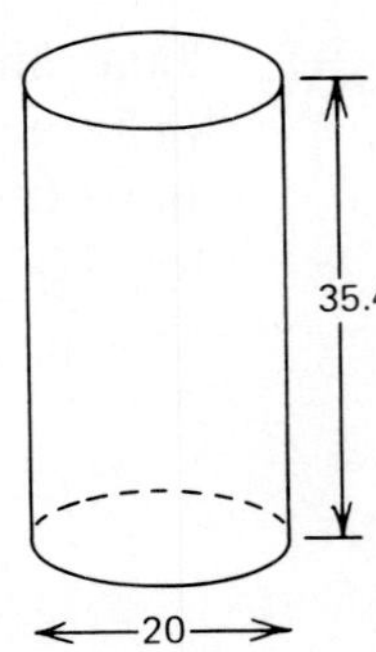

(c)

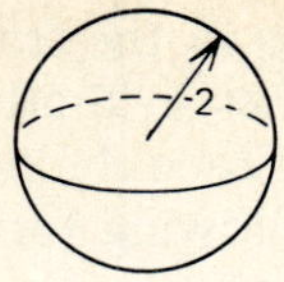

(d)

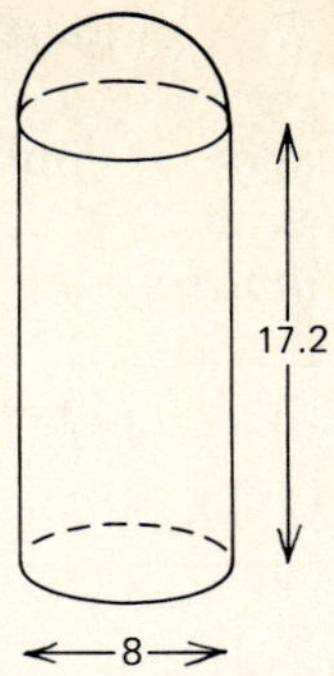

ANSWERS **(a)** 15 **(b)** 11,115.6 **(c)** 33.493 **(d)** 998.1013

Volume is the measure of capacity within a solid. It is measured in **cubic units**: cubic centimeters, cubic inches, cubic yards, and so on. Each of these is a cube with a side of one centimeter, one inch, one yard, and so on. Below we see sketched two such cubic units: one cubic centimeter (abbreviated 1 cc, 1 cu cm, or 1 cm^3) and one cubic inch (abbreviated 1 cu in. or 1 $in.^3$).

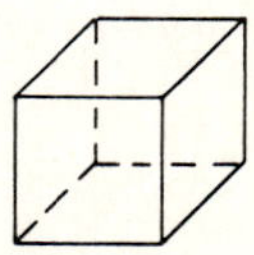

1 cubic centimeter
(1cc, 1 cu cm, or 1 cm^3)

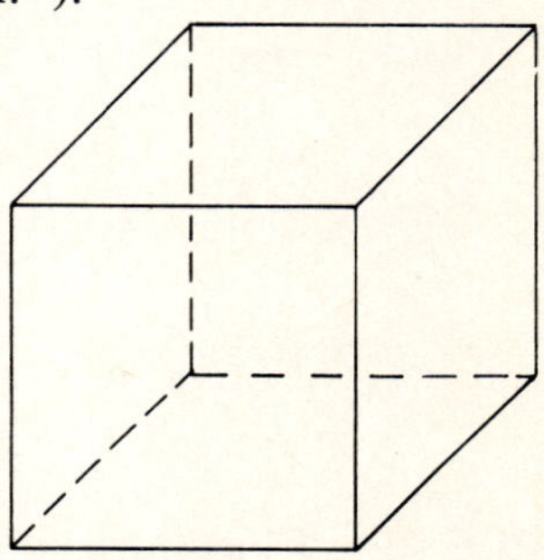

1 cubic inch
(1 cu in. or 1 $in.^3$)

Two Cubic Units

The first solid that we study is the **rectangular box**. This solid has rectangles for its faces.

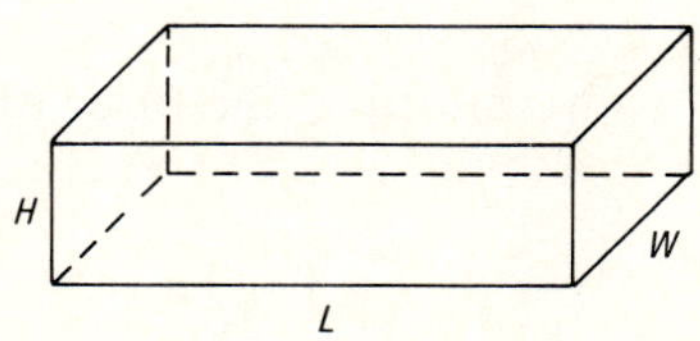

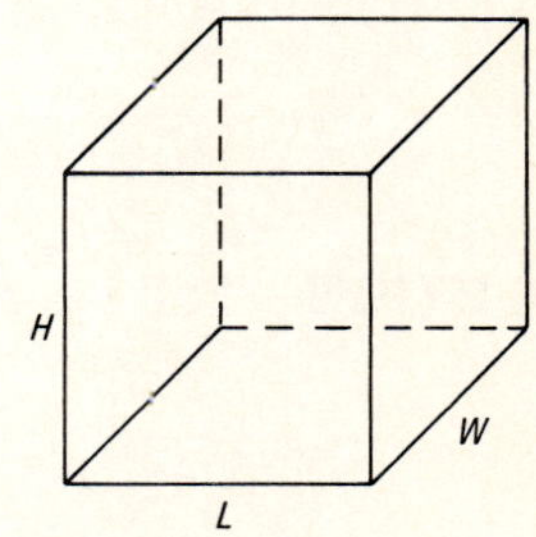

Typical Rectangular Boxes

The three dimensions of a rectangular box are **length (*L*)**, **width (*W*)**, and **height (*H*)**. Consider the 5-by-3-by-2 box shown.

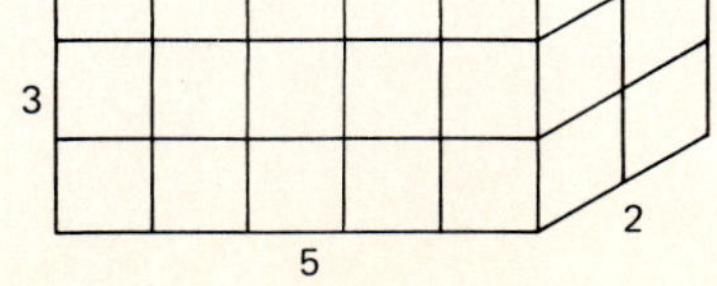

We can see that there are 15 (= 5 X 3) cubes in the front. Since the back row is the same as the front, there are 15 more cubes in the back. This is a total of 30 cubes. Thus we say that the volume is 30. Since the product of all three dimensions 5 X 3 X 2 = 30, this suggests the following formula for the volume: We *multiply the length by the width by the height.*

$$V_{\text{box}} = L \cdot W \cdot H$$

where L is the length, W is the width, and H is the height.

EXAMPLE 44 The following volumes of rectangular boxes are computed by multiplying the length by the width by the height.

(a) $V = 2 \times 5 \times 4 = 40$

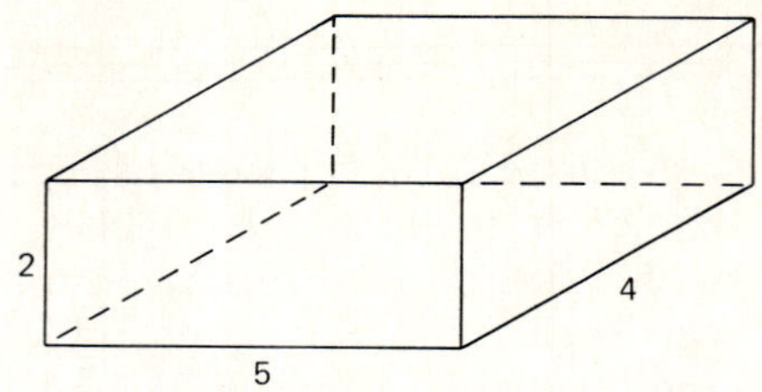

(b) $V = 2 \times 4 \times 3 = 24$

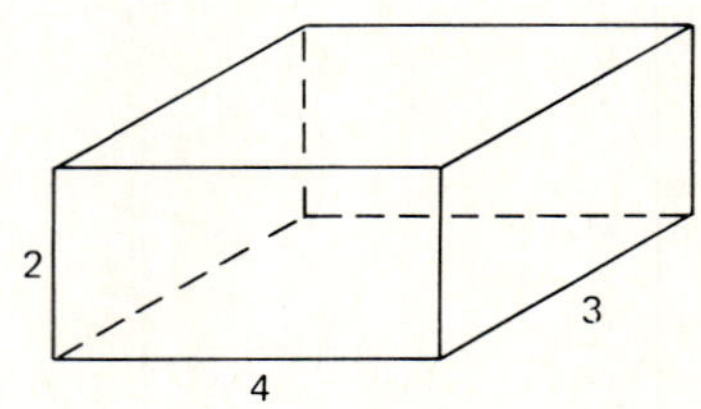

EXAMPLE 45 Find the volume of the box shown.

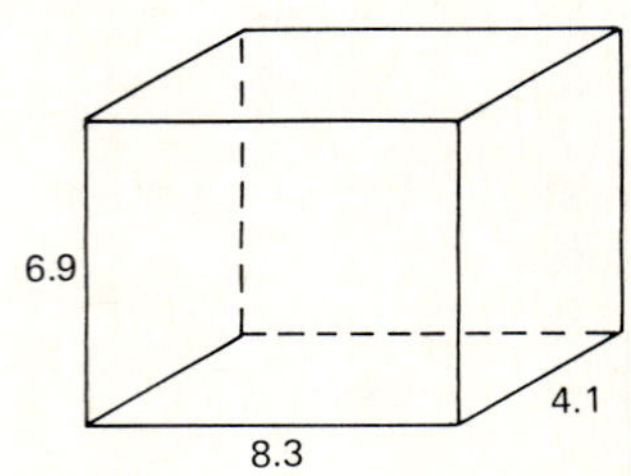

SOLUTION This is a rectangular box, so we multiply the three dimensions together.

$$A_{\text{box}} = L \cdot W \cdot H$$
$$= (8.3) \cdot (4.1) \cdot (6.9) \longrightarrow$$
$$= 234.807 \longleftarrow$$

```
     8.3
  X  4.1
  ------
      83
    33 2
  ------
   34.03

   34.03
 X   6.9
 -------
  30 627
  204 18
 -------
 234.807
```

Just as the square is a special rectangle, the **cube** is a special rectangular box with all the sides equal (length = width = height).

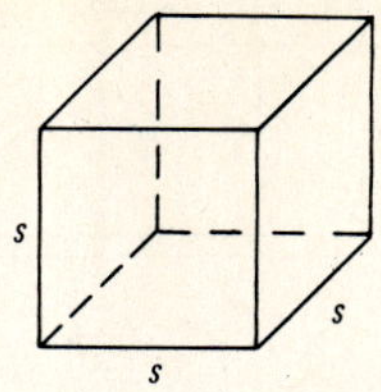

Typical Cube

If the sides of a cube are all s, then the volume is the product $s \cdot s \cdot s$, which is also written s^3 (read "s cubed").

$$V_{\text{cube}} = s \cdot s \cdot s = s^3$$

where s is the side of a cube.

EXAMPLE 46 We compute the volume of each cube below as *the length of a side times itself times itself again*.

(a) $V = 2^3 = 2 \cdot 2 \cdot 2 = 8$

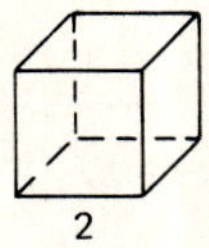

(b) $V = 10^3 = 10 \cdot 10 \cdot 10 = 1000$

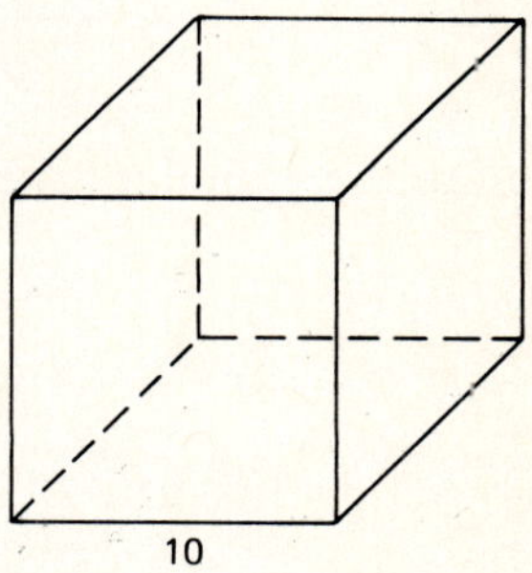

PROBLEM SET Compute the following volumes.

See Examples 44 to 46

1.

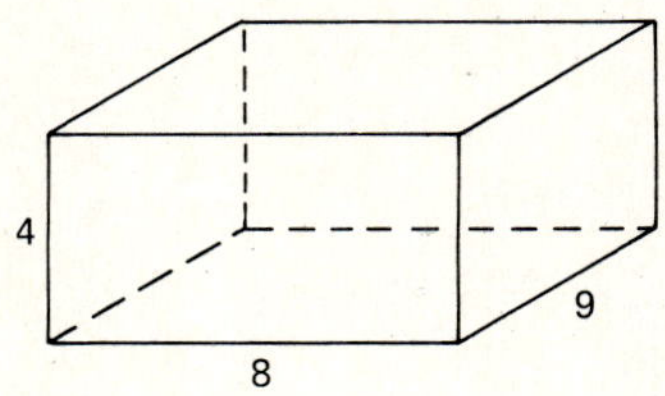

2.

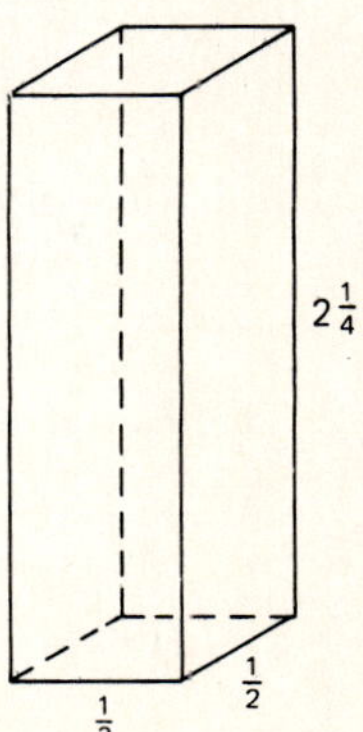

3. A rectangular box: 2.3 by 2 by 1.4.

4. A cube: side = 6.

5. A cube: side = 2.1.

6. A cube: side = $\frac{1}{2}$.

7. Tom is laying a rectangular sidewalk $\frac{1}{2}$ foot deep, 5 feet wide, and 12 feet long. Find the volume.

ANSWERS 1. 288 2. $\frac{3}{8}$ 3. 6.44 4. 216 5. 9.261 6. $\frac{1}{8}$ 7. 30

The **cylinder** is a solid with a circle for a base, as shown.

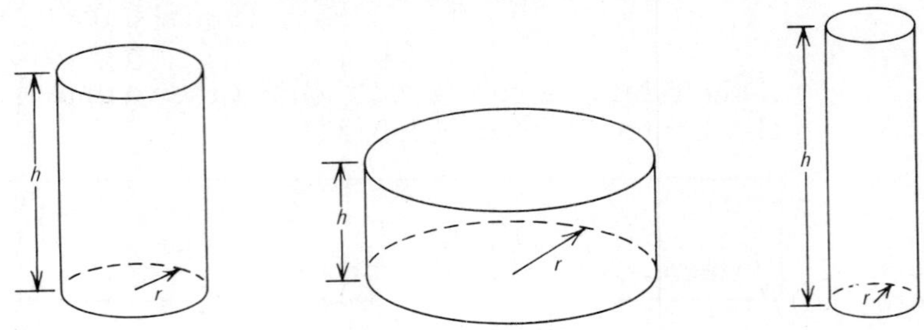

Typical Cylinders

The volume of a cylinder is the *area of the base times the height*. Recall from Section 5.3 that the area of a circle is 3.14 · r · r. Thus we get the volume formula for a cylinder.

$$V_{\text{cylinder}} = \pi \cdot r^2 \cdot h = 3.14 \cdot r \cdot r \cdot h$$

where r is the radius of the base, h is the height, and π is approximately 3.14.

EXAMPLE 47 We use the cylinder formula on the figure shown.

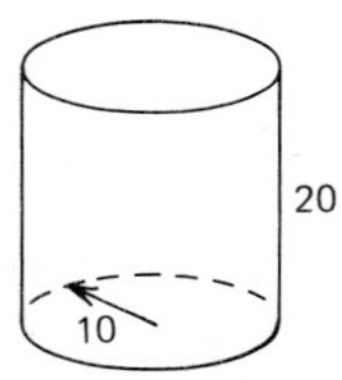

$$\begin{aligned} V_{\text{cylinder}} &= \pi \cdot r^2 \cdot h \\ &= (3.14) \cdot (10) \cdot (10) \cdot (20) \\ &= 6280 \end{aligned}$$

EXAMPLE 48 Find the volume of the cylinder shown.

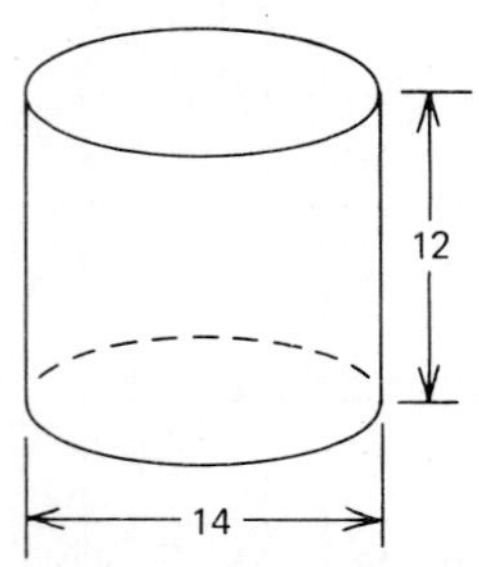

SOLUTION We are given the diameter 14, so the radius is $r = \frac{14}{2} = 7$.

$$V_{\text{cylinder}} = \pi \cdot r^2 \cdot h$$

$$= (3.14) \cdot (7) \cdot (7) \cdot (12) \longrightarrow$$

$$= 1846.32 \longleftarrow$$

$$\begin{array}{r} 7 \\ \times\ 7 \\ \hline 49 \end{array} \qquad \begin{array}{r} 49 \\ \times\ 3.14 \\ \hline 1\ 96 \\ 4\ 9\ \ \\ 147\ \ \ \ \\ \hline 153.86 \end{array}$$

$$\begin{array}{r} 153.86 \\ \times\ \ \ \ 12 \\ \hline 307\ 72 \\ 1538\ 6\ \ \\ \hline 1846.32 \end{array}$$

In working problems of this sort, be sure to find the radius before computing the area.

PROBLEM SET Find the volume of each of the following cylinders.

See Examples 47 and 48

1. 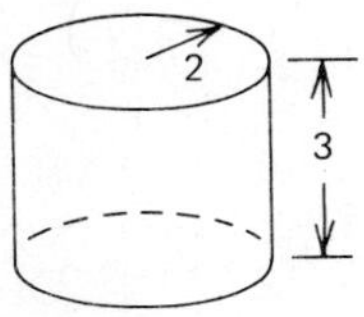

2. 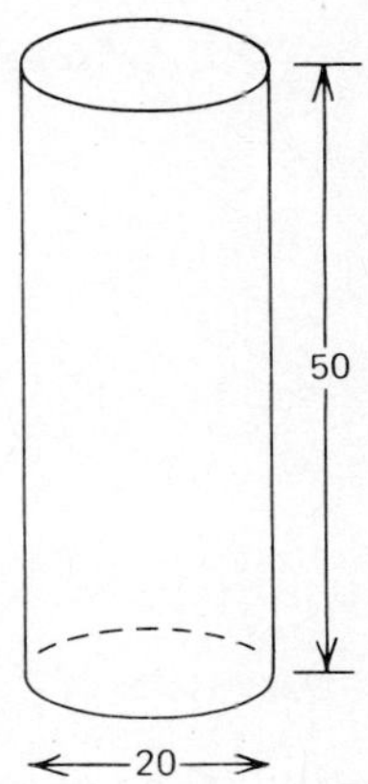

3. A cylinder: height = 7, radius = 7.
4. A cylinder: height = 11, radius = 6.
5. A cylinder: height = 14, diameter = 18.
6. 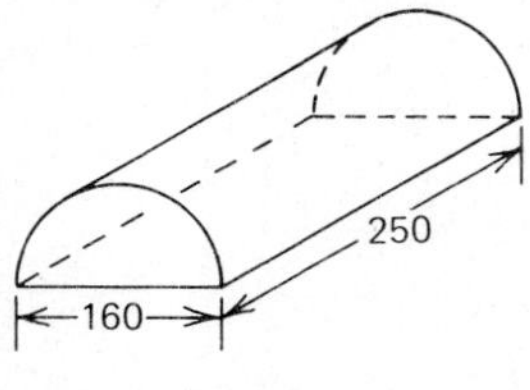

ANSWERS 1. 37.68 2. 15,700 3. 1077.02 4. 1243.44 5. 3560.76 6. 2,512,000

The last geometric figure that we discuss is the **sphere**, which is the shape of a ball.

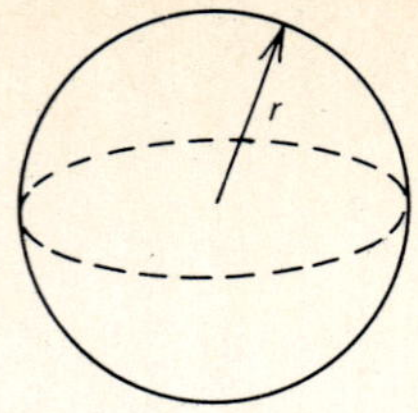

Typical Sphere

The formula for the volume of a sphere is as follows.

$$V_{\text{sphere}} = \frac{4 \cdot \pi \cdot r^3}{3} = \frac{4 \cdot 3.14 \cdot r \cdot r \cdot r}{3}$$

where r is the radius of the sphere.

EXAMPLE 49 Find the volume of the sphere shown.

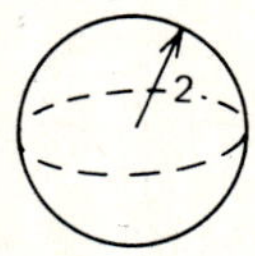

SOLUTION We use the sphere formula.

$$V_{\text{sphere}} = \frac{4 \cdot \pi \cdot r^3}{3}$$

$$= \frac{4 \cdot (3.14) \cdot (2) \cdot (2) \cdot (2)}{3}$$

$$= \frac{4 \cdot 3.14 \cdot 8}{3} \longrightarrow$$

$$= 33.49 \longleftarrow$$

$3.14 \times 8 = 25.12$ $25.12 \times 4 = 100.48$

$100.48 \div 3 = 33.49$

EXAMPLE 50 Find the volume of the sphere shown.

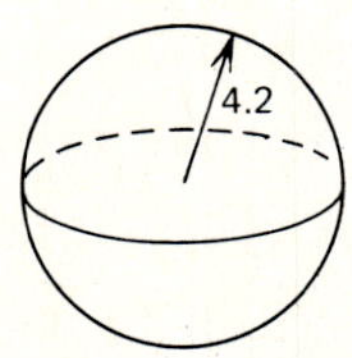

SOLUTION We use the sphere formula with $r = 4.2$.

$$V_{\text{sphere}} = \frac{4 \cdot \pi \cdot r^3}{3}$$

$$= \frac{4 \cdot (3.14) \cdot (4.2) \cdot (4.2) \cdot (4.2)}{3} \longrightarrow$$

$$= 310.18176 \longleftarrow$$

4.2×4.2: 84, 16 8, = 17.64

17.64×4.2: 3 528, 70 56, = 74.088

74.088×3.14: 2 96352, 7 4088, 222 264, = 232.63632

$232.63632 \times 4 = 930.54528$

$930.54528 \div 3 = 310.18176$

EXAMPLE 51 Find the volume of the hemisphere shown.

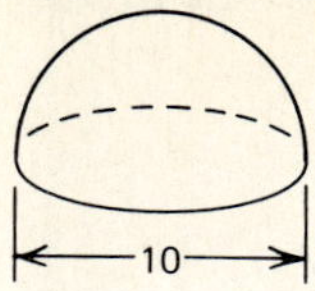

SOLUTION We first compute the volume of an entire sphere with $r = \dfrac{10}{2} = 5$ and divide by 2.

$$V_{\text{hemisphere}} = \frac{1}{2} V_{\text{sphere}}$$

$$= \frac{1}{2} \cdot \frac{4 \cdot \pi \cdot r^3}{3}$$

$$= \frac{1}{2} \cdot \frac{4 \cdot (3.14) \cdot (5) \cdot (5) \cdot (5)}{3} \longrightarrow$$

$$= 261.66 \longleftarrow$$

$5 \times 5 = 25$	$25 \times 5 = 125$
125 × 3.14 5 00 12 5 375 392.50	392.50 × 4 1570.00
$1570.00 \div 3 = 523.33$	$523.33 \div 2 = 261.66$

PROBLEM SET Find the volume of each of the following figures.

See Examples 49 to 51

1. 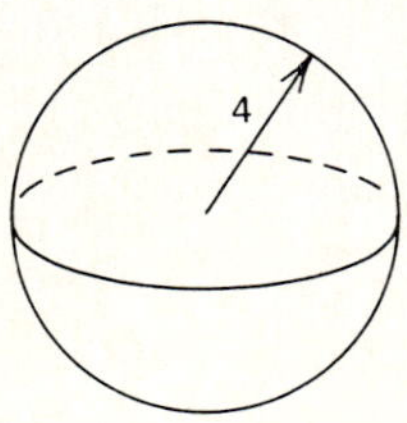

2. 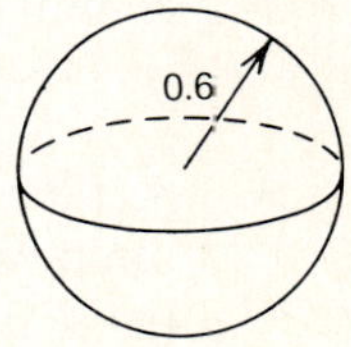

3. A sphere: radius = 10.
4. A sphere: diameter = 16.
5. A hemisphere: radius = 30.
6. A hemisphere: diameter = 100.
7. Ed has a load of black top soil delivered to his yard. The pile forms a hemisphere with a diameter of 2 meters. What is the volume of the dirt pile?
8. The earth has a radius of about 4000 miles. Assuming that the earth is a sphere, what is its volume?
9. Find the volume of the figure shown at the left. (Compute the volumes of the hemisphere and the cylinder separately, and add them.)

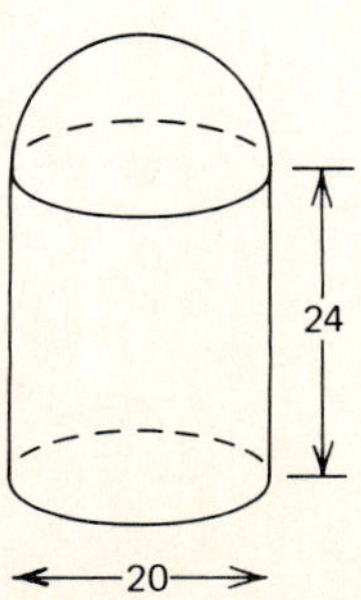

10. Find the volume of the figure shown below. (Subtract the volume of the scooped-out hemisphere from the volume of the cylinder.)

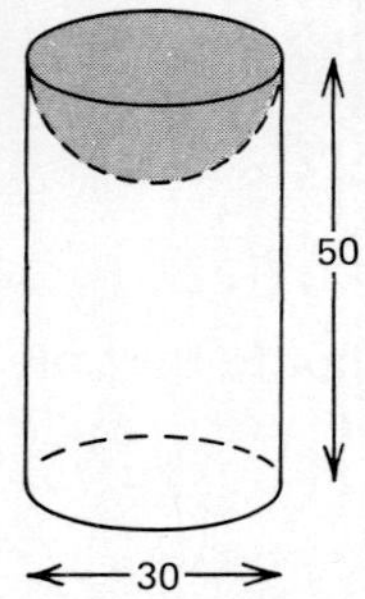

ANSWERS 1. 267.9467 2. 0.90432 3. 4186.667 4. 2143.5733
5. 56,520 6. 261,666.7 7. 2.09 8. 267,946,666,667
9. 9629.33 10. 28,260

Miscellaneous Exercises 5.4

Compute the following volumes.

See Examples 44 and 45

1. A rectangular box: 5 by 10 by 7.
2. A rectangular box: 1.1 by 1.5 by 1.6.

See Example 46

3. A cube: side = 5.
4. A cube: side = $1\frac{1}{3}$.

See Examples 47 and 48

5. A cylinder: height = 20, diameter = 8.
6. A cylinder: height = 6, radius = 6.

See Examples 49 to 51

7. A sphere: radius = 10.
8. A hemisphere: radius = 2.1.

Technical applications

9. A piece of sheet metal is cut as shown (left). It is folded up to form the open box on the right.
   (a) Find the area of the sheet.
   (b) Find the volume of the box.

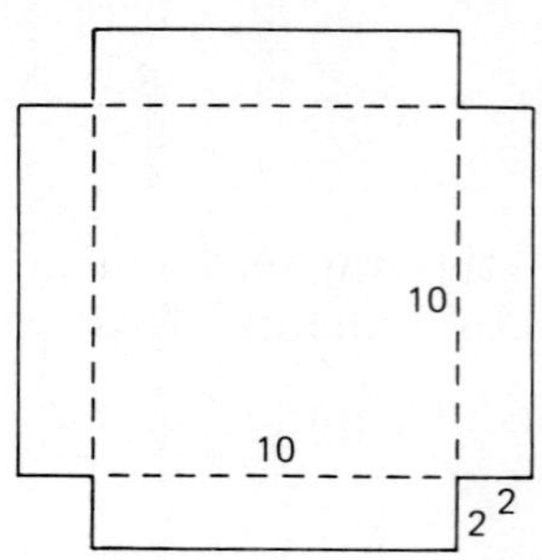

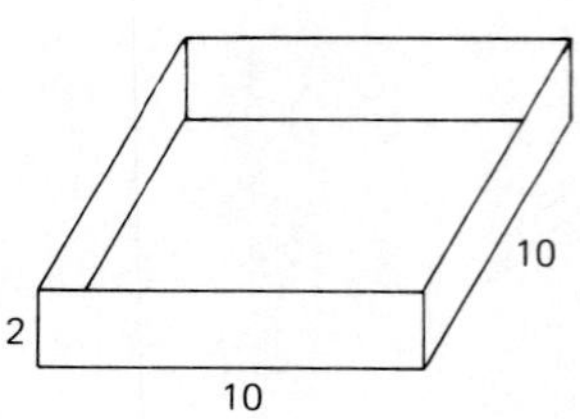

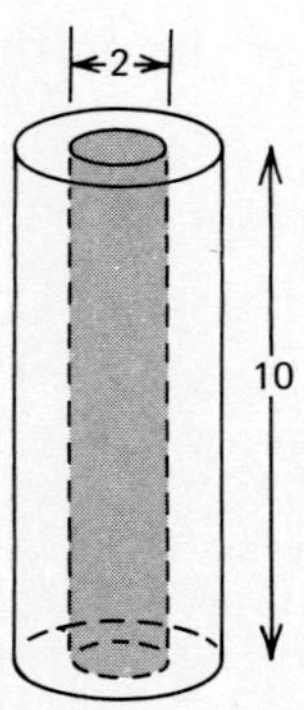

10. A metal bushing is sketched at the left. Compute the volume.

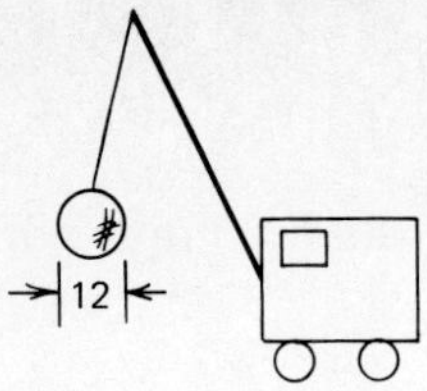

11. The metal wrecking ball shown at the left has a diameter of 12 inches.
   (a) Find the volume of the ball.
   (b) If cast iron weighs 0.26 pound per cubic inch, find the weight of the ball.

Health application

12. A spherical balloon is used to measure a patient's air capacity of the lungs. What is the volume of the air if the balloon is inflated to 10 centimeters in diameter?

Engineering application

13. A semicylindrical tunnel (the opening is a semicircle) is to be drilled through a mountain. The tunnel has a 30-foot diameter and is 1200 feet long. How many cubic feet of dirt and rock will be drilled from the tunnel?

Business applications

14. The Yum Yum Food Company is redesigning its corn-flakes box as shown. Find the volume of the new box.

15. The YumYum Food Company is also redesigning its soup cans as shown at the left. Find the volume of the new can.

Consumer applications

16. Karen receives a load of topsoil for her lawn. It takes the shape of a hemisphere of diameter 10 feet. How many cubic feet is this?

17. The Mendez family is moving. The moving company provides them with cardboard boxes that are 30-inch cubes. Find the volume.

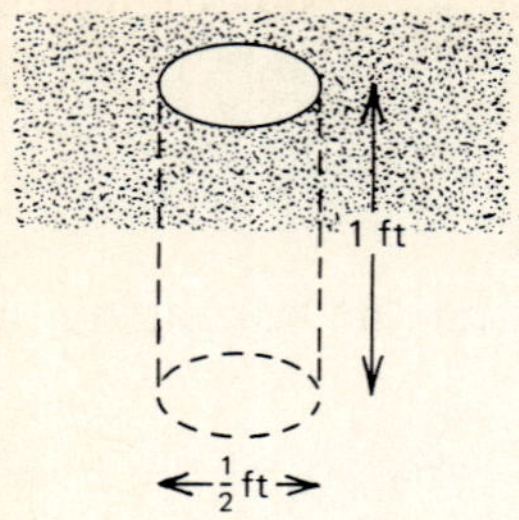

18. Dick is digging holes (cylindrically shaped, as shown at the left) into which he will pour concrete to anchor fence posts.
(a) Find the volume of one hole.
(b) Find the volume of 10 holes.
(c) If a bag of concrete makes 2 cubic feet, how many bags will he need?

Sports application

19. Below are the dimensions of the balls (puck) used in the major sports. Find their volumes.
(a) Baseball: diameter = 2.9 inches
(b) Basketball: diameter = 9.5 inches
(c) Hockey puck: diameter = 3 inches; height = 1 inch
(d) Tennis ball: diameter = 2.5 inches

Now complete Post-Test 1 for Section 5.4 on page T77.

CHAPTER 5 SUMMARY

Important Words and Phrases

angle (5.2)
area (5.3)
base (5.3)
big base (5.3)
center of a circle (5.2)
centi- (5.1)
circle (5.2, 5.3)
circumference (5.2)
cube (5.4)
cubic units (5.4)
cylinder (5.4)
degrees (5.2)
diameter (5.2)
English system (5.1)
height (5.3)
kilo- (5.1)
legs (5.2)
length (5.1, 5.2)
liquid capacity (5.1)
little base (5.3)
measurements (5.1)
metric system (5.1)
micro- (5.1)
milli- (5.1)
multiplying-by-1 trick (5.1)
perimeter (5.2)
π (pi) (5.2, 5.3, 5.4)
polygon (5.2)
protractor (5.2)
radius (5.2)
rectangle (5.2, 5.3)
rectangular box (5.4)
right angle (5.2)
right triangle (5.3)
sides of an angle (5.2)
sphere (5.4)
square (5.2, 5.3)
square units (5.3)
temperature (5.1)
time (5.1)
trapezoid (5.3)
triangle (5.3)
unit fraction (5.1)
vertex (5.2)
volume (5.4)
weight (5.1)
width (5.2)

Important Ideas

(5.1) Understanding the English system of measurement.
Converting units within the English system.
Understanding the metric system.
Converting within the metric system.
Converting between the English and metric systems.

(5.2) Measuring angles.
Computing perimeters of polygons.
Computing circumferences.
Computing perimeters of combination figures.

(5.3) Computing the areas of:
(a) Rectangles.
(b) Squares.
(c) Triangles.
(d) Trapezoids.
(e) Circles.
(f) Combination figures.

(5.4) Computing the volumes of:
(a) Rectangular boxes.
(b) Cubes.
(c) Cylinders.
(d) Spheres.
(e) Combination figures.

CHAPTER 5 REVIEW TEST

Choose the correct (or closest) answer.

1. The approximate distance from the seat of your chair to the floor (in centimeters) is:
(a) 10 **(b)** 45 **(c)** 200 **(d)** 450 **(e)** 1000

2. Eight pounds is how many ounces?
(a) $\frac{1}{2}$ **(b)** $\frac{2}{3}$ **(c)** 98 **(d)** 120 **(e)** 128

3. Eight miles is how many kilometers?
(a) 1.6 **(b)** 4.8 **(c)** 12.8 **(d)** 52.2 **(e)** 116.4

4. What angle is one-fourth of a complete revolution?
(a) 90° **(b)** 120° **(c)** 60° **(d)** 36° **(e)** 1440°

5. What is the approximate measure of angle A?
(a) 12° **(b)** 43° **(c)** 95° **(d)** 139° **(e)** 175°

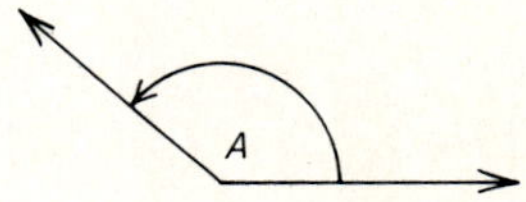

6. Find the perimeter.
(a) 14.25 **(b)** 12.7 **(c)** 12.25 **(d)** 13.7 **(e)** 13.25

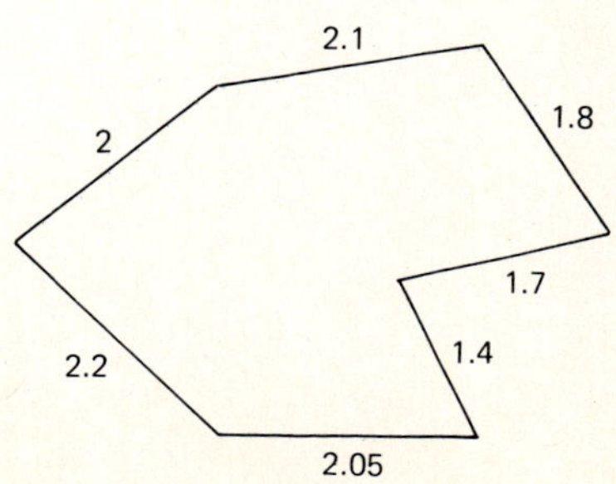

7. Find the perimeter.

(a) $7\frac{1}{5}$ (b) $8\frac{1}{5}$ (c) 7 (d) $5\frac{1}{2}$ (e) 11

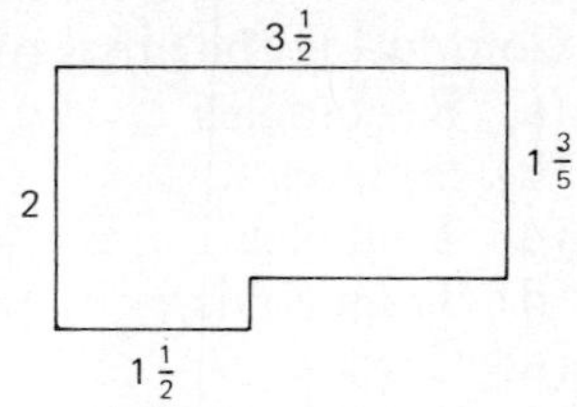

8. Find the circumference of a circle with diameter = 12.
(a) 75.36 (b) 18.84 (c) 37.68 (d) 113.04 (e) 452.16

9. Find the area of a rectangle $2\frac{1}{4}$ by $1\frac{2}{3}$.

(a) $3\frac{11}{12}$ (b) $2\frac{1}{6}$ (c) $3\frac{3}{4}$ (d) $7\frac{5}{6}$ (e) $3\frac{1}{6}$

10. Find the area of a trapezoid with big base = 7, little base = 4, and height = 5.6.
(a) 43.2 (b) 30.8 (c) 61.6 (d) 156.8 (e) 78.4

11. Find the area of a circle with radius = 9.
(a) 1017.36 (b) 244.54 (c) 56.52 (d) 28.26 (e) 254.34

12. Find the area.
(a) 757 (b) 443 (c) 537.2 (d) 286 (e) 246.75

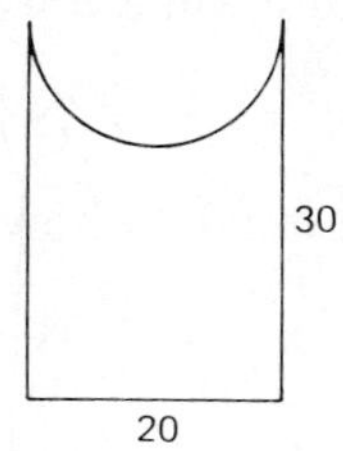

13. Find the volume of a rectangular box, $\frac{3}{4}$ by $1\frac{5}{7}$ by $\frac{2}{3}$.

(a) $1\frac{5}{14}$ (b) $1\frac{10}{84}$ (c) $3\frac{11}{84}$ (d) $1\frac{2}{7}$ (e) $\frac{6}{7}$

14. Find the volume of a cylinder with height = 6 and radius = 4.
(a) 290.34 (b) 150.72 (c) 75.36 (d) 301.44 (e) 678.24

15. Find the volume of a hemisphere with radius = 8.
(a) 134.0 (b) 1071.8 (c) 803.84 (d) 2143.6 (e) 268.0

Now complete Post-Test 1 for Chapter 5 on page T79.

6

INTRODUCTION TO ALGEBRA

Before you begin this chapter, complete the Chapter 6 Pre-Test on page T83.

6.1 ORDER OF OPERATIONS AND THE REAL NUMBERS

SECTION OBJECTIVES After completing this unit, you will be able to:

1. Simplify an expression with many operations.
2. Understand the number line as a picture for the real numbers.
3. Find the opposite of a number.
4. Find the absolute value of a number.

TYPICAL PROBLEMS You will be able to work problems like the ones below.

(a) Simplify $4 + 5 \cdot 6$.

(b) Simplify $2 \cdot (5 + 1)^2$.

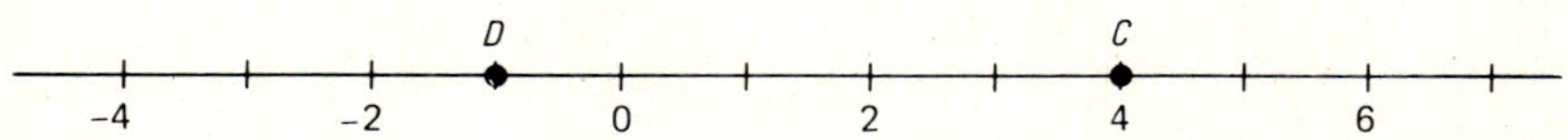

(c) Give the coordinate of point C.

(d) Give the coordinate of point D.

(e) Give the opposite of 7.

(f) Give the opposite of -4.

(g) Give the absolute value of 3.

(h) Give the absolute value of -2.

ANSWERS **(a)** 34 **(b)** 72 **(c)** 4 **(d)** -1 **(e)** -7 **(f)** 4 **(g)** 3 **(h)** 2

The operations x^2 and x^3 are examples of **exponents**. They get their names from geometry. The area of a square with side x is $x \cdot x$ or x^2. The volume of a cube with side x is $x \cdot x \cdot x$ is x^3. Thus we call these "x squared" and "x cubed."

In English, we use punctuation to make expressions more exact. For instance,

Judy said Ethel is tacky

might be taken in two ways:

(a) Judy said, "Ethel is tacky."

(b) "Judy," said Ethel, "is tacky."

It makes a difference how the sentence is punctuated. Similarly,

$$20 - 5 - 3$$

might have two meanings:

(a) $(20 - 5) - 3 = 15 - 3 = 12$

(b) $20 - (5 - 3) = 20 - 2 = 18$

The parentheses act as the quotation marks did. In addition to the operation symbols, we have **grouping symbols,** such as parentheses ( ), brackets [ ], and the fraction bar ———. The symbols are used to make an expression exact, so that there will not be two possible meanings.

Mathematicians have agreed on a **standard order** in which all operations will be done. It can be nicely remembered by the phrase, "*P*lease *E*xcuse *M*y *D*ear *A*unt *S*ally (**PEMDAS**)."

> Standard order of operation:
>
> 1. *Parentheses* (or other grouping symbols): Do all operations within parentheses or brackets, or above and below fraction bars.
> 2. *Exponents:* Do all squaring and cubing from left to right.
> 3. *Multiplication and Division:* From left to right.
> 4. *Addition and Subtraction:* From left to right.

EXAMPLE 1 Simplify $2 + 3 \cdot 4$.

SOLUTION There are no parentheses or exponents, so we do the multiplication first, then the addition.

$$2 + 3 \cdot 4 = 2 + 12 \qquad \boxed{\text{Multiply}}$$

$$= 14 \qquad \boxed{\text{Add}}$$

YES	NO
$2 + 3 \cdot 4 = 2 + 12 = 14$	~~$2 + 3 \cdot 4 = 5 \cdot 4 = 20$~~

We do multiplication before addition.

EXAMPLE 2 Simplify $2(5 - 2)^2$.

SOLUTION We follow the standard order: parentheses, exponent, and multiplication.

$$2(5 - 2)^2 = 2 \cdot 3^2 \qquad \boxed{\text{Parentheses}}$$

$= 2 \cdot 9$ — Square

$= 18$ — Multiply

EXAMPLE 3 Simplify $\dfrac{5^2 - 4}{2^3 - 1}$.

SOLUTION Here we have a fraction bar, which gets top priority. Therefore, we simplify the expressions above and below the fraction bar first.

$$\frac{5^2 - 4}{2^3 - 1} = \frac{25 - 4}{8 - 1}$$

Simplify above and below fraction bars:
1. Exponents
2. Subtractions

$$= \frac{21}{7}$$

$$= 3$$

Divide

EXAMPLE 4 Simplify $3 + 2 \cdot [7 - (5 - 3)^2]$.

SOLUTION Here we have parentheses within brackets. In this case we work out from the inside parentheses to the outside brackets.

$$3 + 2 \cdot [7 - (5 - 3)^2] = 3 + 2 \cdot [7 - 2^2]$$

Inside parentheses

$$= 3 + 2 \cdot [7 - 4]$$

$$= 3 + 2 \cdot [3]$$

Outer brackets:
1. Exponent
2. Subtract

$$= 3 + 6$$

Multiply

$$= 9$$

Add

PROBLEM SET Simplify the following expressions using the standard order of operation.

See Examples 1 to 4

1. $3 + 5 \cdot 2$
2. $4 + 3 \cdot 5$
3. $10 - 2 \cdot 3$

4. $12 - 6 \cdot 2$ 5. $2 + \frac{8}{4}$ 6. $5 - \frac{9}{3}$

7. $2(5 - 3)$ 8. $5(6 + 1)$ 9. $3(2^2 + 1)$

10. $4(10 - 2^3)$ 11. $4(5 - 1)^2$ 12. $3(2 + 3)^2$

13. $\frac{4 + 2}{3 - 1}$ 14. $\frac{10 + 4}{3 + 4}$ 15. $\frac{3 \cdot 6}{7 + 2}$

16. $\frac{20 + 4}{2 \cdot 3}$ 17. $\frac{3^2 + 1}{7 - 2}$ 18. $\frac{21 + 4}{2^3 - 3}$

19. $2[3 + (2 + 2)^2]$ 20. $3[10 - (4 - 1)^2]$ 21. $4[(1 + 2)^3 - 20]$

22. $3[(3 - 1)^2 + 2]$ 23. $\frac{3 - (7 - 6)}{4 - (6 - 4)}$ 24. $\frac{10 - (2 + 2)}{8 - (4 + 1)}$

ANSWERS **1.** 13 **2.** 19 **3.** 4 **4.** 0 **5.** 4 **6.** 2 **7.** 4 **8.** 35
9. 15 **10.** 8 **11.** 64 **12.** 75 **13.** 3 **14.** 2 **15.** 2
16. 4 **17.** 2 **18.** 5 **19.** 38 **20.** 3 **21.** 28 **22.** 18
23. 1 **24.** 2

The most convenient way to picture numbers is with a **number line**.

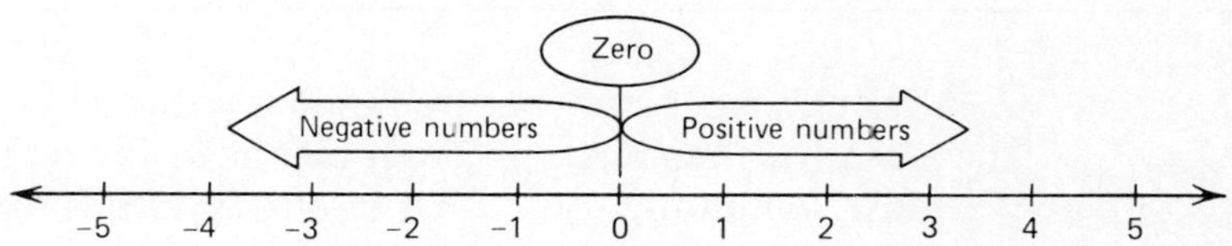

The line continues without end in both directions. The point 0 is called the **origin**. Associated with each point of the line is a number (called its **coordinate**), and the set of all these numbers is called the **real numbers** (also called signed numbers).

The points to the right of 0 are the **positive real numbers**. These are written with a positive sign (+), or more often no sign at all. For instance, +4 (or just 4) is a positive number.

The points to the left of 0 are called the **negative real numbers**. These are written with a negative sign (-). For instance, -3 is a negative number.

EXAMPLE 5 Give the coordinates of the points on the number line.

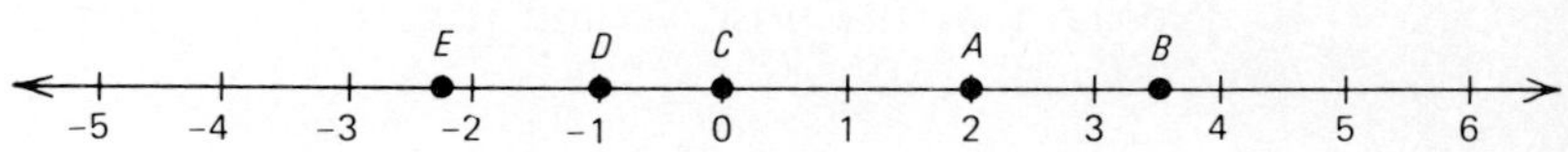

SOLUTION In some cases we will have to estimate the coordinates.

$$A = 2 \qquad B = 3\frac{1}{2}$$

$$C = 0 \qquad D = -1$$

$$E = -2\frac{1}{4}$$

PROBLEM SET Estimate the coordinates of the points on the number line.

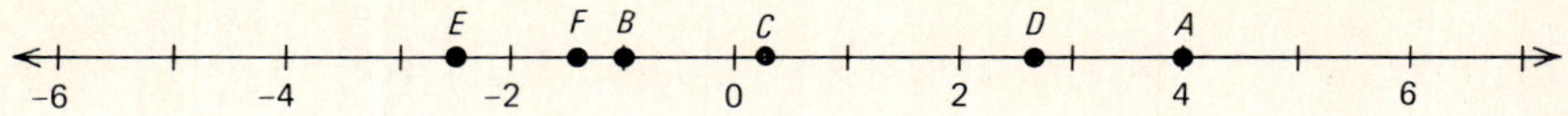

See Example 5

1. Point A 2. Point B 3. Point C
4. Point D 5. Point E 6. Point F

Locate the following numbers on a number line.

7. $G = 5$ 8. $H = -2$ 9. $I = \frac{1}{2}$
10. $J = 3.2$ 11. $K = -3\frac{1}{2}$ 12. $L = -0.7$

ANSWERS 1. 4 2. -1 3. $\frac{1}{4}$ 4. 2.7 5. $-2\frac{1}{2}$ 6. -1.3

7–12.

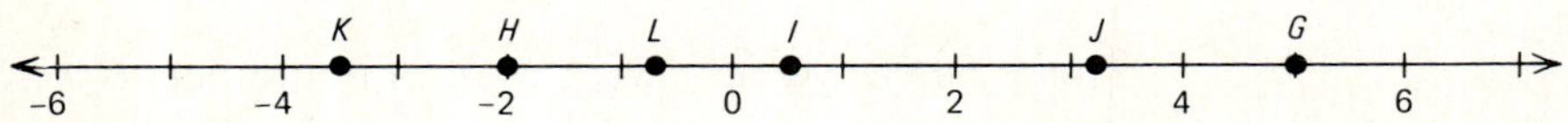

Although most students are familiar with the positive numbers from arithmetic, the negative numbers may be a new idea for many students. Negative numbers are most commonly used to represent opposites for positive quantities.

EXAMPLE 6 The following table shows how negative numbers are used to contrast with positive quantities.

Positives: Phrase	Positives: Number	Negatives: Phrase	Negatives: Number
\$100 in *assets*	+100	\$100 in *debt*	-100
3000 ft *above* sea level	+3000	3000 ft *below* sea level	-3000
20 sec *after* blast-off	+20	20 sec *before* blast-off	-20
5-yard *gain*	+5	5-yard *loss*	-5
10°C *above* freezing	+10	10°C *below* freezing	-10

Notice that the positive and negative numbers come in pairs, such as 5 and -5, or 100 and -100.

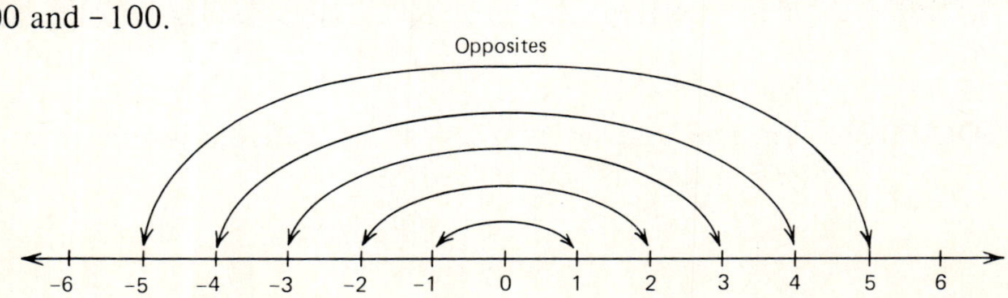

The illustration shows that every number (except 0) has an **opposite** on the other side of 0. We use the notation $-x$ to stand for the opposite of x. (This is also called the **additive inverse**.)

EXAMPLE 7 The following are examples of numbers and their opposites.

Number (x)	Opposite ($-x$)
8	-8
2	-2
-6	6
-5.2	5.2
0	0
$\frac{1}{2}$	$-\frac{1}{2}$
$-\pi$	π

Notice that 0 is its own opposite.

Is $-x$ a negative number? Maybe.

1. Suppose that $x = 7$; then $-x = -7$, which is negative.
2. Suppose that $x = -4$; then $-x$ is the opposite of -4. This is 4. In other words, $-x = -(-4) = 4$, which is positive.

In general, we have

$$-(-x) = x$$

Notice that each pair of opposites, such as 4 and -4, has the same distance, 4, from 0. This distance is always zero or positive and is called the **absolute value**. We write the absolute value of x as $|x|$. See the illustration.

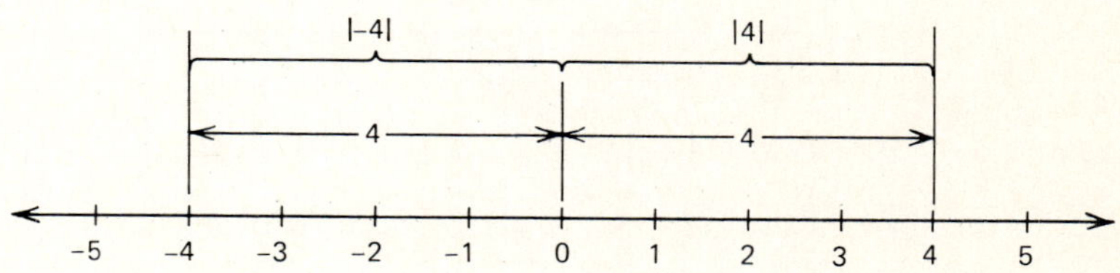

EXAMPLE 8 The following are examples of absolute value, the distance from 0.

(a) $|7| = 7$

(b) $|-8| = 8$

(c) $|14| = 14$

(d) $|0| = 0$

(e) $-|-2| = -2$ (since $|-2| = 2$)

PROBLEM SET Complete the following table.

See Examples 7 and 8

	Number	Opposite	Absolute Value
1.	5		
2.	-2		
3.	-10		
4.	0		
5.		-1	
6.		9	
7.		$\frac{1}{2}$	
8.		$-\frac{1}{4}$	

ANSWERS 1. $-5; 5$ 2. $2; 2$ 3. $10; 10$ 4. $0; 0$ 5. $1; 1$ 6. $-9; 9$
7. $-\frac{1}{2}; \frac{1}{2}$ 8. $\frac{1}{4}; \frac{1}{4}$

Miscellaneous Exercises 6.1

Simplify the following to a single number.

See Examples 1 to 4

1. $3 + 4 \cdot 2$ 2. $12 - 2 \cdot 5$ 3. $5(4 + 3)$ 4. $3(11 - 3^2)$

5. $\dfrac{5^2 + 3}{2 \cdot 2}$ 6. $\dfrac{5 \cdot 6}{8 - 5}$ 7. $3[2 + (1 + 2)^2]$ 8. $\dfrac{12 - (5 - 1)}{5 + (9 - 6)}$

Estimate the coordinates of the points on the number line.

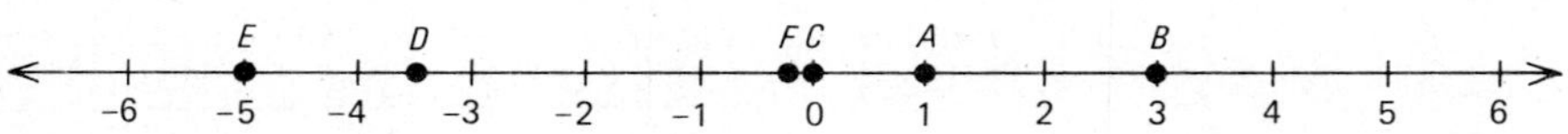

See Example 5

9. Point A 10. Point B 11. Point C
12. Point D 13. Point E 14. Point F

On a number line, locate the following points.

See Example 5

15. $A = 2$ 16. $B = 1\frac{1}{2}$ 17. $C = 3\frac{1}{3}$

18. $D = -1$ 19. $E = -3.7$ 20. $F = -2\frac{1}{4}$

Replace the following phrases by an appropriate positive or negative number.

See Example 6

21. 55 miles per hour forward
22. A \$15 loss at poker

23. A \$20 price increase

24. A 6-pound weight gain

25. Running a movie backward for 4 minutes

26. A $2\frac{1}{4}$-point drop in a stock price

27. A 200-Btu heat loss

28. 20°C above freezing

29. A \$4000 profit from a house sale

30. A \$200 debt

Complete the following table.

See Examples 7 and 8

	Number	Opposite	Absolute Value
31.	18		
32.	-3		
33.	-11		
34.	14		
35.		-2	
36.		10	
37.		9	
38.		-7	

Business applications

39. Locate the following profits on the number line: P_1 = \$10,000; P_2 = -\$15,000; P_3 = \$25,000; P_4 = -\$3000.

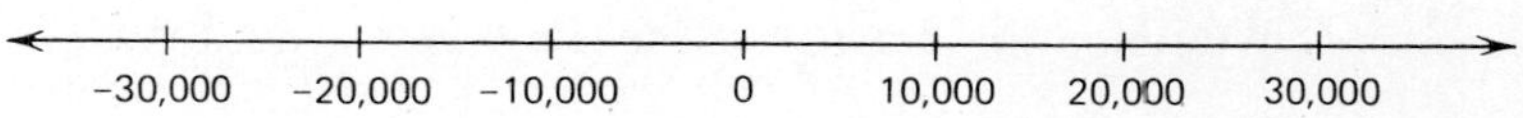

40. Compute the total cost (TC) to produce 100 widgets, which is given by

$$TC = 100^2 + 2000(100) + 200{,}000$$

41. Compute the balance (B) of an investment given by

$$B = 5000(1.07)^2$$

Health application

42. Compute the likelihood (L) of at least 1 out of 3 people contracting a certain disease. This is given by

$$L = 1 - (0.8)^3$$

Psychology application

43. Locate the following moods (-3 = depressed; 0 = neutral; +3 = very happy) on a number line: A(Alice) = +2; B(Bob) = -3; C(Carol) = 0; D(Dave) = 1.

Now complete Post-Test 1 for Section 6.1 on page T85.

6.2 ADDITION AND SUBTRACTION OF REAL NUMBERS

SECTION OBJECTIVES After completing this unit, you will be able to:

1. Add real numbers.
2. Subtract real numbers.

TYPICAL PROBLEMS You will be able to work problems like the ones below.

(a) 3 + (-5) (b) -2 + 7 + (-10) (c) 4 + (-2 + 8) + (-3)

(d) -3 - (9) (e) -8 - (-10) (f) -5 - 8 - (4 - 5) + 7

ANSWERS (a) -2 (b) -5 (c) 7 (d) -12 (e) 2 (f) -5

We can use the number line to help us picture the **addition** of real numbers.

EXAMPLE 9 Add 4 + 3.

SOLUTION This may be a simple problem, but it is still a good place to start. These are both positive numbers, so we can imagine them as assets. If we have \$4 in assets and \$3 in assets, together we have \$7 in assets. On a number line, this appears as follows.

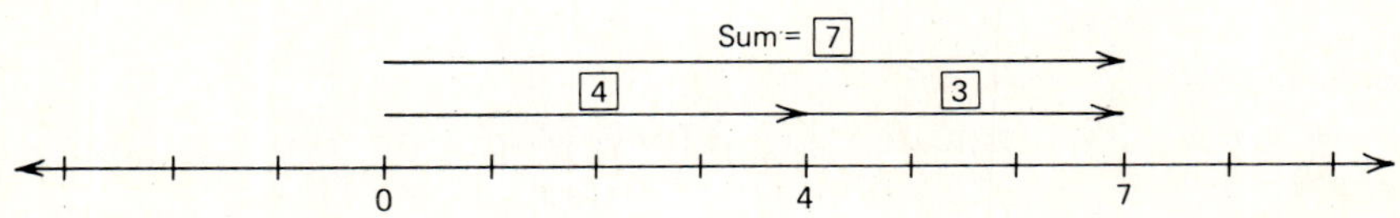

We place the 3-arrow after the 4-arrow, and get a 7-arrow. Thus 4 + 3 = 7.

EXAMPLE 10 Add -2 + (-3).

SOLUTION This time the numbers are both negative, so we can imagine them as debts. If we have a \$2 debt and a \$3 debt, together we have a \$5 debt.

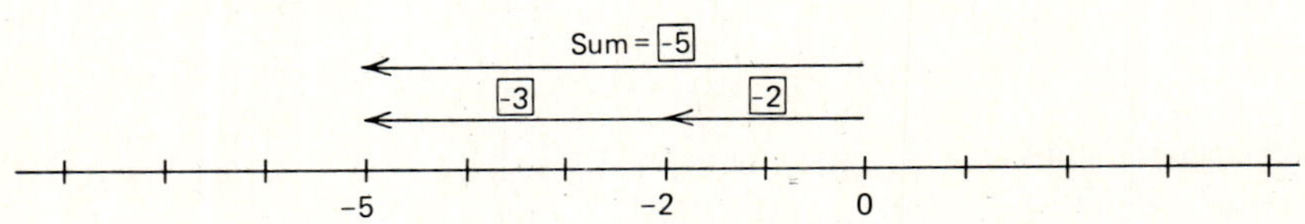

On the number line, we see that a -3-arrow and a -2-arrow produce a -5-arrow. Thus -2 + (-3) = -5.

These two examples suggest the following rule.

To add two real numbers with the *same* sign:
1. Add their absolute values.
2. Give the answer the common sign.

EXAMPLE 11 The following additions are of real numbers with the same sign.

(a) $5 + 7 = 12$
(b) $(+3) + (+11) = +14$
(c) $-4 + (-9) = -13$
(d) $-10 + (-2) = -12$
(e) $-20 + (-30) = -50$

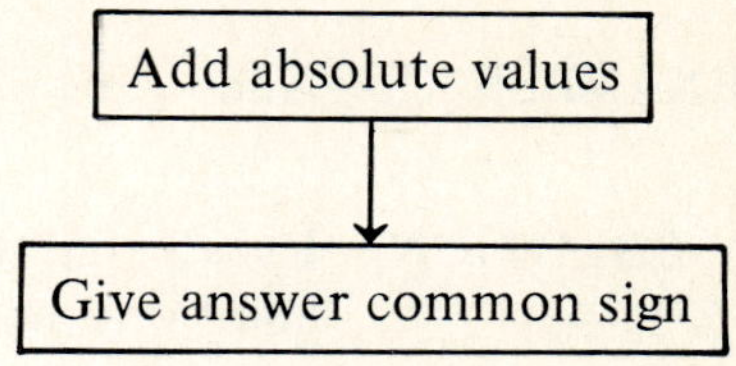

EXAMPLE 12 Add $7 + (-2)$.

SOLUTION These numbers have opposite signs. Suppose that we have a \$7 asset and a \$2 debt; then our net is \$5 in assets.

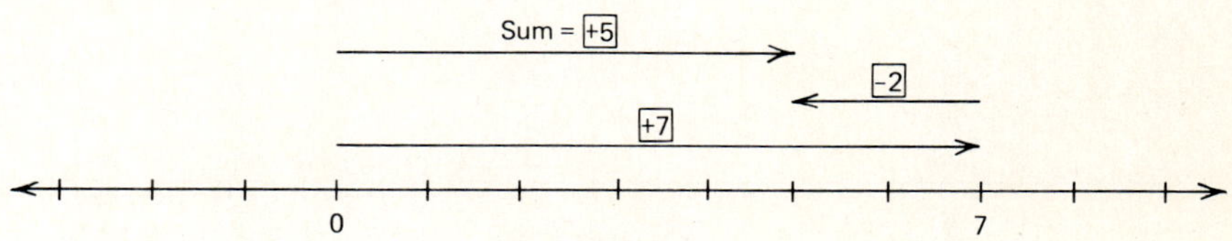

On the number line, if we attach a -2-arrow to the end of a 7-arrow, we end up with a 5-arrow. Thus $7 + (-2) = 5$.

EXAMPLE 13 Add $-6 + 4$.

SOLUTION Again, we have opposite signs. Suppose that we have a \$6 debt and \$4 assets; then together, we have a net of \$2 debt.

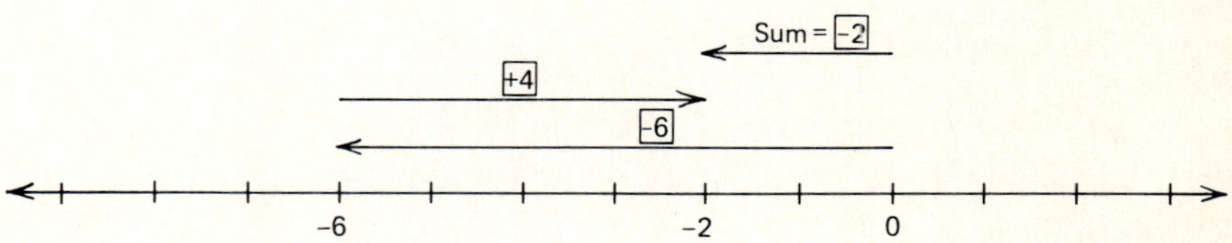

On the number line, we see that attaching a 4-arrow to a -6-arrow gives a -2-arrow; thus $-6 + 4 = -2$.

These examples suggest our next rule.

To add two real numbers of *opposite* signs:
1. Subtract the absolute values (larger − smaller).
2. Give the answer the sign of the number with the larger absolute value.

EXAMPLE 14 The following additions are of real numbers with opposite signs.

(a) $2 + (-9) = -7$
(b) $6 + (-1) = 5$
(c) $-7 + 4 = -3$
(d) $-2 + 10 = 8$
(e) $-20 + 15 = -5$

Subtract absolute values → Take sign of larger absolute value

EXAMPLE 15 Add $-3 + (-7 + 9) + (-8)$.

SOLUTION Recall that we perform the operation within the parentheses first. Then we do additions, left to right.

$$-3 + (-7 + 9) + (-8) = -3 + 2 + (-8)$$ Add: $-7 + 9 = 2$

$$= -1 + (-8)$$ Add: $-3 + 2 = -1$

$$= -9$$ Add: $-1 + (-8) = -9$

PROBLEM SET Find each of the following sums.

See Examples 9 to 15

1. $2 + (-9)$	2. $11 + 3$	3. $14 + (-5)$
4. $-10 + (-6)$	5. $-6 + 15$	6. $-8 + 7$
7. $9 + (-10)$	8. $8 + (-5)$	9. $12 + (+6)$
10. $-4 + 13$	11. $-7 + (-9)$	12. $14 + (-11)$
13. $-20 + (-15)$	14. $30 + (-17)$	15. $20 + (+16)$
16. $17 + (-19)$	17. $-15 + (-3)$	18. $-17 + 13$
19. $2 + 3 + 6$	20. $2 + 5 + (-8)$	21. $5 + (-2) + (-6)$
22. $-4 + (-2) + 9$	23. $-6 + (-9) + (-3)$	24. $-2 + (-5) + (-1)$
25. $-4 + 3 + (-8)$	26. $-5 + 10 + (-3)$	27. $-6 + 11 + 2$
28. $-8 + 2 + 3$	29. $12 + (-6) + 1$	30. $2 + (-5) + 1$
31. $-2 + (-6 + 8) + 3$	32. $-4 + 7 + (-8 + 10)$	33. $4 + (-2 + 5) + (-3)$
34. $-6 + [3 + (-8)] + (-4)$		

ANSWERS 1. -7 2. 14 3. 9 4. -16 5. 9 6. -1 7. -1 8. 3 9. 18 10. 9 11. -16 12. 3 13. -35 14. 13 15. 36 16. -2 17. -18 18. -4 19. 11 20. -1 21. -3 22. 3 23. -18 24. -8 25. -9 26. 2 27. 7 28. -3 29. 7 30. -2 31. 3 32. 5 33. 4 34. -15

We now show how **subtraction** is the opposite of addition.

EXAMPLE 16 Subtract 7 - 4.

SOLUTION Consider this as a temperature problem: if it is 4° one hour and 7° the next hour, what is the temperature jump?

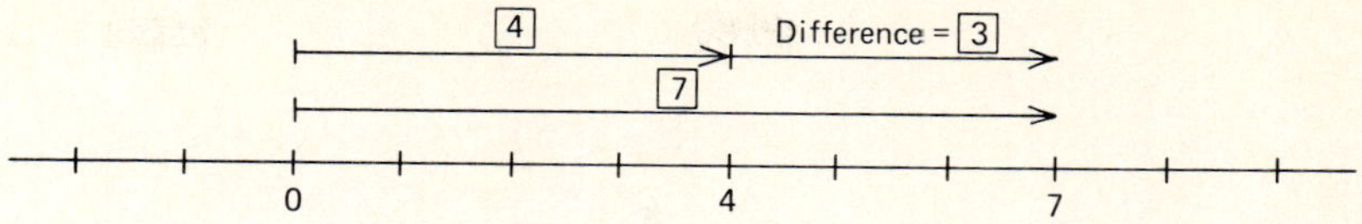

We see on the number line that the difference between 7° and 4° is 3°. This is also the result that we get by adding 7 + (-4).

EXAMPLE 17 Subtract 5 - (-3).

SOLUTION Consider this as a temperature problem: if it is -3° one hour and 5° the next hour, what is the temperature jump?

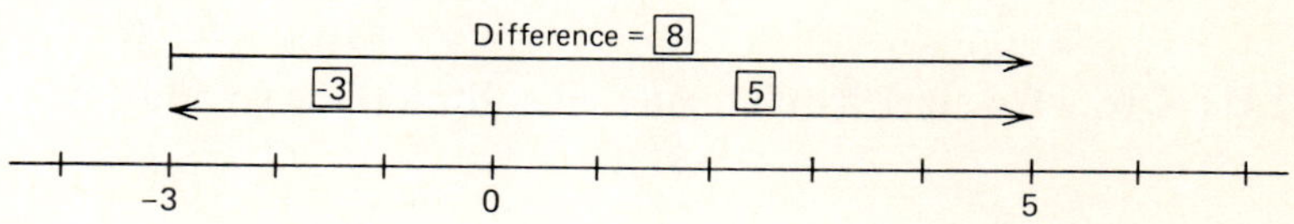

We see on the number line that the difference is 8°. Thus 5 - (-3) = 8. Also, 5 + (+3) = 8.

These examples suggest the next rule.

To subtract real numbers, add the opposite of the second number.

As a formula, this is written as follows.

$$x - y = x + (-y)$$

Another way of stating the subtraction rule is that we *change the sign of the second number and add.*

EXAMPLE 18 The following subtractions use the fact that subtraction means *add the opposite of the second number.*

(a) $7 - 3 = 7 + (-3) = 4$
(b) $5 - 13 = 5 + (-13) = -8$
(c) $-3 - 9 = -3 + (-9) = -12$
(d) $-5 - (-11) = -5 + (+11) = 6$
(e) $-7 - (-1) = -7 + (+1) = -6$

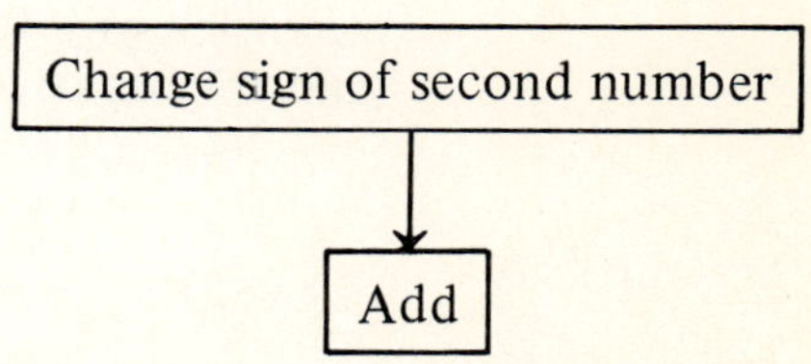

EXAMPLE 19 Evaluate $15 - (x - y)$ for $x = -2$ and $y = 3$.

SOLUTION We first substitute and then do the subtraction within the parentheses.

$15 - (x - y) = 15 - (-2 - 3)$	Substitute $x = -2, y = 3$
$= 15 - [-2 + (-3)]$	Change sign
$= 15 - (-5)$	Add within brackets
$= 15 + (+5)$	Change sign
$= 20$	Add

EXAMPLE 20 Simplify $3 + (-5) - 7 - (-8) + (-4)$.

SOLUTION We start by changing all subtractions to additions of opposites.

$3 + (-5) - 7 - (-8) + (-4) = \underbrace{3 + (-5)} + (-7) + (+8) + (-4)$	Subtraction means addition of opposites
$= \underbrace{-2 + (-7)} + (+8) + (-4)$	Add left to right: $3 + (-5) = -2$
$= \underbrace{-9 + (+8)} + (-4)$	Add: $-2 + (-7) = -9$
$= \underbrace{-1 + (-4)}$	Add: $-9 + (8) = -1$
$= -5$	Add: $-1 + (-4) = -5$

Another way to simplify a problem such as this is to group the positives together and the negatives together, and then add.

$3 + (-5) - 7 - (-8) + (-4) = 3 + (-5) + (-7) + (+8) + (-4)$	Subtraction means addition of opposites
$= [3 + 8] + [-5 + (-7) + (-4)]$	Group positives, group negatives

$= 11 + (-16)$ — Simplify each group

$= -5$ — Simplify

PROBLEM SET Simplify the following.

See Examples 16 to 20

1. $3 - 10$	**2.** $5 - 7$	**3.** $-3 - 5$
4. $10 - (-12)$	**5.** $8 - 3$	**6.** $4 - (-7)$
7. $-9 - (-3)$	**8.** $-6 - 7$	**9.** $-5 - (-8)$
10. $10 - 18$	**11.** $-4 - (-9)$	**12.** $9 - (-2)$
13. $3 - 9$	**14.** $-5 - (-2)$	**15.** $-7 - 10$
16. $-11 - 13$	**17.** $-9 - (-15)$	**18.** $8 - (-21)$
19. $3 - (8 - 1)$	**20.** $4 - (7 - 9)$	**21.** $-4 - (7 - 3)$
22. $10 - (2 - 6)$	**23.** $-8 - [3 - (-5)]$	**24.** $5 - [-8 - (-7)]$
25. $-10 - [-6 - (-8)]$	**26.** $-5 - [9 - (-8)]$	**27.** $-11 - [-3 - 7]$

28. $-12 - [-5 - (-6)]$

29. $-7 - 5 + 2 - 3 - (-6)$

30. $-8 - (-7) - 6 + 4 - 2$

31. $-3 - (-6) + 10 - 4 - (-3)$

32. $10 + (-4) - 5 - (-2) + 3$

33. $7 - 9 + (-3) - (-5) - 10$

34. $-1 - (-5) - 6 - 8 + (-4)$

ANSWERS **1.** -7 **2.** -2 **3.** -8 **4.** 22 **5.** 5 **6.** 11 **7.** -6 **8.** -13 **9.** 3 **10.** -8 **11.** 5 **12.** 11 **13.** -6 **14.** -3 **15.** -17 **16.** -24 **17.** 6 **18.** 29 **19.** -4 **20.** 6 **21.** -8 **22.** 14 **23.** -16 **24.** 6 **25.** -12 **26.** -22 **27.** -1 **28.** -13 **29.** -7 **30.** -5 **31.** 12 **32.** 6 **33.** -10 **34.** -14

Miscellaneous Exercises 6.2

Simplify the following.

See Examples 9 to 15

1. $-4 + 10$	**2.** $-5 + (-12)$
3. $8 + (-13)$	**4.** $-2 + (-21)$
5. $3 + (-5) + 7$	**6.** $-7 + 9 + (-10)$
7. $-11 + (-8) + (-6) + 1$	**8.** $-9 + (-15) + (-20) + 8 + (-1)$

See Examples 16 to 20

9. 4 - 10

10. -3 - 8

11. -7 - (-11)

12. 8 - (-13)

13. 2 - 5 - 9

14. 3 - 5 - (-12)

15. -5 - (-2 + 7)

16. -10 - [-4 - (-5)]

17. -3 - 4 + (-2) - (-9) + (-1) + 7

Business applications

18. A business lists its assets (+) and debts (-) as:

$10,000, -$7000, $2500, $3000, -$4500

Find the company's net worth.

19. In business, we have the formula

change in profit = change in revenue - change in cost

Use this formula to complete the following table.

Change in Profit	Change in Revenue	Change in Cost
?	20,000	15,000
?	30,000	38,000
?	-12,000	-17,000
?	-8,000	2,000

Health applications

20. A patient's weight gains (and losses) are recorded as

+2, -1, -1.5, +1, -2.5, -0.5, +1, -1.5

What is the patient's net gain (or loss)?

21. The pressure exerted on the lungs is measured in a unit called *centimeters of water* (which we will not explain). For instance, for a normal person, *inspiration* (breathing in) has a pressure of about -30 centimeters of water, and *expiration* (breathing out) has a pressure of about +25 centimeters of water.

A very ill patient may need the help of an artificial respirator to breathe. This machine provides the difference between the needed pressure and the patient's own ability (needed - ability). Complete the following table.

	Inspiration Pressure			Expiration Pressure		
Patient	Needed	Ability	Respirator	Needed	Ability	Respirator
A	-30	-17	?	+25	+12	?
B	-40	-23	?	+30	+14	?
C	-35	-12	?	+28	+17	?

Science application

22. The change of temperature in a substance is given by

temperature change = final temperature - original temperature

Use this formula to complete the following table.

Temperature Change	Final Temperature	Original Temperature
?	18°C	2°C
?	-14°C	27°C
?	-2°C	-9°C
?	5°C	-13°C

Ecology application

23. Over the period of one week, the daily changes in the air pollution (from lead) from a factory (in micrograms per cubic meter) were:

$$+3, \quad -10, \quad +5, \quad +6, \quad -2, \quad -11, \quad -13$$

Find the net change over the week.

Chemical application

24. In chemistry, elements and compounds have valences, which add as real numbers do. Find the net valences of:
(a) Cr_2O_7 if the valence of Cr_2 is +12 and the valence of O_7 is -14.
(b) Fe_2O_3 if the valence of Fe_2 is +6 and the valence of O_3 is -6.

Now complete Post-Test 1 for Section 6.2 on page T87.

6.3 MULTIPLICATION AND DIVISION OF REAL NUMBERS

SECTION OBJECTIVES

After completing this unit, you will be able to:

1. Multiply real numbers.
2. Divide real numbers.
3. Simplify expressions with more than one operation.

TYPICAL PROBLEMS

You will be able to work problems like the ones below.

(a) $(-5)(-10)$ (b) $7(-3)$ (c) $9 + (-4)(-2)$

(d) $\dfrac{-24}{-4}$ (e) $\dfrac{100}{-2}$ (f) $\dfrac{(-2)(-6) + (-1)(-3)}{1-6}$

ANSWERS (a) 50 (b) -21 (c) 17 (d) 6 (e) -50 (f) -3

We will look at (but not prove) the rules for **multiplying** real numbers. Recall from arithmetic that multiplying by a whole number can be considered as *repeated addition*. Similarly, we can consider multiplying by a negative integer as *repeated subtraction*. The following table shows all the possibilities for multiplying positive and negative numbers.

Factors	Meaning	Product
$3 \cdot 4$	Add 4 three times $= 4 + 4 + 4 = 12$	12
$3 \cdot (-4)$	Add (-4) three times $= (-4) + (-4) + (-4) = -12$	-12
$(-3) \cdot 4$	Subtract 4 three times $= -4 - 4 - 4$ $= (-4) + (-4) + (-4) = -12$	-12
$(-3) \cdot (-4)$	Subtract (-4) three times $= -(-4) - (-4) - (-4)$ $= 4 + 4 + 4 = 12$	12

Notice that all the products have an absolute value of 12; only the signs are different. When we multiplied two positives or two negatives, we got a *positive* number. When we multiplied one positive and one negative, we got a *negative* product. This suggests the following general rule.

> To multiply two real numbers:
>
> **1.** Multiply their absolute values.
>
> **2a.** If the numbers have the *same* sign, the product is *positive*.
>
> **2b.** If the numbers have *opposite* signs, the product is *negative*.

EXAMPLE 21 The following multiplications are pairs of real numbers with the *same* sign. The products are all *positive*.

(a) $2 \cdot 5 = 10$

(b) $3 \cdot 7 = 21$

(c) $(-4)(-6) = 24$

(d) $(-10)(-3) = 30$

(e) $\left(-\frac{1}{2}\right)\left(-\frac{1}{3}\right) = \frac{1}{6}$

EXAMPLE 22 The following multiplications are pairs of real numbers with *opposite* signs. The products are all *negative*.

(a) $(-5)(7) = -35$

(b) $(-4)(10) = -40$

(c) $(6)(-3) = -18$

(d) $(8)(-2) = -16$

(e) $\left(-\frac{2}{3}\right)\left(\frac{5}{7}\right) = -\frac{10}{21}$

PROBLEM SET

Find the following products.

See Examples 21 and 22

1. $2 \cdot 7$	2. $-2(7)$	3. $4(-6)$	4. $-3(-7)$
5. $-4(5)$	6. $10(-4)$	7. $5(6)$	8. $6(-4)$
9. $-5(-8)$	10. $(-8)(-6)$	11. $-4(12)$	12. $9(-3)$
13. $-11(-4)$	14. $-3(-5)$	15. $5(10)$	16. $-7(5)$
17. $\frac{1}{3}\left(-\frac{1}{4}\right)$	18. $-\frac{1}{2}\left(\frac{1}{5}\right)$		

ANSWERS 1. 14 2. −14 3. −24 4. 21 5. −20 6. −40 7. 30 8. −24 9. 40 10. 48 11. −48 12. −27 13. 44 14. 15 15. 50 16. −35 17. $\frac{-1}{12}$ 18. $\frac{-1}{10}$

Let us now state some simple, but important, properties for multiplication.

> ***PROPERTY*** For any real number x,
>
> **(a)** $1 \cdot x = x \cdot 1 = x$
> **(b)** $0 \cdot x = x \cdot 0 = 0$
> **(c)** $(-1)x = x(-1) = -x$

EXAMPLE 23 The following examples show the three properties stated above.

(a) $1 \cdot 7 = 7$
(b) $(-4) \cdot 1 = -4$
(c) $0 \cdot (-5) = 0$
(d) $15 \cdot 0 = 0$
(e) $(-1) \cdot \frac{1}{2} = -\frac{1}{2}$
(f) $(-6)(-1) = 6$

PROBLEM SET

Find the following products.

See Example 23

1. $6 \cdot 0$	2. $(1)9$	3. $(-8)(-1)$
4. $-7(1)$	5. $0\left(\frac{1}{2}\right)$	6. $-1(14)$

ANSWERS 1. 0 2. 9 3. 8 4. −7 5. 0 6. −14

Division is the opposite of multiplication. Recall from arithmetic that

$$10 \div 2 = \boxed{?} \qquad \text{means} \qquad 2 \cdot \boxed{?} = 10 \qquad (\text{Thus, } \boxed{?} = 5.)$$

Division has the same meaning with real numbers. For instance,

$$-10 \div 2 = \frac{-10}{2} = \boxed{?} \qquad \text{means} \qquad 2 \cdot \boxed{?} = -10$$

We can see that $\boxed{?} = -5$. Using this thinking, we can derive the rules for division that are very similar to the rules for multiplication.

> To divide two real numbers:
>
> **1.** Divide their absolute values.
>
> **2a.** If the numbers have the *same* sign, the quotient is *positive*.
>
> **2b.** If the numbers have *opposite* signs, the quotient is *negative*.

EXAMPLE 24 The following divisions are pairs of numbers with the *same* sign. The quotients are all *positive*.

(a) $\frac{10}{2} = 5$

(b) $\frac{-12}{-6} = 2$

(c) $\frac{-20}{-4} = 5$

(d) $\frac{36}{9} = 4$

(e) $\frac{-100}{-1} = 100$

EXAMPLE 25 The following divisions are pairs of numbers with *opposite* signs. The quotients are all *negative*.

(a) $\frac{-12}{2} = -6$

(b) $\frac{-25}{5} = -5$

(c) $\frac{21}{-3} = -7$

(d) $\frac{30}{-10} = -3$

(e) $\frac{18}{-1} = -18$

Special mention should be given to the number 0. *We can never divide by zero.*

YES	NO
$\frac{0}{7} = 0$	~~$\frac{7}{0} = 0$~~

We cannot divide by zero. The fraction $\frac{7}{0}$ makes no sense.

YES	NO
$\frac{6}{6} = 1$	~~$\frac{0}{0} = 1$~~

The fraction $\frac{0}{0}$ makes no sense.

PROBLEM SET Divide the following numbers.

See Examples 24, 25, and the Yes/No boxes

1. $\frac{4}{-2}$ 2. $\frac{-6}{3}$ 3. $\frac{-10}{-2}$ 4. $\frac{-26}{-13}$ 5. $\frac{30}{15}$

6. $\frac{18}{-6}$ 7. $\frac{32}{-8}$ 8. $\frac{-16}{-2}$ 9. $\frac{-20}{-10}$ 10. $\frac{0}{-5}$

11. $\frac{30}{-5}$ 12. $\frac{14}{7}$ 13. $\frac{14}{0}$ 14. $\frac{-20}{4}$ 15. $\frac{-90}{-3}$

16. $\frac{-42}{6}$ 17. $\frac{0}{-6}$ 18. $\frac{100}{-2}$ 19. $\frac{-50}{-50}$ 20. $\frac{0}{0}$

ANSWERS **1.** -2 **2.** -2 **3.** 5 **4.** 2 **5.** 2 **6.** -3 **7.** -4 **8.** 8
9. 2 **10.** 0 **11.** -6 **12.** 2 **13.** Not defined **14.** -5
15. 30 **16.** -7 **17.** 0 **18.** -50 **19.** 1 **20.** Not defined

Recall that the standard order of operation is: parentheses, exponents, multiplication, division, addition, and subtraction (*P*lease *E*xcuse *M*y *D*ear *A*unt *S*ally).

EXAMPLE 26 Simplify $-3(-2 - 5) - 7(-4 + 6)$.

SOLUTION We follow the standard order for operation: parentheses, multiplication, addition, and subtraction. First, we change all subtractions to additions of opposites.

$$-3(-2 - 5) - 7(-4 + 6) = -3[-2 + (-5)] + (-7)(-4 + 6)$$

Subtraction is addition of opposite

$= -3[-7] + (-7)(2)$	Parentheses
$= 21 + (-14)$	Multiply
$= 7$	Add

EXAMPLE 27 Simplify $\dfrac{(5)(-4) - (4)(-2)}{2(4 - 5)}$.

SOLUTION We first simplify above and below the fraction bar; then we divide.

$\dfrac{(5)(-4) - (4)(-2)}{2(4 - 5)} = \dfrac{-20 - (-8)}{2(-1)}$	Multiply above and below fraction bar
$= \dfrac{-20 + (+8)}{-2}$	Change sign, add
$= \dfrac{-12}{-2}$	Add
$= 6$	Divide

PROBLEM SET Simplify the following expressions using the standard order of operation.

See Examples 26 and 27

1. $-2 + 3 \cdot 4$
2. $4 - (-3)(5)$
3. $(-2)(3) - (5)(-1)$
4. $(-3)(4) + (-2)(-7)$
5. $-5(6 - 10)$
6. $2(-3 - 5)$
7. $10 - 4(2 - 5)$
8. $-3 + (-4)(1 - 8)$
9. $-2(1 - 6) - 4(-2 + 8)$
10. $5(3 - 5) - 4(-2 - 6)$
11. $12 + \dfrac{-6}{3}$
12. $-4 - \dfrac{8}{-4}$
13. $\dfrac{3(-6)}{9}$
14. $\dfrac{-30}{(-2)(-5)}$
15. $\dfrac{(-10)(10)}{(-2)(-25)}$
16. $\dfrac{(-8)(-8)}{(-2)(-4)}$

17. $\dfrac{2-10}{1-3}$

18. $\dfrac{-6+20}{2-9}$

19. $\dfrac{2(-3)+4(-1)}{1-6}$

20. $\dfrac{-6(-2)-4(-5)}{-2+6}$

ANSWERS 1. 10 2. 19 3. −1 4. 2 5. 20 6. −16 7. 22 8. 25 9. −14 10. 22 11. 10 12. −2 13. −2 14. −3 15. −2 16. 8 17. 4 18. −2 19. 2 20. 8

Miscellaneous Exercises 6.3

Simplify the following expressions.

See Examples 21 and 22

1. $(-3)(4)$
2. $(-6)(-8)$
3. $(-2)(50)$
4. $7(12)$
5. $10(-4)$
6. $-\dfrac{1}{2}\left(-\dfrac{1}{4}\right)$

See Example 23

7. $1(-41)$
8. $0(2356)$
9. $(-10)(-1)$

See Examples 24 and 25

10. $\dfrac{10}{-2}$
11. $\dfrac{-12}{-4}$
12. $\dfrac{0}{5}$
13. $\dfrac{-150}{3}$
14. $\dfrac{-8}{-8}$
15. $\dfrac{-9}{0}$

See Examples 26 and 27

16. $-5+(-3)(4)$
17. $-3(2-6)+(-7)(-2)$
18. $\dfrac{(-3)(-8)}{(-2)(4)}$
19. $\dfrac{(-5)(2)-(-3)(-6)}{3-10}$

Chemistry application

20. In chemistry, the valence of an ion of an element is the product of the valence of the element and the number of atoms. Complete the following table.

Element (valence)	Ion	Net Valence
Cr (+6)	Cr_2	$2(+6)=+12$
O (−2)	O_7	$7(-2)=-14$
H (+1)	H_3	?
N (−3)	N_1	?
O (−2)	O_3	?
C (+4)	C_2	?

Psychology application

21. The z-score for an IQ is given by the formula

$$z=\frac{\text{IQ}-100}{15}$$

Find the z-score for the following cases:
(a) IQ = 130
(b) IQ = 85
(c) IQ = 55
(d) IQ = 100

Business applications

22. The extra profit (P) in producing more (or fewer) items is the *marginal profit* (MP) times the number of extra units produced (U). Written as a formula, this is

$$P = (MP)(U)$$

Use the formula to complete the following table.

Extra Profit (P)	Marginal Profit (MP)	Extra Units (U)
?	50	100
?	-80	200
?	200	-30
?	-500	-20

23. *Elasticity* (e) measures the effect of price changes on consumer demand. The formula is

$$e = \frac{-P(\Delta Q)}{Q(\Delta P)}$$

where P is the price, Q is the demand, ΔP is the change in price, and ΔQ is the change in demand. Complete the following table.

e	P	Q	ΔP	ΔQ
?	10	200	+1	-40
?	40	800	-2	+20
?	200	60,000	+10	-300
?	100	20,000	+5	-5000

Electricity application

24. According to Ohm's law,

$$i = \frac{V}{R}$$

where i is the current in a simple circuit, V is the voltage, and R is the resistance. Complete the following table.

i	V	R
?	2000	200
?	-300	100
?	-80	20

Now complete Post-Test 1 for Section 6.3 on page T89.

6.4 THE DISTRIBUTIVE LAW AND SIMPLIFYING EXPRESSIONS

SECTION OBJECTIVES

After completing this unit, you will be able to:

1. Use the distributive law.
2. Simplify certain algebraic expressions.

TYPICAL PROBLEMS You will be able to work problems like the ones below.

(a) Simplify $5(x + y)$. (b) Simplify $3(a - b - 7)$.

(c) Simplify $6x + 5 + 8$. (d) Simplify $10a - 7a + 8$.

(e) Simplify $5t + 4 - 2t - 10 - 6t + 9$. (f) Simplify $12 - 3t - 5 + 2t + 1 + 15 - 9t$.

ANSWERS (a) $5x + 5y$ (b) $3a - 3b - 21$ (c) $6x + 13$ (d) $3a + 8$ (e) $-3t + 3$ (f) $23 - 10t$

One of the most important properties of the real numbers is the **distributive law**. This property involves both addition and multiplication. It can be stated as follows.

> ***DISTRIBUTIVE LAW*** For any real numbers a, b, and c,
>
> $$a(b + c) = a \cdot b + a \cdot c$$
>
> $$a(b - c) = a \cdot b - a \cdot c$$

We can think of the outside number "distributing" over the inside numbers. For instance,

$$\boxed{5}\,(2 + 9) = \boxed{5} \cdot 2 + \boxed{5} \cdot 9$$
$$= 10 + 45 = 55$$

Of course, we could have added within the parentheses first and gotten $5 \cdot (11) = 55$, the same answer.

EXAMPLE 28 The following are examples of the distributive law. Notice that we start using letters along with the numbers. It is standard in algebra to use letters to stand for unknown numbers. The term $5 \cdot x$ (or $5x$) means five times some number called x.

(a) $3(p + q) = 3p + 3q$

(b) $7(x - 2) = 7x - 14$

(c) $10(n + 4) = 10n + 40$

(d) $-2(a + b - 6) = -2a - 2b + 12$

(e) $-5(x - y + 2) = -5x + 5y - 10$

PROBLEM SET Use the distributive law to simplify the following.

See Example 28

1. $5(a + b)$ 2. $2(m - n)$ 3. $10(x + y - z)$

4. $3(p + 4)$ 5. $7(q - 6)$ 6. $-4(x - 5)$

7. $8(a + b - 3)$ 8. $10(3 - s + t)$ 9. $-2(u - v - 11)$

ANSWERS 1. $5a + 5b$ 2. $2m - 2n$ 3. $10x + 10y - 10z$ 4. $3p + 12$ 5. $7q - 42$ 6. $-4x + 20$ 7. $8a + 8b - 24$ 8. $30 - 10s + 10t$ 9. $-2u + 2v + 22$

In any expression, a number or a product of a number and a variable is called a **term**. We simplify an expression by **combining like terms**, or terms with the same variable. For instance, $4a$, $7a$, $12a$, and $-2a$ are like terms, because they all have the same variable, a. The term x can also be thought of as $1 \cdot x$. We can use the distributive law to help us combine like terms.

EXAMPLE 29 The following use the distributive law ("in reverse") to combine the like terms in the following expressions.

(a) $2a + 7a = (2 + 7)a = 9a$

(b) $3x + 4x + 5x = (3 + 4 + 5)x = 12x$

(c) $11n - 8n = (11 - 8)n = 3n$

(d) $5y + 7y - 4y + y = (5 + 7 - 4 + 1)y = 9y$

PROBLEM SET Simplify the following expressions.

See Example 29

1. $3x + 6x$
2. $7a + 8a$
3. $13t - 6t$
4. $14r - 2r$
5. $2s + 3s + 8s$
6. $6y + 8y + y$
7. $6x - 5x + 9x$
8. $p - 5p + 3p - 7p$
9. $9n - n - 5n + 3n$

ANSWERS 1. $9x$ 2. $15a$ 3. $7t$ 4. $12r$ 5. $13s$ 6. $15y$ 7. $10x$ 8. $-8p$ 9. $6n$

EXAMPLE 30 Simplify the expression $2x + 5 + 8$.

SOLUTION We combine the like terms, 5 and 8.

$$2x + 5 + 8 = 2x + 13$$ Combine terms

EXAMPLE 31 Simplify the expression $2x + 5x + 6$.

SOLUTION We use the distributive law to combine the like terms, $2x$ and $5x$.

$$2x + 5x + 6 = (2 + 5)x + 6$$ Distributive law ↓

$= 7x + 6$	Simplify

Often, we must group the like terms together first.

EXAMPLE 32 Simplify the expression $8x - 3 - 2x + 4 + 5x + 1$.

SOLUTION We first rewrite all the subtractions as additions of opposites (change the sign and add). Then we regroup: all the x-terms together, and all the numbers together. Finally, we simplify.

$8x - 3 - 2x + 4 + 5x + 1 = 8x + (-3) + (-2x) + 4 + 5x + 1$	Change signs and add
$= [8x + (-2x) + 5x] + [-3 + 4 + 1]$	Regroup
$= [(8 + (-2) + 5)x] + [-3 + 4 + 1]$	Distributive law
$= 11x + 2$	Simplify

Notice that we group all the x-terms together and all the single numbers together.

EXAMPLE 33 Simplify the expression $T = -4 - 3x + 6x + 2 + 4x - 10 - 5x$.

SOLUTION We first rewrite all the subtractions as additions of opposites; then we regroup and simplify.

$T = -4 + (-3x) + 6x + 2 + 4x + (-10) + (-5x)$	Change signs and add
$= [-3x + 6x + 4x + (-5x)] + [-4 + 2 + (-10)]$	Regroup
$= [(-3 + 6 + 4 + (-5))x] + [-4 + 2 + (-10)]$	Distributive law
$= 2x - 12$	Simplify

PROBLEM SET Simplify the following expressions.

See Examples 30 to 33

1. $7a + 5 + 3$
2. $8x - 4 + 6$
3. $6r + 3 - 2 - 8$
4. $-4x - 9 + 10 - 3$
5. $4p + 7 - 10 + 4$
6. $6y - 3 - 4 - 8$
7. $6x + 3x - 2$
8. $7x - 8x + 2$
9. $10a - 3a - 12$
10. $8u - 11u + 3$
11. $g - 3g + 8g - 5$
12. $6x - 7x - 9x + 7$
13. $7x + 3 + 2x + 7$
14. $4x + 2 + 3x + 7$
15. $8a - 7 - 2a - 3$
16. $4x - 2 - 9x + 3$
17. $4u - 10 - u + 3 - 5u$
18. $7r + 11 - 5r + 17 - 4r$
19. $3 - x - 7 + 4x - 12$
20. $7 - x - 3x - 4 + 8x - 3$

ANSWERS **1.** $7a + 8$ **2.** $8x + 2$ **3.** $6r - 7$ **4.** $-4x - 2$ **5.** $4p + 1$ **6.** $6y - 15$ **7.** $9x - 2$ **8.** $-x + 2$ **9.** $7a - 12$ **10.** $-3u + 3$ **11.** $6g - 5$ **12.** $-10x + 7$ **13.** $9x + 10$ **14.** $7x + 9$ **15.** $6a - 10$ **16.** $-5x + 1$ **17.** $-2u - 7$ **18.** $-2r + 28$ **19.** $3x - 16$ **20.** $4x$

Miscellaneous Exercises 6.4

Simplify the following expressions.

See Example 28

1. $7(a + b)$
2. $8(x - y)$
3. $5(p + q + 4)$
4. $10(n - m - 3)$

See Example 29

5. $3x + 8x$
6. $9a - 2a$
7. $2p + 8p + 7p$
8. $5t - 3t - 7t - t$

See Examples 30 and 31

9. $4m + 3 + 12$
10. $9t - 4 + 3$
11. $7y + 5 - 9 + 10$
12. $5r + 4r + 2$
13. $9t + t - 5t - 3$
14. $x + 2x + 3x + 4x + 5$

See Examples 32 and 33

15. $3z + 5 + 5z - 2$
16. $9u - 8 - u + 3$
17. $4t - 7t + 2 - t + 1 - 6t$
18. $10n - n - 5 + 2n + 3n + 12$

Now complete Post-Test 1 for Section 6.4 on page T91.

6.5 SOLVING EQUATIONS

SECTION OBJECTIVES After completing this unit, you will be able to solve simple equations.

TYPICAL PROBLEMS You will be able to work problems like the ones below.

Solve for the unknown letter.

(a) $n + 3 = 10$ **(b)** $x - 5 = 12$

(c) $6 \cdot t = 54$ **(d)** $\frac{u}{5} = 7$

(e) $7x + 3 = 11x - 5$ **(f)** $3x + 4x - 3 = 2x - 8$

(g) $2(x + 1) = 4x - 8$

ANSWERS **(a)** 7 **(b)** 17 **(c)** 9 **(d)** 35 **(e)** 2 **(f)** −1 **(g)** 5

The equation $2 + 3 = 5$ is true.

The equation $2 + 5 = 6$ is false.

How about the equation $2 + x = 7$? True or false? The answer depends on x: If $x = 5$, it is true; if $x \neq 5$, it is false. The values that make such equations true are called the **solutions** (or the **solution set**). We spend the rest of the chapter finding the solutions for equations.

We have already found solutions to equations in Chapters 1 through 4. To do this, we used the facts that addition and subtraction are opposite operations, and multiplication and division are opposite operations. Let us now restate this in a slightly more formal way.

> ***ADDITION PROPERTY OF EQUALITY*** Let A, B, and C be algebraic expressions. If $A = B$, then $A + C = B + C$.

In words, we can *add the same number to both sides of an equation, and the results are still equal.* Since subtracting and adding the opposite are the same, we can also *subtract the same number from both sides of an equation, and the results will still be equal.*

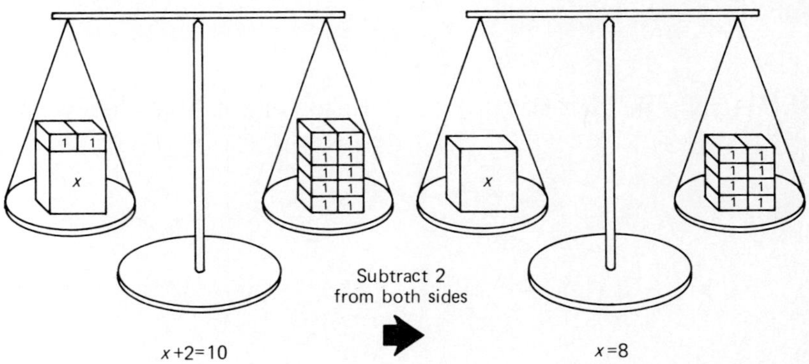

The illustration shows the equation $x + 2 = 10$ on the left. As a scale, it balances. If we remove (subtract) two units from both sides of the scales, it will still balance. But now we see that the unknown block, x, is 8.

EXAMPLE 34 Solve $x - 4 = 11$ for x.

SOLUTION Here we wish to get x alone. Since 4 is subtracted from x, we add 4 to both sides. This will produce $-4 + 4$ on the left, which will cancel, leaving x alone.

$x - 4 = 11$	Given
$x - 4 \boxed{+ 4} = 11 \boxed{+ 4}$	Add 4 to both sides
$x = 15$	Since $-4 + 4 = 0$

The solution is 15. We can check this: $15 - 4 = 11$.

EXAMPLE 35 Solve $a + 6 = 20$, for a.

SOLUTION Here, a is not alone. We can remove the 6 that is added to a by subtracting 6 from both sides.

$a + 6 = 20$	Given
$a + 6 \boxed{- 6} = 20 \boxed{- 6}$	Subtract 6 from both sides
$a = 14$	Since $6 - 6 = 0$

The solution is 14. We can check this: $14 + 6 = 20$.

EXAMPLE 36 Solve $5x - 4x - 7 = 12$.

SOLUTION We first simplify the left-side expression; then we solve.

$5x - 4x - 7 = 12$	Given
$x - 7 = 12$	Combine terms
$x - 7 \boxed{+ 7} = 12 \boxed{+ 7}$	Add 7 to both sides
$x = 19$	Since $-7 + 7 = 0$

The solution is 19.

EXAMPLE 37 Solve $8t - 3 = 7t + 5$.

SOLUTION We use the addition property to move all the t-terms to the left side of the equation, and all the number terms to the right side.

$8t - 3 = 7t + 5$	Given
$8t - 3 \boxed{-7t} = 7t + 5 \boxed{-7t}$	Subtract $7t$ from both sides
$t - 3 = 5$	Simplify
$t - 3 \boxed{+3} = 5 \boxed{+3}$	Add 3 to both sides
$t = 8$	Simplify

The solution is $t = 8$. We can check this by substituting $t = 8$ into both sides of the original equation.

Left side $= 8(8) - 3 = 64 - 3 = 61$

Right side $= 7(8) + 5 = 56 + 5 = 61$ *Checks.*

PROBLEM SET Solve the following equations using the addition property of equality.

See Examples 34 to 37

1. $x - 3 = 12$
2. $a - 6 = 10$
3. $t - 8 = 2$
4. $x - 4 = -8$
5. $y - 2 = -1$
6. $u - 10 = 21$
7. $x + 7 = 11$
8. $a + 2 = 5$
9. $r + 8 = 5$
10. $y + 7 = 1$
11. $x + 12 = -2$
12. $t + 4 = -6$
13. $3x - 2x - 7 = 2$
14. $8x - 7x + 2 = 5$
15. $10t + 3 - 9t = 6$
16. $4y - 5 - 3y = 7$
17. $16a - 2 - 15a = -4$
18. $2a + 5 - a = -6$
19. $5x = 4x + 6$
20. $9a = 8 + 8a$
21. $10t - 1 = 9t + 8$
22. $2y - 5 = y - 8$
23. $6u + 5 = 5u - 3$
24. $11x + 7 = 10x - 18$

ANSWERS 1. 15 2. 16 3. 10 4. −4 5. 1 6. 31 7. 4 8. 3 9. −3 10. −6 11. −14 12. −10 13. 9 14. 3 15. 3 16. 12 17. −2 18. −11 19. 6 20. 8 21. 9 22. −3 23. −8 24. −25

Just as we can add the same number to both sides of an equation, we can *multiply both sides of an equation by the same nonzero number, and the results will still be equal.*

> ***MULTIPLICATION PROPERTY OF EQUALITY*** Let A, B, and C be algebraic expressions ($C \neq 0$). If $A = B$, then $AC = BC$.

EXAMPLE 38 Solve $6x = 18$.

SOLUTION To get the x alone, we multiply both sides by $\frac{1}{6}$ to remove the 6 $\left[\text{since } \frac{1}{6}(6) = 1\right]$.

$6x = 18$	Given
$\boxed{\frac{1}{6}}(6x) = \boxed{\frac{1}{6}}(18)$	Multiply both sides by $\frac{1}{6}$
$x = 3$	Simplify

In Example 38, it is reasonable to have divided both sides by 6. Since multiplying both sides by $\frac{1}{C}$ is the same as dividing by C, we can say that we can *divide both sides of an equation by the same nonzero number, and the results will also be equal.*

EXAMPLE 39 Solve $-4a = 100$.

SOLUTION Instead of multiplying both sides by $\frac{-1}{4}$, we divide both sides by -4. This gives the same result.

$-4a = 100$	Given
$\frac{-4a}{\boxed{-4}} = \frac{100}{\boxed{-4}}$	Divide both sides by -4
$a = -25$	Simplify

EXAMPLE 40 Solve $\frac{k}{5} = 7$.

SOLUTION Here we multiply both sides by 5 to get k alone.

$\frac{k}{5} = 7$	Given
$\boxed{5}\left(\frac{k}{5}\right) = \boxed{5}(7)$	↓ Multiply both sides by 5
$k = 35$	↓ Simplify

EXAMPLE 41 Solve $-x = 12$.

SOLUTION This almost looks like an answer, but it is not. We want x, *not* $-x$. Think of $-x$ as $-1(x)$, so that multiplying by -1 will produce x.

$-x = 12$	Given
$\boxed{-1}(-x) = \boxed{-1}(12)$	↓ Multiply both sides by -1
$x = -12$	↓ Simplify

PROBLEM SET Solve the following equations using the multiplication property of equality.

See Examples 38 to 41

1. $2x = 10$
2. $3a = 15$
3. $5x = 20$
4. $7t = 21$
5. $-6a = 42$
6. $-10x = -80$
7. $3k = -24$
8. $-4m = 16$
9. $-6x = -6$
10. $3y = -12$
11. $\frac{x}{4} = 3$
12. $\frac{a}{6} = 5$
13. $\frac{r}{5} = 12$
14. $\frac{m}{10} = -2$
15. $\frac{x}{-2} = 8$
16. $\frac{z}{4} = -5$
17. $\frac{k}{-3} = -2$
18. $\frac{x}{-6} = -1$
19. $-x = 4$
20. $-a = 2$
21. $-t = -7$
22. $-x = -8$

23. $-m = \frac{3}{5}$ 24. $-k = \frac{-4}{3}$

ANSWERS 1. 5 2. 5 3. 4 4. 3 5. -7 6. 8 7. -8 8. -4
9. 1 10. -4 11. 12 12. 30 13. 60 14. -20 15. -16
16. -20 17. 6 18. 6 19. -4 20. -2 21. 7 22. 8
23. $-\frac{3}{5}$ 24. $\frac{4}{3}$

We can put these properties all together to produce a general procedure for solving equations of this type.

> To solve simple equations:
>
> 1. Use the distributive law to remove all parentheses and brackets.
> 2. Simplify the left and right sides of the equation by combining like terms.
> 3. Move all terms with the unknown to one side of the equation.
> 4. Move all of the number terms to the other side of the equation.
> 5. Multiply or divide to get the unknown by itself.
> 6. Check by substituting the solution into the original equation.
>
> (If a step is not needed, skip it.)

EXAMPLE 42 Solve $8x + 2x - 3 = 7$.

SOLUTION We first simplify the expression on the left; then we get x alone and solve.

$$8x + 2x - 3 = 7$$

$$10x - 3 = 7$$

$$10x - 3 \boxed{+3} = 7 \boxed{+3}$$

$$10x = 10$$

$$\frac{10x}{\boxed{10}} = \frac{10}{\boxed{10}}$$

$$x = 1$$

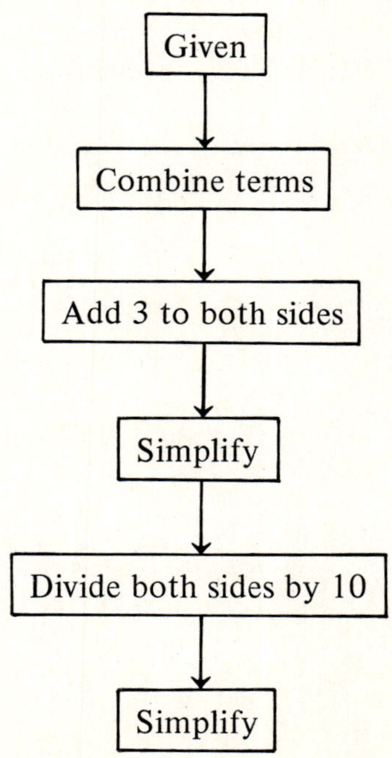

EXAMPLE 43 Solve $8y + 1 = 5y - 5$.

SOLUTION We use the addition and subtraction property to get all the y-terms together on the left and all the numbers on the right.

$8y + 1 = 5y - 5$	Given
$8y + 1 \boxed{-5y} = 5y - 5 \boxed{-5y}$	Subtract $5y$ from both sides
$3y + 1 = -5$	Simplify each side
$3y + 1 \boxed{-1} = -5 \boxed{-1}$	Subtract 1 from both sides
$3y = -6$	Simplify
$\dfrac{3y}{\boxed{3}} = \dfrac{-6}{\boxed{3}}$	Divide both sides by 3
$y = -2$	Simplify

Let us check this answer by substituting $y = -2$ into the right and left sides of the original equation. If they are the same, $y = -2$ is our solution.

$$\textit{Left side} = 8(-2) + 1 = -16 + 1 = -15$$
$$\textit{Right side} = 5(-2) - 5 = -10 - 5 = -15$$

The left side equals the right side; thus the solution is $y = -2$. *Checks.*

EXAMPLE 44 Solve $2(x + 5) = 5x + 16$.

SOLUTION We first remove the parentheses. Then we get the x-terms on the left and the number terms on the right.

$2(x + 5) = 5x + 16$	Given
$2x + 10 = 5x + 16$	Distributive law
$2x + 10 \boxed{-5x} = 5x + 16 \boxed{-5x}$	Subtract $5x$ from both sides

$-3x + 10 = 16$	Simplify
$-3x + 10 \boxed{-10} = 16 \boxed{-10}$	Subtract 10 from both sides
$-3x = 6$	Simplify
$\dfrac{-3x}{\boxed{-3}} = \dfrac{6}{\boxed{-3}}$	Divide both sides by -3
$x = -2$	Simplify

We can check this by substituting.

Left side $= 2(-2 + 5) = 2(3) = 6$

Right side $= 5(-2) + 16 = -10 + 16 = 6$ *Checks.*

PROBLEM SET Solve the following equations and check the solutions.

See Examples 42 to 44

1.	$4x + 2x - 1 = 5$	2.	$3a + 2a + 4 = 14$
3.	$2x + 5 + 6x = -3$	4.	$10k - 4 - 3k = 10$
5.	$3m - 5m + 2 = 8$	6.	$z - 6z - 3 = 2$
7.	$8x - 1 = 3x + 9$	8.	$4a - 7 = a + 8$
9.	$4t - 6 = t - 3$	10.	$3z - 8 = 5z - 2$
11.	$6x - 5 = 4x + 3$	12.	$10k + 3 = 7k - 9$
13.	$4x + 7 = 9x + 2$	14.	$6a - 4 = 9a + 8$
15.	$2(x + 5) = 7x - 5$	16.	$6(x - 3) = 4x + 2$
17.	$6(2a - 1) = 10a + 4$	18.	$4(3k - 5) = 7k - 35$
19.	$3(5x - 2) = 11x + 34$	20.	$10(3 - m) = -4m - 6$
21.	$5(6 - y) = -12y + 2$	22.	$4(6 - 2x) = 12x + 4$

ANSWERS 1. 1 2. 2 3. −1 4. 2 5. −3 6. −1 7. 2 8. 5
9. 1 10. −3 11. 4 12. −4 13. 1 14. −4 15. 3
16. 10 17. 5 18. −3 19. 10 20. 6 21. −4 22. 1

Miscellaneous Exercises 6.5

Solve the following equations.

See Examples 34 and 35

1.	$x - 5 = 12$	2.	$a - 4 = 20$
3.	$t + 7 = 9$	4.	$z + 20 = 14$

See Examples 36 and 37

5. $6y - 5y - 2 = 7$
6. $10a - 9a + 6 = 8$
7. $4r - 3r + 3 = 8$
8. $5x - 8 - 4x = -1$

See Examples 38 to 41

9. $4t = 36$
10. $3x = -30$
11. $-5z = 25$
12. $-2w = -12$
13. $\frac{n}{6} = 3$
14. $\frac{k}{-2} = -4$
15. $-a = 5$
16. $-x = -9$

See Examples 42 to 44

17. $5x + 2x - 3 = 11$
18. $6a - a + 3 = 18$
19. $5x - 7 = 2x + 5$
20. $9a + 7 = 5a - 5$
21. $3(z - 2) = z + 6$
22. $7(n + 5) = -10n + 1$

Business applications

23. The relation between price (P) and demand (Q) for a certain item is given by

$$Q = 500(10 - P)$$

Find P if $Q = 2000$.

24. In Illinois, the income tax (T) that a family pays is given by

$$T = 0.025(I - 1000E)$$

where I is income and E is the number of exemptions. Find I if $T = \$300$ and $E = 3$.

25. The formula for profit (P) is given by

$$P = R - C$$

where R is the revenue and C is the cost. Find R if $P = \$20{,}000$ and $C = \$100{,}000$.

26. The average cost (A) to produce Q items is given by

$$A = \frac{C}{Q}$$

where C is the total cost. Find the total cost C if $A = \$10$ and $Q = 500$ items.

27. The relation between the demand (Q) for an item and its price (P) is given by

$$Q + 1000P = 6000$$

Find P if $Q = 2000$ items.

Health application

28. Young's formula for a child's dosage (C) of medicine is given by

$$C = \left(\frac{A}{A + 12}\right) D$$

where A is the child's age and D is the adult dosage. Find D if $C = 50$ and $A = 6$ years.

Consumer application

29. The cost (C) to operate a certain automobile for one year is given by

$$C = 1200 + 0.08m$$

where m is the number of miles driven in one year. Find m if $C = \$2000$.

Science applications

30. The relation between Fahrenheit (F) temperature and Celsius (C) is given by

$$F = \frac{9}{5}C + 32$$

Find C if $F = 68°$.

31. The relation between Fahrenheit (F) and Celsius (C) can also be written

$$C = \frac{5}{9}(F - 32)$$

Find F if $C = 30°$.

Now complete Post-Test 1 for Section 6.5 on page T93.

6.6 SOLVING PROBLEMS

SECTION OBJECTIVE After completing this unit, you will be able to solve stated problems.

TYPICAL PROBLEMS You will be able to work problems like the ones below.

(a) One number is 5 more than another number. Their sum is 39. Find the numbers.

(b) The sum of two consecutive integers is 47. Find the numbers.

(c) In a recent election, Klein received 130 more votes than Peterson. If 1000 total votes were cast for the two candidates, how many votes did each get?

(d) The perimeter of a square is 76. What is the length of a side?

ANSWERS **(a)** 17 and 22 **(b)** 23 and 24 **(c)** Klein: 565; Peterson: 435 **(d)** Length = 19

In Sections 4.3 and 4.4, we saw how to solve certain percentage and proportion problems by translating an English statement into a solvable equation. Many algebra problems start as word problems. These are English sentences that can be translated into algebraic symbols and equations. Table 1 is a table of **translations** that we can use in these word problems.

To solve word problems:

1. Read (and reread) the problem carefully.
2. Identify the unknown (or unknowns) by a letter (or letters).
3. Translate the English information into an algebraic equation.
4. Solve the equation.
5. Check the solution.

Table 1

English	Algebra
plus, more than, added to, increased by, sum of, total of	$+$
minus, less than, decreased by, subtracted from, difference of, fewer	$-$
times, of, product of, multiplied by	$\cdot$
divided by, per, ratio of, quotient of	$\div$
is, are, was, should be, represents	$=$
what, some number	x
consecutive integers	$x, x + 1, x + 2$
consecutive odd (or even) integers	$x, x + 2, x + 4$

EXAMPLE 45 One number is 3 more than another number. Their sum is 41. Find the numbers.

SOLUTION We have two numbers here. We can call one of them x. Since the other is 3 *more than* x, it is $x + 3$.

$$\begin{cases} \text{Let } x = \text{one number} \\ x + 3 = \text{other number} \end{cases}$$

Their sum is 41

$$\underbrace{x + (x + 3)} = 41$$

$$2x + 3 = 41$$

$$2x + 3 \boxed{-3} = 41 \boxed{-3}$$

$$2x = 38$$

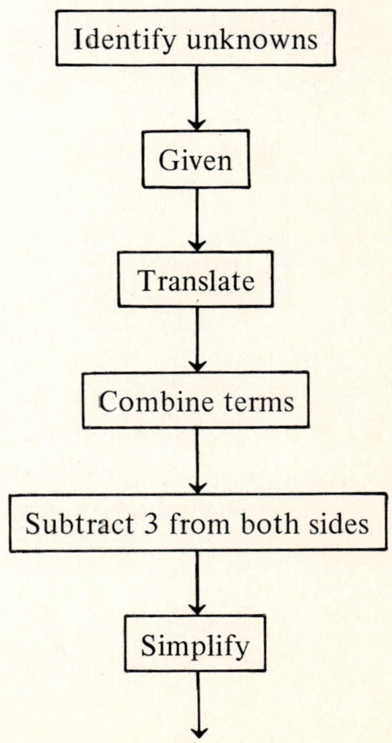

$\dfrac{2x}{\boxed{2}} = \dfrac{38}{\boxed{2}}$	Divide both sides by 2
$x = 19$	Simplify
$x + 3 = 22$	Find other unknown

The numbers are 19 and 22. We can check: 22 is 3 more than 19, and their sum is 41.

EXAMPLE 46 The sum of two consecutive even integers is 70. Find the numbers.

SOLUTION We have two numbers. We call one of them x. Since they are consecutive even numbers, they are two apart. Thus the other number is $x + 2$.

$\begin{cases} \text{Let } x = \text{one number} \\ x + 2 = \text{next even number} \end{cases}$	Identify unknowns
Sum is 70	Given
$\underbrace{x + (x + 2)} = 70$	Translate
$2x + 2 = 70$	Simplify
$2x + 2 \boxed{-\ 2} = 70 \boxed{-\ 2}$	Subtract 2 from both sides
$2x = 68$	Simplify
$\dfrac{2x}{\boxed{2}} = \dfrac{68}{\boxed{2}}$	Divide both sides by 2
$x = 34$	Simplify
$x + 2 = 36$	Find other number

The numbers are 34 and 36. We can see that their sum is 70.

PROBLEM SET

See Examples 45 and 46

1. One number is 8 more than another number. If their sum is 42, find the numbers.
2. One number is 12 less than another number. If their sum is 90, find the numbers.
3. One number is 5 times another number. If their difference is 28, find the numbers.
4. One number is one-third of another number. If their sum is 48, find the numbers.
5. The sum of two consecutive integers is 57. Find the numbers.
6. The sum of two consecutive odd integers is 76. Find the numbers.

ANSWERS 1. 17 and 25 2. 39 and 51 3. 35 and 7 4. 12 and 36 5. 28 and 29 6. 37 and 39

EXAMPLE 47 Larry and Nancy earn \$23,300 together. Larry earns \$900 more than Nancy. How much does each of them earn?

SOLUTION If Nancy's salary is x, then Larry's is 900 more, or $x + 900$. The total is 23,300.

$$\begin{cases} \text{Let } x = \text{Nancy's salary} \\ x + 900 = \text{Larry's salary} \end{cases}$$

Total is 23,300

$$\underbrace{x + (x + 900)} = 23{,}300$$

$$2x + 900 = 23{,}300$$

$$2x + 900 \boxed{-\ 900} = 23{,}300 \boxed{-\ 900}$$

$$2x = 22{,}400$$

$$\frac{2x}{\boxed{2}} = \frac{22{,}400}{\boxed{2}}$$

$$x = 11{,}200$$

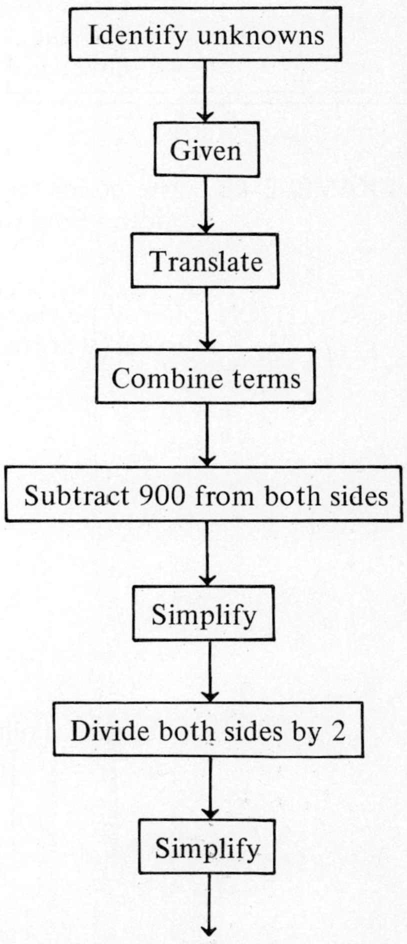

$$x + 900 = 12{,}100$$

Find other unknown

Therefore, Nancy's salary is \$11,200, and Larry's is \$12,100. Together, this checks as a \$23,300 total.

PROBLEM SET

See Example 47

1. In a certain condominium, there are twice as many two-bedroom units as there are three-bedroom units. There are a total of 147 units. How many of each are there?
2. In another condominium, there are the same number of one- and two-bedroom units, and 40 fewer three-bedroom units. If there are 350 total units, how many of each are there?
3. Bob has 1200 feet of recording tape. He cuts it into two pieces: one is 150 feet longer than the other. How long is each piece?
4. The front four of the Tigers' defense is made of four very heavy players. Jones is 10 pounds heavier than Smith, who is 5 pounds heavier than Wilson, who is 10 pounds heavier than Green. Together, they weigh 1010 pounds. How much does each of them weigh?

ANSWERS

1. Two-bedroom: 98; three-bedroom: 49
2. One-bedroom: 130; two-bedroom: 130; three-bedroom: 90
3. 525 and 675
4. Green: 240; Wilson: 250; Smith: 255; Jones: 265

EXAMPLE 48 The perimeter of a rectangle is 26. The length is one more than three times the width. Find the dimensions.

SOLUTION Let W be the width. (It is usually easier to start with W as the smaller number.) Recall that the perimeter of a rectangle is $P = 2L + 2W$.

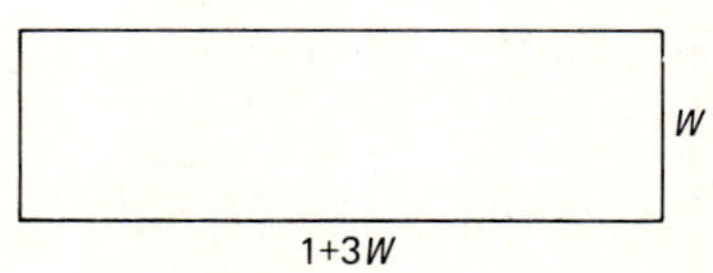

Let W = width

Identify unknown

Length is one more than three times width

$$L = 1 + 3 \cdot W$$

Given

Translate

Perimeter of the rectangle is 26	Given
$2L + 2W = 26$	Translate
$2(1 + 3W) + 2W = 26$	Substitute $L = 1 + 3W$
$2 + 6W + 2W = 26$	Distributive law
$2 + 8W = 26$	Combine terms
$8W = 24$	Subtract 2 from both sides
$W = 3$	Divide both sides by 8
$L = 1 + 3W = 10$	Find other dimension

The dimensions are 3 and 10. The perimeter is $2(3) + 2(10) = 26$.

PROBLEM SET

See Example 48

1. The perimeter of a square is 92. Find the length of a side.
2. The lengths of the sides of a triangle are three consecutive odd integers. The perimeter is 87. Find the length of each side.
3. The width of a rectangle is 6 less than the length. The perimeter is 56. Find the length and width.
4. The length of a rectangle is 2 less than five times the width. The perimeter is 56. Find the length and the width.

ANSWERS 1. 23 2. 27, 29, and 31 3. Width: 11; length: 17
4. Width: 5; length: 23

Miscellaneous Exercises 6.6

See Examples 45 and 46

1. One number is 11 more than another number. Their sum is 81. Find the numbers.
2. One number is 7 less than another number. Their sum is 75. Find the numbers.

See Example 47

3. One number is 6 times another number. Their sum is 56. Find the numbers.
4. The sum of two consecutive integers is 105. Find the numbers.
5. The sum of two consecutive even integers is 98. Find the numbers.
6. Becky is three years older than Lisa. The sum of their ages is 59. How old is each woman?
7. One movie runs 10 minutes less than twice the time of another movie. The total time is 290 minutes. What is the length of each movie?
8. Ann, Bob, and Chuck form a partnership. Ann puts up \$3000 more than Bob, who puts up \$4000 more than Chuck. Their total assets are \$38,000. How much did each put up?

See Example 48

9. The perimeter of a certain square is 108. What is the length of each side?
10. The length of a certain rectangle is 1 less than 4 times the width. The perimeter is 68. Find the length and width.

Now complete Post-Test 1 for Section 6.6 on page T95.

CHAPTER 6 SUMMARY

Important Words and Phrases

absolute value (6.1)
addition of real numbers (6.2)
addition property of equality (6.5)
additive inverse (6.1)
combining like terms (6.4)
coordinate (6.1)
distributive law (6.4)
division of real numbers (6.3)
exponents (6.1)
grouping symbols (6.1)
like terms (6.4)
multiplication of real numbers (6.3)
multiplication property of equality (6.5)
negative real numbers (6.1)
number line (6.1)
opposite (6.1)
order of operation (6.1)
origin (6.1)
PEMDAS (6.1)
positive real numbers (6.1)
real numbers (6.1)
solution (6.5)
standard order (6.1)
subtraction of real numbers (6.2)
term (6.4)
translation (6.6)

Important Ideas

(6.1) Knowing the standard order of operation.
Using the number line.
Finding the opposite of a number.
Finding the absolute value of a number.

(6.2) Adding real numbers.
Subtracting real numbers.

(6.3) Multiplying real numbers.
Dividing real numbers.
Simplifying expressions with more than one operation.

(6.4) Using the distributive law.
Simplifying expressions with like terms.

(6.5) Solving simple equations.

(6.6) Solving stated problems.

CHAPTER 6 REVIEW TEST

Choose the correct answer.

1. Simplify $\dfrac{11+5}{5-1}$.
   (a) 5 (b) 4 (c) $\dfrac{8}{3}$ (d) 2 (e) 10

2. What is the coordinate of point D?

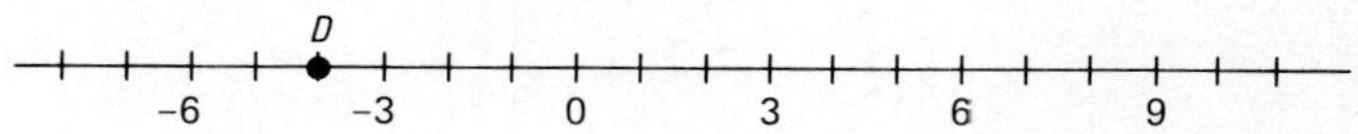

   (a) −2 (b) −1 (c) −3 (d) 4 (e) −4

3. What is $-(-9)$?
   (a) 9 (b) −9 (c) ±9 (d) −3 (e) 3

4. Simplify $-4+3$.
   (a) −7 (b) 1 (c) −1 (d) 7 (e) −12

5. Simplify $-2+(-3+4)+7$.
   (a) −16 (b) 16 (c) 4 (d) 6 (e) 8

6. Simplify $-4-(-1)$.
   (a) 3 (b) 5 (c) −3 (d) −5 (e) −4

7. Simplify $-3-6-(-1)+(-4)+5$.
   (a) −7 (b) −9 (c) −1 (d) −3 (e) −13

8. Simplify $(-3)(-7)$.
   (a) 10 (b) −10 (c) −4 (d) −21 (e) 21

9. Simplify $3+(-2)(5)$.
   (a) 13 (b) −7 (c) 5 (d) −5 (e) 0

10. Simplify $\dfrac{-20}{-4}$.
    (a) −16 (b) −24 (c) −5 (d) 5 (e) 24

11. Simplify $\dfrac{(-1)(-2)(-3)(-5)}{(-1)(-10)}$.
    (a) −3 (b) 3 (c) −6 (d) 6 (e) 0

12. Simplify $5(x-3)$.
    (a) $5x-5$ (b) $5x-3$ (c) $3x-5$ (d) $5x+15$ (e) $5x-15$

13. Simplify $4t-5+8t+3$.
    (a) $12t-2$ (b) $12t-8$ (c) $12t+8$ (d) $4t+2$ (e) $-4t+2$

14. Simplify $3(x-1)+2(x+3)-1$.
    (a) $5x+1$ (b) $5x+4$ (c) $5x+2$ (d) $x+1$ (e) $x+4$

15. Solve $-3x=30$.
    (a) −33 (b) 33 (c) 27 (d) −10 (e) 10

16. Solve $4t+3t+2=16$.
    (a) 3 (b) 4 (c) 2 (d) 5 (e) 1

17. Solve $5(a-1)=2a+7$.
    (a) 4 (b) 3 (c) 2 (d) −4 (e) −2

18. The sum of two consecutive integers is 53. What is the *smaller* number?
(a) 25 **(b)** 28 **(c)** 27 **(d)** 29 **(e)** 26

19. Ralph is 3 years older than Alice. If their total ages are 95 years, what is *Ralph's* age?
(a) 47 **(b)** 49 **(c)** 48 **(d)** 46 **(e)** 50

20. The length of a rectangle is 10 more than twice the width. The perimeter is 68. What is the *length?*
(a) 24 **(b)** 28 **(c)** 22 **(d)** 26 **(e)** 30

Now complete Post-Test 1 for Chapter 6 on page T97.

7

THE HAND CALCULATOR

Before you begin this chapter, complete the Chapter 7 Pre-Test on page T101.

7.1 ELEMENTARY HAND CALCULATOR OPERATIONS

SECTION OBJECTIVES

After completing this unit, you will be able to:

1. Perform elementary operations on a hand calculator.
2. Use a hand calculator to help work application problems like the ones seen earlier in the text.

TYPICAL PROBLEMS

You will be able to work problems like the ones below.

Use a hand calculator to work the following problems.

(a) 37.1 + 8.07 + 5.9 **(b)** 2000 − 73.15

(c) 1.07 × 42.9 × 17.3 **(d)** 42,925 ÷ 0.72

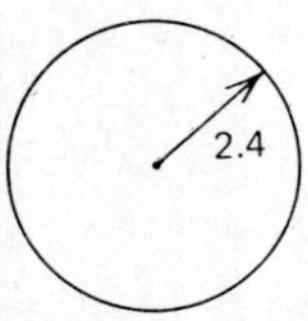

(e) Write $\frac{2}{11}$ as a three-place decimal.

(f) Seventeen is what percent of 53?

(g) Compute the area of the circle at the left.

ANSWERS **(a)** 51.07 **(b)** 1926.85 **(c)** 794.1219 **(d)** 59,618.056 **(e)** 0.182 **(f)** 32% **(g)** 18.0864

Now that you understand the basic arithmetic operations and how to apply them, you are ready to learn how to use the hand calculator to its fullest advantage.

Figure 7.1 shows a sketch of a typical hand calculator that can be purchased for \$6 to \$16. In this section, we discuss the following keys and their uses.

1. [C]. This key *clears* the entire display and sets the display to 0. It is best to clear the machine before every problem.
2. [CE]. This key (*clear entry*) erases a mistaken entry but not the other numbers in the problem.
3. [0], [1], [2], [3], [4], [5], [6], [7], [8], [9]. These are the usual *digits.*
4. [·]. This is the *decimal point*.

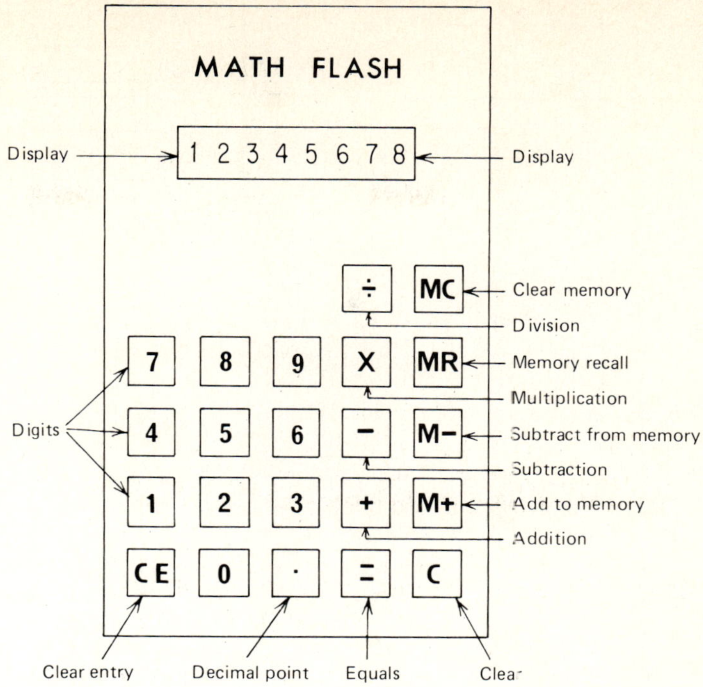

Figure 7.1

Typical Hand Calculator

5. ÷, ×, +, −. These are the usual *arithmetic operations:* division, multiplication, addition, and subtraction.
6. =. The *equal sign* is usually used to give a final answer.

As we go through our examples, we show the following three columns.

1. PUNCH: These are the keys that you actually punch on the calculator.
2. DISPLAY: This shows you what is actually being displayed on the machine after the last punch.
3. MEANING: In this column, we comment on each step.

EXAMPLE 1 Display 52.079.

SOLUTION We simply punch the digits and decimal point just as they appear.

PUNCH	DISPLAY	MEANING
C	0.	Clear
5 2 · 0 7 9	52.079	Enter 52.079

EXAMPLE 2 Add 3 + 5.

SOLUTION This is a simple warm-up problem.

PUNCH	DISPLAY	MEANING
[C]	0.	Clear
[3]	3.	3
[+][5]	5.	Plus 5
[=]	8.	Answer

Notice that we simply press the keys in the usual order: [3][+][5][=], and the answer appears as 8.

EXAMPLE 3 Add 3.19 + 0.7 + 15.258.

SOLUTION We enter and add each of these decimals.

PUNCH	DISPLAY	MEANING
[C]	0.	Clear
[3][·][1][9]	3.19	3.19
[+][·][7]	0.7	Plus 0.7
[+][1][5][·][2][5][8]	15.258	Plus 15.258
[=]	19.148	Answer

EXAMPLE 4 Subtract 2000 - 731.37.

SOLUTION We now use the [-] key.

PUNCH	DISPLAY	MEANING
[C]	0.	Clear
[2][0][0][0]	2000.	2000
[-][7][3][1][·][3][7]	731.37	Minus 731.37
[=]	1268.63	Answer

EXAMPLE 5 Use a hand calculator to balance the following checkbook:

June 1	Balance from May	$ 214.79
June 3	Best Foods	-45.23
June 5	Guzzle Gas Co.	-24.71
June 6	Deposit	+371.42
June 8	Goniff Motors	-103.48
June 10	Shmatte Clothes Co.	-38.91
	Balance	?

SOLUTION We enter the balance from May and subtract the checks and add the deposits.

PUNCH	DISPLAY	MEANING
[C]	0.	Clear
[2][1][4][·][7][9]	214.79	Enter balance
[-][4][5][·][2][3]	45.23	Subtract food bill
[-][2][4][·][7][1]	24.71	Subtract gas bill
[+][3][7][1][·][4][2]	371.42	Add deposit
[-][1][0][3][·][4][8]	103.48	Subtract car payment
[-][3][8][·][9][1]	38.91	Subtract clothes bill
[=]	373.88	Balance

EXAMPLE 6 Suppose that you hit the wrong key halfway through a problem. Do we have to start all over? No, we use the [CE] key. For example, add 17.2 + 18.05 + 2.5 + 8.3.

PUNCH	DISPLAY	MEANING
[C]	0.	Clear
[1][7][·][2]	17.2	Enter 17.2
[+][1][8][·][0][5]	18.05	Plus 18.05
[+][2][·][4]	2.4	Oops! Should be 2.5
[CE]	0.	Erase 2.4
[2][·][5]	2.5	Enter 2.5
[+][8][·][3]	8.3	Plus 8.3
[=]	46.05	Answer

Notice that we did not have to start over. After the mistake, we punched [CE] to erase the mistake and then entered the correct number. (*Note:* On some calculators, the clear-entry operation is done by punching a [CE/C] key once; to clear, you punch [CE/C] twice.)

PROBLEM SET Use a hand calculator to perform the following additions and subtractions.

See Examples 1 to 6

1. 2 + 5
2. 3 + 7
3. 8 - 5
4. 12 - 7
5. 2.1 + 8.4
6. 1.9 - 1.7
7. 41,782 + 9,872 + 30,189 + 19,727
8. 2.5 + 8.41 + 6.536 + 12.03
9. 4.02 + 10.59 + 17.82 + 8.41
10. 7182 - 5285
11. 4.237 - 1.5893
12. 0.00372 - 0.000977

13.		14.		15.	
	478.33		6000		82.071
	+ 25.75		− 751.25		− 15.6
	− 447.91		− 821.39		+ 23.43
	− 15.23		− 1266.53		− 80.39

ANSWERS **1.** 7 **2.** 10 **3.** 3 **4.** 5 **5.** 10.5 **6.** 0.2 **7.** 101,570 **8.** 29.476 **9.** 40.84 **10.** 1897 **11.** 2.6477 **12.** 0.002743 **13.** 40.94 **14.** 3160.83 **15.** 9.511

Multiplication and division are just as easy as addition and subtraction. When dividing, however, we often get more decimal places than we need; in these cases, we round the numbers off (review Section 3.1).

EXAMPLE 7 Multiply 6 × 7.

SOLUTION This is a short problem to warm up.

PUNCH	DISPLAY	MEANING
[C]	0.	Clear
[6]	6.	6
[×] [7]	7.	Times 7
[=]	42.	Answer

EXAMPLE 8 Multiply 259 × 82.07 × 0.0021.

SOLUTION We use the [×] key to multiply these numbers.

PUNCH	DISPLAY	MEANING
[C]	0.	Clear
[2] [5] [9]	259.	Enter 259
[×] [8] [2] [·] [0] [7]	82.07	Times 82.07
[×] [·] [0] [0] [2] [1]	0.0021	Times 0.0021
[=]	44.637873	Answer

EXAMPLE 9 Divide 731 ÷ 19.2.

SOLUTION We use the [÷] key. Also, we will round off the final answer.

PUNCH	DISPLAY	MEANING
[C]	0.	Clear
[7] [3] [1]	731.	Enter 731
[÷] [1] [9] [·] [2]	19.2	Divided by 19.2
[=]	38.072916	Answer

Since the original numbers, 731 and 19.2, do not have much decimal-place accuracy, it is not reasonable to expect the answer to have six places of accuracy. Therefore, we round the answer to the nearest tenth (one decimal place).

38.072916	Determine place to be rounded
↑	↓
38.0[7]2916	Look one place right
↑	↓
38.1	Round up since 7 is 5 or more

Now we say that the answer is 38.1.

EXAMPLE 10 Write $\frac{3}{17}$ as a three-place decimal.

SOLUTION Recall that $\frac{3}{17}$ means $3 \div 17$. Therefore, we can use the [÷] key to convert fractions to decimals. We then round off our machine answer to three decimal places.

PUNCH	DISPLAY	MEANING
[C]	0.	Clear
[3]	3.	Numerator
[÷] [1] [7]	17.	Divided by denominator
[=]	0.1764705	The fraction $\frac{3}{17}$

We now round this off to three places.

0.1764705	Determine place to be rounded
↑	↓
0.176[4]705	Look one place right
↑	↓
0.176	Round down, since 4 is less than 5

Sometimes, we have to simplify a fraction such as

$$\frac{(7.3)\,(18.2)}{(9.05)\,(0.21)\,(3.41)}$$

that only has multiplications in the numerator and denominator. We have a simple trick for computing these fractions on a hand calculator.

> To simplify a fraction containing only multiplications (no additions or subtractions):
>
> 1. Multiply all terms in the numerator.
> 2. Divide by each term in the denominator.

As a formula this might be written as follows.

$$\frac{A \cdot B \cdot C}{D \cdot E \cdot F} = A \cdot B \cdot C \div D \div E \div F$$

EXAMPLE 11 Simplify $\dfrac{(7.3)(18.2)}{(9.05)(0.21)(3.41)}$.

SOLUTION We use the procedure stated above.

PUNCH	DISPLAY	MEANING
[C]	0.	Clear
[7] [·] [3]	7.3	Multiply numerator terms
[×] [1] [8] [·] [2]	18.2	
[÷] [9] [·] [0] [5]	9.05	Divide by each denominator term
[÷] [·] [2] [1]	0.21	
[÷] [3] [·] [4] [1]	3.41	
[=]	20.500854	Answer

As in the preceding problems, we round the answer off to 20.5.

PROBLEM SET Use a hand calculator to perform the following operations.

See Examples 7 to 9

1. 4 × 9
2. 8 × 11
3. 32 ÷ 4
4. 60 ÷ 12
5. 1.2 × 7
6. 8 ÷ 0.2
7. 62.7 × 4.92
8. 8.07 × 5400
9. 0.031 × 87.3
10. 0.041 × 0.009 × 79.7
11. 1.065 × 1.065 × 1.065
12. 17.3 ÷ 5.1
13. 60,000 ÷ 13
14. 25,000 ÷ 38.75
15. 0.731 ÷ 0.025
16. $2.1\overline{)75.33}$
17. $0.53\overline{)0.00259}$

Use a hand calculator to write each of the following fractions as three-place decimals.

See Examples 10 and 11

18. $\frac{2}{9}$

19. $\frac{3}{11}$

20. $\frac{52}{219}$

21. $\frac{(42)(81)}{102}$

22. $\frac{(7.3)(8.5)}{(6.1)(2.54)}$

23. $\frac{(1.79)(253)(415)}{(6.3)(0.23)}$

ANSWERS **1.** 36 **2.** 88 **3.** 8 **4.** 5 **5.** 8.4 **6.** 40 **7.** 308.484 **8.** 43,578 **9.** 2.7063 **10.** 0.0294093 **11.** 1.2079496 **12.** 3.3921568 **13.** 4615.3846 **14.** 645.16129 **15.** 29.24 **16.** 35.871428 **17.** 0.0048867 **18.** 0.222 **19.** 0.273 **20.** 0.237 **21.** 33.353 **22.** 4.005 **23.** 129,703.96

We are now going to work a few application problems using the hand calculator to do the arithmetic. (*Remember:* You still have to set up the problem; the calculator does only the arithmetic.)

EXAMPLE 12 Kim drove 272.8 miles on 12.3 gallons of gasoline. How many miles per gallon is this?

SOLUTION Recall that *per* usually means division; thus we find the mpg by dividing the number of miles by the number of gallons. Therefore,

$$\text{mpg} = \frac{272.8}{12.3}$$

PUNCH	DISPLAY	MEANING
[C]	0.	Clear
[2][7][2][·][8]	272.8	Miles
[÷][1][2][·][3]	12.3	Divided by gallons
[=]	22.178862	mpg (answer)

Thus the mpg is about 22.2 miles per gallon.

EXAMPLE 13 A recent election had the following results.

Mr. Kandelman	1,272,754
Ms. Wolff	981,632
Mr. Altman	52,787
Others	9,813

What percent of the vote did Kandelman get?

SOLUTION Recall from Chapter 4 that we must first find the base (or total) in order to compute percents.

PUNCH	DISPLAY	MEANING
C	0.	Clear
1 2 7 2 7 5 4	1272754.	Kandelman
+ 9 8 1 6 3 2	981632.	Wolff
+ 5 2 7 8 7	52787.	Altman
+ 9 8 1 3	9813.	Others
=	2316986.	Total (base)

Now, we can use the $A = P \cdot B$ formula (review Section 4.3, page 201). Here $A = 1{,}272{,}754$ (amount for Kandelman), $B = 2{,}316{,}986$ (base), and P is the unknown. Substituting into $A = P \cdot B$, we get

$$A = P \cdot B$$

$$1{,}272{,}754 = P \cdot 2{,}316{,}986$$

$$\frac{1{,}272{,}754}{2{,}316{,}986} = P$$

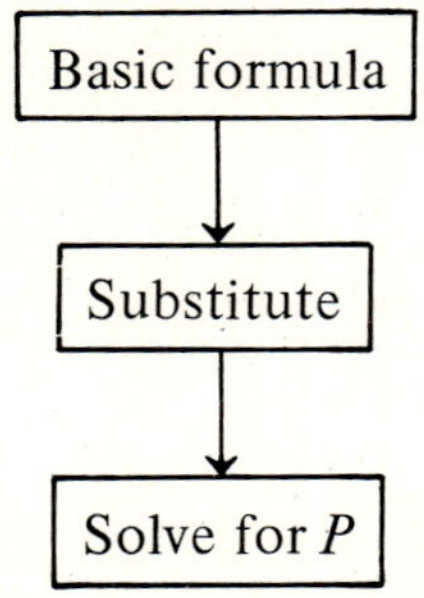

PUNCH	DISPLAY	MEANING
C	0.	Clear
1 2 7 2 7 5 4	1272754.	Kandelman
÷ 2 3 1 6 9 8 6	2316986.	Divide by total
=	0.5493145	Answer

To get a percent, we move the decimal point two places to the right. Thus $P = 54.93145\%$. Since the percent of votes is usually rounded to the nearest tenth of a percent, we get 54.9%.

EXAMPLE 14 Find the volume of the sphere.

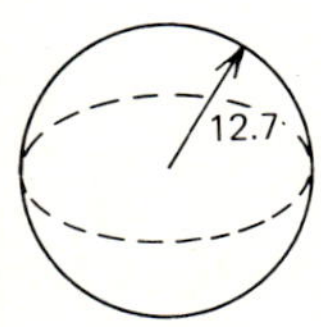

SOLUTION Recall from Chapter 5 (page 288) that the volume of a sphere is given by

$$V = \frac{4\pi r^3}{3}$$

Substituting $r = 12.7$ and $\pi = 3.14$, we get

$$V = \frac{4(3.14)\,(12.7)\,(12.7)\,(12.7)}{3}$$

Now we compute this on a hand calculator.

PUNCH	DISPLAY	MEANING
[C]	0.	Clear
[4]	4.	4
[×][3][·][1][4]	3.14	Times π
[×][1][2][·][7]	12.7	
[×][1][2][·][7]	12.7	Times r cubed
[×][1][2][·][7]	12.7	
[÷][3]	3.	Divide by 3
[=]	8575.8966	Answer

PROBLEM SET The following problems are typical of those seen throughout the text. Solve them with a hand calculator.

See Examples 12 to 14

1. A certain town has a population of 29,532 people. The town is 13.7 square miles. How many people per square miles is this?
2. Below are the week's sales figures for Sam Arbuckle. What is the total for the week?

Monday	\$528.73
Tuesday	479.38
Wednesday	502.54
Thursday	517.82
Friday	498.78

3. The heat needed to raise the temperature of a 2.95-pound bar of silver by 3.35°F is given by 2.95 × 3.35 × 0.056. Compute this.
4. The Bensens have \$2173.43 in the bank at 7.25% interest. How much interest will they earn in one year?
5. A recent election had the following results.

Washington	75,179
Adams	71,038
Jefferson	9,783
Others	1,876

What percent of the vote did each candidate (and "others") receive?

6. The Smiths want to clear \$37,100 on the sale of their house after a $7\frac{1}{2}\%$ commission. What must the selling price be?
7. What is the area of a rectangle 27.9 by 19.3?
8. What is the area of a circle with radius 52.73?
9. What is the volume of a cylinder with height = 71.3 and diameter = 68.2?
10. How many inches is 74.95 centimeters?

ANSWERS 1. 2155.6204 2. \$2527.25 3. 0.55342 4. \$157.57
5. Washington: 47.6%; Adams: 45.0%; Jefferson: 6.2%; others: 1.2%
6. \$40,108.11 7. 538.47 8. 8730.6221 9. 260,332.22
10. 29.51 inches

Miscellaneous Exercises 7.1

Use a hand calculator to perform the indicated operations.

See Examples 1 to 6

1. $2 + 7$
2. $17 - 8$
3. $2781 + 7385 + 19834$

$$\begin{array}{r} 2781 \\ 7385 \\ +19834 \\ \hline \end{array}$$

4. $$\begin{array}{r} 210817 \\ -\ 94659 \\ \hline \end{array}$$

5. $7.18 + 8.2 + 15.3$
6. $97.1 - 76.034$

See Examples 7 to 9

7. 5×8
8. $48 \div 4$
9. $$\begin{array}{r} 7814 \\ \times\ 983 \\ \hline \end{array}$$
10. $217\overline{)814683}$
11. 7.03×12.417
12. $283.17 \div 0.16$

See Examples 10 and 11

13. $\dfrac{7}{19}$
14. $\dfrac{27}{153}$
15. $\dfrac{(7.3)(6.9)}{(17.8)(0.8)}$
16. $\dfrac{(14)(0.375)}{(7.91)(2500)(0.817)}$

Business applications

17. A company's gross revenue was \$34,069,572. Its expenses (costs) were \$29,583,952. Find the net profit (revenue − cost).
18. The present value of \$1 (in 10 years, 8% interest) is 0.46319. Compute the present value of \$345,958 by calculating $0.46319 \times 345{,}958$.
19. The ABC Conglomerate controls 45,359 out of 218,573 proxies in a smaller company. What percent is this?

Health applications

20. A patient's systolic blood pressure dropped from 115 to 83. What percent is this drop?
21. A man has $4\frac{1}{2}$-inch-diameter leg. A nurse wants to wrap the leg with a bandage.
   (a) What is the circumference of the leg?
   (b) How much bandage is required to wrap the leg 25 times?

Science applications

22. The diameter of the earth is about 7927 miles.
   (a) What is the radius of the earth?
   (b) What is the circumference of the earth?
   (c) What is the volume of the earth?
   (d) What is the surface area of the earth? (Use the formula for the surface area of a sphere, $S = 4\pi r^2$.)
23. A substance has a mass of 23.56 grams and a volume of 0.683 cubic centimeter Compute the density (which is mass/volume).

Technical applications

24. In a scale drawing, 0.0012 centimeter represents 1 meter. If a wall is actually 14.8 meters, what would its size be in a scale drawing?
25. A bolt has a $\frac{5}{8}$-inch-diameter head.
   (a) What is the radius of the head?
   (b) Write the radius as a four-place decimal.
   (c) Compute the area of the head.

Consumer applications

26. One month, Laura keeps the following records for gas mileage.

Miles	Gallons
312.9	16.8
290.4	15.6
337.8	17.4
302.5	16.1

(a) Find the total number of miles driven.
(b) Find the total number of gallons used.
(c) Compute the average miles per gallon.

27. For a 35-year-old man, 5-year-term life insurance costs \$6.55 per \$1000 coverage. How much will it cost for \$55,000 of coverage?

28. The Smedleys put \$2345 in the bank at $7\frac{1}{4}\%$ interest.

(a) How much interest do they earn in the first year?
(b) What is their new balance after one year (original deposit + interest)?
(c) How much interest does the new balance earn in the second year?
(d) What is the new balance after two years?

29. Balance the following checkbook.

Apr. 1	Balance from March	\$ 214.69
Apr. 2	Deposit	+421.05
Apr. 4	Rent	-255.00
Apr. 4	Electricity	-31.31
Apr. 5	Phone Co.	-19.23
Apr. 6	Groceries	-42.69
Apr. 9	Clothes	-15.43
Apr. 10	Gasoline Co.	-23.79
Apr. 12	Cash	-25.00
	Balance	?

Now complete Post-Test 1 for Section 7.1 on page T103.

7.2 THE MEMORY AND EXPONENTS

SECTION OBJECTIVES

After completing this unit, you will be able to:

1. Use the memory in a calculator to do complex computations.
2. Compute exponents.

TYPICAL PROBLEMS

You will be able to work problems like the ones below.

Use a hand calculator to compute the following.

(a) (2)(7) + (4)(8)

(b) (2.1)(7.5) + (8.2)(4.9) - (2.7)(0.03)

(c) $\frac{1}{3} + \frac{2}{7} + \frac{5}{8}$

(d) $\dfrac{6}{11} - \dfrac{3}{13}$

(e) $\dfrac{(3)\,(7.9)}{2.9} + \dfrac{(5.2)\,(7.3)}{(1.8)\,(4.3)}$

(f) 1.2^8

(g) $400 \cdot (0.92)^{10}$

ANSWERS **(a)** 46 **(b)** 55.849 **(c)** 1.2440476 **(d)** 0.31468531 **(e)** 13.076807 **(f)** 4.299817 **(g)** 173.75528

Many problems require complex calculations. This is where the memory is used. Here is a brief description of the memory and the memory keys.

1. *Memory.* Your calculator should have a memory. The memory works as follows. Have you ever worked a problem and written the answer on paper (or said to yourself), "Remember the number, 7.463. I'll need it later." This is exactly what the memory does: It remembers one number while you are working with other numbers.
2. [MC]. This is the *memory clear* key. This key says to the calculator, "Erase or forget whatever number you were remembering. Start over at 0."
3. [M+]. This key *adds* to the memory. Suppose that the calculator is remembering the number 7 in memory. If we punch [4] and then [M+], we are adding 4 to the 7 in the memory; the memory is now 11. Also, when the memory is cleared or zero, punching [M+] enters a number into the memory. For example, if there is a 4.56 on the display and we punch [M+], 4.56 will be stored in the memory.
4. [M−]. This key *subtracts* from the memory. Suppose that 20 is in the memory. If 12 is on the display and we punch [M−], we are subtracting 12 from the 20 in the memory; the memory is now 8.
5. [MR]. This is the *memory recall*. This key tells us what is in the memory. After a long problem, it is like the [=] key.

EXAMPLE 15 Simplify $4 \cdot 8 + 3 \cdot 7$ on a calculator.

SOLUTION Recall from Chapter 6 that multiplication has priority over addition; in other words, we multiply first and then add. Thus we get

$$4 \cdot 8 + 3 \cdot 7 = 32 + 21 = 53$$

Now let us try to work this in the order that it appears. On most calculators, we get the following.

PUNCH	DISPLAY	MEANING
[C]	[0.]	Clear
[4] [×] [8]	[8.]	4 times 8

PUNCH	DISPLAY	MEANING
[+]	32.	Plus
[3][X][7]	7.	3 times 7
[=]	245.	*Wrong answer!*

What happened? We know that the answer should be 53, but we got 245. Many calculators (but not all) do the operations in the order that they are given instead of by algebraic priority.

To do a problem like this, we have to use the memory. Every time we complete a multiplication, we use [M+] to store the product in the memory.

PUNCH	DISPLAY	MEANING
[C][MC]	0.	Clear display and memory
[4][X][8][=]	32.	4 times 8 = 32
[M+]	32.	Store in memory
[3][X][7][=]	21.	3 times 7 = 21
[M+]	21.	Add to the memory
[MR]	53.	Recall answer

Notice that we did each multiplication separately: We multiplied 4 X 8 = 32; then we added this to the memory with [M+]; then we multiplied 3 X 7 = 21; we then added this to the memory with [M+]; finally, we use [MR] to recall the total from the memory.

EXAMPLE 16 Simplify (2.7)(7.8) + (31.7)(0.4) - (8.1)(0.7).

SOLUTION We compute each product separately. Then we use [M+] to add to the memory and [M-] to subtract from the memory.

PUNCH	DISPLAY	MEANING
[C][MC]	0.	Clear display and memory
[2][·][7][X][7][·][8][=]	21.06	2.7 times 7.8 = 21.06
[M+]	21.06	Store in memory
[3][1][·][7][X][·][4][=]	12.68	31.7 times 0.4 = 12.68
[M+]	12.68	Add to memory
[8][·][1][X][·][7][=]	5.67	8.1 times 0.7 = 5.67
[M-]	5.67	Subtract from memory
[MR]	28.07	Recall answer

EXAMPLE 17 Use a hand calculator to compute the total cost of the following party items.

Item	Quantity	Cost per Unit
Paper plates	3 dozen	\$1.49 per dozen
Paper cups	3 dozen	\$1.29 per dozen
Potato chips	3 bags	\$0.89 per bag
Doritos	4 bags	\$1.59 per bag
Dip	2 tubs	\$0.89 per tub
Beer	12 six-packs	\$2.29 per six-pack
Soda pop	5 six-packs	\$1.89 per six-pack
Cole slaw	6 pounds	\$0.85 per pound

SOLUTION If we were working this with pencil and paper, we would multiply each item's quantity by its price. We would then write down each product and total them.

This is exactly what we are doing with the memory. We compute each product and store (or add) it to the memory with the [M+] key.

PUNCH	DISPLAY	MEANING
[C] [MC]	0.	Clear display and memory
[3] [×] [1] [·] [4] [9] [=]	4.47	Plates cost \$4.47
[M+]	4.47	Store in memory
[3] [×] [1] [·] [2] [9] [=]	3.87	Cups cost \$3.87
[M+]	3.87	Add to memory
[3] [×] [·] [8] [9] [=]	2.67	Chips cost \$2.67
[M+]	2.67	Add to memory
[4] [×] [1] [·] [5] [9] [=]	6.36	Doritos cost \$6.36
[M+]	6.36	Add to memory
[2] [×] [·] [8] [9] [=]	1.78	Dip costs \$1.78
[M+]	1.78	Add to memory
[1] [2] [×] [2] [·] [2] [9] [=]	27.48	Beer costs \$27.48
[M+]	27.48	Add to memory
[5] [×] [1] [·] [8] [9] [=]	9.45	Pop costs \$9.45
[M+]	9.45	Add to memory
[6] [×] [·] [8] [5] [=]	5.10	Cole slaw costs \$5.10
[M+]	5.10	Add to memory
[MR]	61.18	Total cost

Notice how we computed this total in blocks: quantity times price, equals, and store in the memory. We then used [MR] to recall the total from the memory.

PROBLEM SET Use the memory on a hand calculator to compute each of the following.

See Examples 15 to 17

1. $2(7) + 5(8)$
2. $6(11) + 4(13)$
3. $8(22) - 5(31)$
4. $13(15) - 7(21)$
5. $(21.2)(15.1) + (2)(17.1) + (3)(0.51)$
6. $(8.5)(7) + (11.2)(6) + (0.03)(2500)(2.3)$
7. $(6)(7.8) - (11)(2.3)(13.7) + (3.6)(2.89)(213.6)$
8. $(3)(7.1)(6.5) - (4)(0.3)(1.783) + (8.1)(0.752)$
9. Compute the cost of the following project.

Item	Quantity	Cost per Item
Paneling	15 panels	\$12.50 per panel
Nails	2 pounds	\$0.75 per pound
2 by 4 studs	45 studs	\$1.25 per stud
Corner molding	48 feet	\$0.27 per foot
Floor tile	120 tiles	\$0.89 per tile
Ceiling tiles	28 tiles	\$2.75 per tile
Ceiling grids	216 feet	\$0.12 per foot

ANSWERS **1.** 54 **2.** 118 **3.** 21 **4.** 48 **5.** 355.85 **6.** 299.2 7. 1922.4844 **8.** 142.4016 **9.** \$467.93

Another use of the memory is adding and subtracting fractions. Recall from Chapter 2 that we add and subtract fractions by rewriting them with an LCD. For example,

$$\frac{1}{3} + \frac{4}{7} + \frac{5}{6} = \frac{14}{42} + \frac{24}{42} + \frac{35}{42} = \frac{73}{42}$$

EXAMPLE 18 Add $\frac{1}{3} + \frac{4}{7} + \frac{5}{6}$ on a calculator.

SOLUTION With a calculator, we do not need to find the LCD. Rather, we write each fraction as a decimal (by division). Then we add each of these decimals into the memory.

PUNCH	DISPLAY	MEANING
[C] [MC]	0.	Clear display and memory
[1] [÷] [3] [=]	.33333333	$\frac{1}{3}$
[M+]	.33333333	Store in the memory
[4] [÷] [7] [=]	.57142857	$\frac{4}{7}$
[M+]	.57142857	Add to the memory

PUNCH	DISPLAY	MEANING
5 ÷ 6 =	.83333333	$\frac{5}{6}$
M+	.83333333	Add to the memory
MR	1.7380952	Answer

This answer agrees with the fraction answer that we got: $\frac{73}{42}$. If we write $\frac{73}{42}$ as a decimal, we get 1.7380952.

EXAMPLE 19 Subtract $\frac{23}{41} - \frac{17}{35}$ on a calculator.

SOLUTION This would be a difficult problem to do by finding the LCD. It is much easier to do with a calculator with an M– key.

PUNCH	DISPLAY	MEANING
C MC	0.	Clear display and memory
2 3 ÷ 4 1 =	.56097561	$\frac{23}{41}$
M+	.56097561	Store in memory
1 7 ÷ 3 5 =	.48571429	$\frac{17}{35}$
M–	.48571429	Subtract from memory
MR	.07526132	Answer

EXAMPLE 20 Simplify $\frac{(7)(8.3)}{9.7} + \frac{6.1}{0.9} - \frac{100}{(8.7)(33)}$.

SOLUTION We simplify each fraction separately; then we add it to or subtract it from the memory. We use M+ and M–.

PUNCH	DISPLAY	MEANING
C MC	0.	Clear display and memory
7	7.	First fraction
× 8 · 3	8.3	
÷ 9 · 7	9.7	
=	5.9896907	
M+	5.9896907	Store in memory

PUNCH	DISPLAY	MEANING
[6] [·] [1]	6.1	Second fraction
[÷] [·] [9]	0.9	
[=]	6.7777778	
[M+]	6.7777778	Add to memory
[1] [0] [0]	100.	Third fraction
[÷] [8] [·] [7]	8.7	
[÷] [3] [3]	33.	
[=]	.34831069	
[M−]	.34831069	Subtract from memory
[MR]	12.419158	Answer

PROBLEM SET Add and subtract the following fractions with a hand calculator. Round each answer to the nearest thousandth.

See Examples 18 to 20

1. $\frac{1}{2}+\frac{1}{3}$
2. $\frac{2}{19}+\frac{5}{7}+\frac{4}{5}$
3. $\frac{7}{11}-\frac{8}{17}$
4. $\frac{4}{21}-\frac{7}{53}$
5. $\frac{2}{7}+\frac{7}{18}-\frac{9}{29}$
6. $\frac{14}{47}+\frac{15}{79}-\frac{2}{13}+\frac{53}{103}$
7. $\frac{(0.7)(8)}{9}+\frac{(4)(11)}{8.2}$
8. $\frac{(1.2)(510)}{0.76}-\frac{(4.1)(408)}{19}$
9. $\frac{(3.4)(8.02)}{7.14}+\frac{(4.9)(0.45)}{(5.3)(3.4)}-\frac{3.7}{(8.08)(93.3)}$
10. $\frac{10.79}{8.03}-\frac{121.7}{53.6}+\frac{(62)(47.33)}{(7.05)(0.83)}$

ANSWERS **1.** 0.833 **2.** 1.620 **3.** 0.166 **4.** 0.058 **5.** 0.364 **6.** 0.848 **7.** 5.988 **8.** 717.221 **9.** 3.937 **10.** 500.562

Recall that in Chapter 6 we talked about exponents such as squaring and cubing. **Exponents** are generally repeated multiplications; for instance,

$$2^2 = 2 \times 2 = 4$$
$$2^3 = 2 \times 2 \times 2 = 8$$
$$2^4 = 2 \times 2 \times 2 \times 2 = 16$$
$$2^5 = 2 \times 2 \times 2 \times 2 \times 2 = 32$$

and so on

EXAMPLE 21 Find 2^7.

SOLUTION We can do this repeated operation on a calculator in two ways. First, we can simply multiply 2 X 2 X 2 X 2 X 2 X 2 X 2. There is a second method that works on many calculators that is even faster. It works as follows using the [X] and [=] keys.

PUNCH	DISPLAY	MEANING
[C]	0.	Clear
[2]	2.	Enter 2
[X]	2.	Start multiplying
[=]	4.	2^2
[=]	8.	2^3
[=]	16.	2^4
[=]	32.	2^5
[=]	64.	2^6
[=]	128.	2^7 (answer)

Notice that we punched [2], [X], and then [=] six times.

> To compute an exponent, B^n:
>
> 1. Clear the calculator.
> 2. Enter B.
> 3. Punch [X].
> 4. Punch [=] $(n - 1)$ times.

EXAMPLE 22 Find $(1.7831)^6$.

SOLUTION It would be a bother to multiply 1.7831 as a factor six times. Instead, we enter 1.7831 once, punch [X], and then [=] five times (one less than the exponent 6).

PUNCH	DISPLAY	MEANING
[C]	0.	Clear
[1] [·] [7] [8] [3] [1]	1.7831	Enter 1.7831
[X]	1.7831	Start multiplying
[=]	3.1794456	1.7831^2
[=]	5.6692694	1.7831^3
[=]	10.108874	1.7831^4
[=]	18.025133	1.7831^5
[=]	32.140616	1.7831^6

EXAMPLE 23 Compute $1500 \cdot (1.07)^{20}$.

SOLUTION This computation might come up in a bank interest problem: \$1500 invested at 7% interest for 20 years. We first compute $(1.07)^{20}$; then we multiply by 1500.

PUNCH	DISPLAY	MEANING
C	0.	Clear
1 · 0 7	1.07	Enter 1.07
×	1.07	Start multiplying
=	1.1449	1.07^2
=	1.225043	1.07^3
=	1.310796	1.07^4
⋮	⋮	⋮
=	3.6165265	1.07^{19}
=	3.8696833	1.07^{20}
× 1 5 0 0	1500.	Times 1500
=	5804.5249	Answer

This says that \$1500 invested at 7% interest for 20 years will grow to about \$5805.

PROBLEM SET Compute the following with a hand calculator.

See Examples 21 to 23

1. 3^4
2. 6^3
3. 12^5
4. 9^4
5. $(1.083)^9$
6. $(2.073)^{13}$
7. $(0.876)^{10}$
8. $2000 \cdot (1.05)^4$
9. $6000 \cdot (0.9)^7$
10. The Rodriguez family invests \$3000 at 8% interest for 7 years. This grows to $3000 \cdot (1.08)^7$. Compute this.

ANSWERS **1.** 81 **2.** 216 **3.** 248,832 **4.** 6561 **5.** 2.0495381 **6.** 13,055.422 **7.** 0.2660975 **8.** 2431.0125 **9.** 2869.7814 **10.** 5141.47

Miscellaneous Exercises 7.2

Use a hand calculator to compute each of the following expressions.

See Examples 15 to 17

1. 2(11) + 8(17) + 3(23)
2. 6(8.2) − 5(7.05) + 11(0.37)
3. (0.071)(5123) + (0.387)(1831)
4. (54437)(6.1) − (478)(27.3) + 82(503.6)

See Examples 18 to 20

5. $\frac{1}{7} + \frac{1}{8} + \frac{1}{9}$

6. $1 - \frac{1}{2} + \frac{1}{3} - \frac{1}{4} + \frac{1}{5} - \frac{1}{6} + \frac{1}{7}$

7. $1 + \frac{1}{4} + \frac{1}{9} + \frac{1}{16} + \frac{1}{25} + \frac{1}{36} + \frac{1}{49}$

8. $\frac{1}{2} + \frac{1}{4} + \frac{1}{8} + \frac{1}{16} + \frac{1}{32} + \frac{1}{64} + \frac{1}{128} + \frac{1}{256}$

9. $\frac{27}{41} + \frac{53}{5.1} - \frac{1.8}{0.7}$

10. $\frac{(2.5)(3.2)}{6.9} - \frac{(4.2)(0.7)}{1.3} + \frac{8.4}{9.2} + \frac{0.25}{(6.3)(.021)}$

See Examples 21 to 23

11. 4^8
12. 2^{15}
13. 1.7^9
14. 8.031^4
15. $2000(1.3)^6$
16. $45{,}000(0.9)^5$

Business applications

17. A firm owns a \$300,000 piece of property which is growing in value 12% per year. After 7 years it will be worth $300{,}000(1.12)^7$. Compute this.

18. A firm buys a \$55,000 machine. It depreciates 10% per year. After 6 years, its worth is $55{,}000(0.9)^6$. Compute this.

19. A company must compute the total commission due all its salespeople. Below are the sales and commission rates for its salespeople. Compute the total commission that the company must pay.

Salesperson	Sales	Commission Rate
Sern	250,000	4%
Jones	175,000	5%
Wang	232,000	$3\frac{1}{2}$%
Kownowski	198,000	$4\frac{1}{2}$%
Bigelow	319,000	$5\frac{1}{2}$%

Health application

20. A patient is to receive a number of injections of different medicines. A nurse is interested in the total amount of medicine to be given to the patient. Compute this.

Medicine	Number of Injections	Strength
A	4	0.46
B	5	0.75
C	2	1.2
D	8	0.375

Consumer application

21. Compute the total cost for the following gift.

Item	Quantity	Cost per Unit
Drummers drumming	12	$30 per drummer
Pipers piping	11	$35 per piper
Lords a-leaping	10	$75 per lord
Ladies dancing	9	$40 per lady
Maids a-milking	8	$15 per maid
Swans a-swimming	7	$35 per swan
Geese a-laying	6	$15 per goose
Golden rings	5	$250 per ring
Calling birds	4	$25 per bird
French hens	3	$15 per hen
Turtle doves	2	$12 per dove
Partridge	1	$20 per partridge
Pear tree	1	$45 per tree

Now complete Post-Test 1 for Section 7.2 on page T105.

CHAPTER 7 SUMMARY

Important Words and Phrases

arithmetic operations [+] [−] [×] [÷] (7.1)

clear [C] (7.1)

clear entry [CE] (7.1)

decimal point [·] (7.1)

digits [0] [1] [2] [3] [4] [5] [6] [7] [8] [9] (7.1)

equal sign [=] (7.1)

exponents (7.2)

hand calculator (7.1)

memory (7.2)

memory add [M+] (7.2)

memory clear [MC] (7.2)

memory recall [MR] (7.2)

memory subtract [M−] (7.2)

order of operations (7.2)

Important Ideas

(7.1) Performing elementary operations on a calculator.
Using a calculator to work applications.

(7.2) Using the memory to do complex calculations.
Computing exponents.

Choose the correct (or closest) answer. Use a hand calculator.

1. $703.4 + 382.68 + 1188.4 = ?$
   (a) 2374.43 (b) 2275.49 (c) 2274.48
   (d) 2174.58 (e) 2374.47

2. $205.073 - 194.7 = ?$
   (a) 10.373 (b) 9.373 (c) 10.483
   (d) 10.384 (e) 9.374

3. $(0.0783)(863.52) = ?$
   (a) 66.612 (b) 6.7613 (c) 676.13
   (d) 67.614 (e) 68.361

4. $4.072 \div 0.8671 = ?$
   (a) 0.2129 (b) 2.1294 (c) 46.961
   (d) 4.7962 (e) 4.6961

5. $\dfrac{(1.79)(52)}{4.2} = ?$
   (a) 21.2893 (b) 22.1619 (c) 0.00819
   (d) 390.936 (e) 23.941

6. $(4)(2.1) + (1.1)(29.3) - (8.2)(0.752) = ?$
   (a) 34.4636 (b) 203.153 (c) 204.213
   (d) 32.6832 (e) 38.4032

7. $\dfrac{2}{3} - \dfrac{3}{4} + \dfrac{4}{5} - \dfrac{5}{6} + \dfrac{7}{8} = ?$
   (a) 3.925 (b) 0.85732 (c) 0.67432
   (d) 2.4837 (e) 0.75833

8. $\dfrac{1.7}{(4)(9.1)} - \dfrac{(4.2)(9.8)}{0.6} + \dfrac{7500}{(0.9)(8.3)} = ?$
   (a) 843.9503 (b) 935.463 (c) 1072.66
   (d) 603.685 (e) 923.493

9. $17^5 = ?$
   (a) 83,521 (b) 24,137,569 (c) 1,419,857
   (d) 1,728,393 (e) 283,942

10. $4200 \cdot (1.075)^8 = ?$
   (a) 5415 (b) 6574.03 (c) 8052.40
   (d) 7490.61 (e) 8329.84

Now complete Post-Test 1 for Chapter 7 on page T107.

PRE-TESTS AND POST-TESTS

Name ______________

Section ______________

Date ______________

Score ______________

1

Pre-Test

1. In 289,033, what is the value of the 9? 1. ______________ *(1.1)**

2. In 602,137, what digit is in the ten-thousands' column? 2. ______________ *(1.1)*

3. Write 29,168 in words. 3. ______________________________ *(1.1)*

4. $\begin{array}{r} 472 \\ +985 \\ \hline \end{array}$ 4. ______________ *(1.2)*

5. $\begin{array}{r} 4685 \\ 9183 \\ +\ 766 \\ \hline \end{array}$ 5. ______________ *(1.2)*

6. $\begin{array}{r} 29 \\ 624 \\ 521 \\ +9806 \\ \hline \end{array}$ 6. ______________ *(1.2)*

7. $\begin{array}{r} 824 \\ -269 \\ \hline \end{array}$ 7. ______________ *(1.3)*

8. $\begin{array}{r} 4132 \\ -1397 \\ \hline \end{array}$ 8. ______________ *(1.3)*

9. $\begin{array}{r} 10001 \\ -\ 6835 \\ \hline \end{array}$ 9. ______________ *(1.3)*

10. Find n: $n + 8 = 20$. 10. ______________ *(1.3)*

(over)

*These numbers refer to the sections in the text.

1

11. $\begin{array}{r} 49 \\ \times 63 \\ \hline \end{array}$

11. ______________ (*1.4*)

12. $\begin{array}{r} 247 \\ \times\ 58 \\ \hline \end{array}$

12. ______________ (*1.4*)

13. $\begin{array}{r} 954 \\ \times 162 \\ \hline \end{array}$

13. ______________ (*1.4*)

For problems 14 to 16, divide and show quotient and remainder (if any).

14. $27\overline{)2256}$

14. ______________ (*1.5*)

15. $83\overline{)62582}$

15. ______________ (*1.5*)

16. $657\overline{)325557}$

16. ______________ (*1.5*)

17. Find x: $5 \cdot x = 30$.

17. ______________ (*1.5*)

For problems 18 and 19, factor composites into primes; if number is prime, write "Prime."

18. 110

18. ______________ (*1.6*)

19. 101

19. ______________ (*1.6*)

20. Find the least common multiple (LCM) of 18 and 24.

20. ______________ (*1.6*)

1.1 PLACE VALUE, NAMING WHOLE NUMBERS, AND ROUNDING

Name ____________________

Section ____________________

Date ____________________

Score ____________________

Post-Test 1

1

1. Write 2479 in expanded form.	1.	____________	*(1–3)**
2. Write 405,008 in expanded form.	2.	____________	*(1–3)*
3. In 319, what is the value of the 1?	3.	____________	*(4–6)*
4. In 10,098, what is the value of the 1?	4.	____________	*(4–6)*
5. In 3,456,908, what digit is in the hundred-thousands' place?	5.	____________	*(7–9)*
6. In 309,562, what digit is in the thousands' place?	6.	____________	*(7–9)*
7. Write 492 in words.	7.	____________	*(10–12)*
8. Write 69,125 in words.	8.	____________	*(13–15)*
9. Write 34,025,644 in words.	9.	____________	*(13–15)*
10. Round 23,782 to the nearest thousand.	10.	____________	*(16–18)*

*These numbers refer to the examples in Chapter 1.

1

1.1 PLACE VALUE, NAMING WHOLE NUMBERS, AND ROUNDING

Post-Test 2

Name ______________

Section ______________

Date ______________

Score ______________

---

1. Write 5921 in expanded form. 1. ______________ *(1–3)*

2. Write 30,604 in expanded form. 2. ______________ *(1–3)*

3. In 459, what is the value of the 4? 3. ______________ *(4–6)*

4. In 20,094, what is the value of the 2? 4. ______________ *(4–6)*

5. In 34,958,021, what digit is in the millions' place? 5. ______________ *(7–9)*

6. In 39,087 what digit is in the tens' column? 6. ______________ *(7–9)*

7. Write 706 in words. 7. ______________ *(10–12)*

8. Write 51,817 in words. 8. ______________ *(13–15)*

9. Write 47,003,892 in words. 9. ______________ *(13–15)*

10. Round 158,432 to the nearest ten thousand. 10. ______________ *(16–18)*

Name ______________

Section ______________

Date ______________

Score ______________

Post-Test 1

	Problem		Answer	Objectives
1.	47 +29	1.	______	(20–24)
2.	1642 + 894	2.	______	(20–24)
3.	244 682 + 67	3.	______	(20–24)
4.	15,691 4,822 + 8,497	4.	______	(20–24)
5.	7,503 29,623 48,065 + 7,810	5.	______	(20–24)
6.	6,811 44,536 7,189 +23,647	6.	______	(20–24)
7.	Add 2153 + 14,927 + 9644.	7.	______	(20–24)
8.	Add 14 + 123 + 83 + 206.	8.	______	(20–24)
9.	Add 822 + 1237 + 8451 + 981.	9.	______	(20–24)
10.	For breakfast, Danny has 621 calories. At lunch, he has 776 calories. For supper, he has 849 calories. Later, he has 625 calories worth of pizza. How many calories did he eat this day?	10.	______	(25, 26)

1.2 ADDITION OF WHOLE NUMBERS

Post-Test 2

Name ____________

Section ____________

Date ____________

Score ____________

1. $\begin{array}{r} 69 \\ +52 \\ \hline \end{array}$ — 1. ____________ (*20–24*)

2. $\begin{array}{r} 4035 \\ +\ 986 \\ \hline \end{array}$ — 2. ____________ (*20–24*)

3. $\begin{array}{r} 8607 \\ 22 \\ +\ 538 \\ \hline \end{array}$ — 3. ____________ (*20–24*)

4. $\begin{array}{r} 4932 \\ 23567 \\ +\ 5491 \\ \hline \end{array}$ — 4. ____________ (*20–24*)

5. $\begin{array}{r} 8799 \\ 6197 \\ 901 \\ +3912 \\ \hline \end{array}$ — 5. ____________ (*20–24*)

6. $\begin{array}{r} 8557 \\ 42683 \\ 79104 \\ +\ 5123 \\ \hline \end{array}$ — 6. ____________ (*20–24*)

7. Add 77 + 214 + 397. — 7. ____________ (*20–24*)

8. Add 24 + 152 + 73 + 209. — 8. ____________ (*20–24*)

9. Add 2413 + 20,563 + 9176. — 9. ____________ (*20–24*)

10. One month, Mike made trips of 213, 443, 182, and 309 miles. What was the total mileage for these trips? — 10. ____________ (*25,26*)

1.3 SUBTRACTION OF WHOLE NUMBERS

Name ________________

Section ________________

Date ________________

Score ________________

Post-Test 1

1

Subtract the following whole numbers.

1. $\begin{array}{r} 68 \\ -41 \\ \hline \end{array}$ — 1. ____________ *(28, 29)*

2. $\begin{array}{r} 5782 \\ -1234 \\ \hline \end{array}$ — 2. ____________ *(28, 29)*

3. $\begin{array}{r} 427 \\ -184 \\ \hline \end{array}$ — 3. ____________ *(30–32)*

4. $\begin{array}{r} 2904 \\ -447 \\ \hline \end{array}$ — 4. ____________ *(30–32)*

5. 42,847 - 6769 — 5. ____________ *(30–32)*

6. 93,000,000 - 26,475,000 — 6. ____________ *(30–32)*

7. 27,906,137 - 18,137,053 — 7. ____________ *(30–32)*

8. In a mayoral election, Smith defeated Jones 47,952 to 42,878. What was the margin of victory? — 8. ____________ *(33)*

Find n in the following problems.

9. $n + 23 = 37$ — 9. ____________ *(34, 35)*

10. $n - 12 = 20$ — 10. ____________ *(36, 37)*

1.3 SUBTRACTION OF WHOLE NUMBERS

Name ______________________

Section ______________________

Date ______________________

Score ______________________

Post-Test 2

Subtract the following numbers.

1. $\begin{array}{r} 47 \\ -15 \\ \hline \end{array}$ — 1. ______________ (28, 29)

2. $\begin{array}{r} 487 \\ -235 \\ \hline \end{array}$ — 2. ______________ (28, 29)

3. $\begin{array}{r} 617 \\ -443 \\ \hline \end{array}$ — 3. ______________ (30–32)

4. $\begin{array}{r} 27{,}106 \\ -\ \ 9{,}828 \\ \hline \end{array}$ — 4. ______________ (30–32)

5. 12,496,202 - 9,713,615 — 5. ______________ (30–32)

6. 20,000 - 3492 — 6. ______________ (30–32)

7. 34,981 - 21,029 — 7. ______________ (30–32)

8. A church needs to raise $57,000 to meet its budget. It has raised $34,178. How much more does it need? — 8. ______________ (33)

Find n in the following problems.

9. $n + 17 = 30$ — 9. ______________ (34, 35)

10. $n - 20 = 35$ — 10. ______________ (36, 37)

Name ______________________

Section ______________________

Date ______________________

Score ______________________

Post-Test 1

Multiply the following whole numbers.

1. 7×26 — 1. ______________ (*40–44*)

2. $800 \times 21{,}000$ — 2. ______________ (*40–44*)

3. $17{,}000 \times 4000$ — 3. ______________ (*40–44*)

4. 79×63 — 4. ______________ (*47–50*)

5. 71×458 — 5. ______________ (*47–50*)

6. 817×265 — 6. ______________ (*47–50*)

7. 28×5176 — 7. ______________ (*47–50*)

8. $509 \times 21{,}783$ — 8. ______________ (*47–50*)

9. Estimate $4891 \times 21{,}795$. — 9. ______________ (*45, 46*)

10. A certain cupcake has 160 calories. If Lisa eats two cupcakes per day, how many calories is this in a 365-day year? — 10. ______________ (*51, 52*)

1

1.4 MULTIPLICATION OF WHOLE NUMBERS

Post-Test 2

Name ______________________

Section ______________________

Date ______________________

Score ______________________

Multiply the following whole numbers.

1. 9 X 74 — 1. ______________ (*40–44*)

2. 300 X 49,000 — 2. ______________ (*40–44*)

3. 600 X 7,500,000 — 3. ______________ (*40–44*)

4. 64 X 43 — 4. ______________ (*47–50*)

5. 86 X 751 — 5. ______________ (*47–50*)

6. 427 X 615 — 6. ______________ (*47–50*)

7. 39 X 4273 — 7. ______________ (*47–50*)

8. 7003 X 32,746 — 8. ______________ (*47–50*)

9. Estimate 52,109 X 89,231. — 9. ______________ (*45, 46*)

10. There are about 220,000,000 Americans. The average American eats about 188 pounds of meat per year. How much total meat does America eat in one year? — 10. ______________ (*51, 52*)

Name ______________________

Section ______________________

Date ______________________

Score ______________________

Post-Test 1

Divide the following whole numbers.

1.	$8\overline{)4776}$	1.	______________	(54, 55)
2.	$7\overline{)60352}$	2.	______________	(54, 55)
3.	$43\overline{)4128}$	3.	______________	(56–59)
4.	$57\overline{)39011}$	4.	______________	(56–59)
5.	$195\overline{)170294}$	5.	______________	(56–59)
6.	$588\overline{)542724}$	6.	______________	(56–59)
7.	Divide 335,797 ÷ 49.	7.	______________	(56–59)
8.	Divide 6,000,000 ÷ 750.	8.	______________	(56–59)
9.	Find n: $6 \cdot n = 54$.	9.	______________	(62, 63)
10.	It is 2820 miles from New York to Los Angeles. If a person walked 30 miles a day, how long would it take to walk from New York to Los Angeles?	10.	______________	(60, 61)

1.5 DIVISION OF WHOLE NUMBERS

Name ______________

Section ______________

Date ______________

Score ______________

Post-Test 2

Divide the following whole numbers.

1. $6\overline{)50556}$ — 1. ______________ (54, 55)

2. $9\overline{)14385}$ — 2. ______________ (54, 55)

3. $59\overline{)26347}$ — 3. ______________ (56–59)

4. $83\overline{)500324}$ — 4. ______________ (56–59)

5. $369\overline{)2860857}$ — 5. ______________ (56–59)

6. $703\overline{)2816899}$ — 6. ______________ (56–59)

7. Divide 5,924,124 ÷ 684. — 7. ______________ (56–59)

8. Divide 47,000,000 ÷ 1600. — 8. ______________ (56–59)

9. Find n: $5 \cdot n = 50$. — 9. ______________ (62, 63)

10. It takes 3500 calories to lose 1 pound of body weight. If a cookie is 125 calories, how many fewer cookies must you eat to lose 1 pound? — 10. ______________ (60, 61)

1.6 PRIME FACTORIZATION AND LEAST COMMON MULTIPLE

Name ____________________

Section ____________________

Date ____________________

Score ____________________

Post-Test 1

For each of the following numbers, state if it is a prime number or write it as a product of primes.

1. 36 — 1. ______________ *(66–73)*

2. 45 — 2. ______________ *(66–73)*

3. 67 — 3. ______________ *(66–73)*

4. 77 — 4. ______________ *(66–73)*

5. 120 — 5. ______________ *(66–73)*

Find the least common multiple (LCM) for each of the following sets of numbers.

6. 6 and 10 — 6. ______________ *(74–78)*

7. 12 and 18 — 7. ______________ *(74–78)*

8. 28 and 36 — 8. ______________ *(74–78)*

9. 6, 8, and 12 — 9. ______________ *(74–78)*

10. 12, 15, and 20 — 10. ______________ *(74–78)*

1

1.6 PRIME FACTORIZATION AND LEAST COMMON MULTIPLE

Post-Test 2

Name ______________________

Section ______________________

Date ______________________

Score ______________________

For each of the following numbers, state if it is a prime number or write it as a product of primes.

1. 48 — 1. ______________ *(66–73)*

2. 42 — 2. ______________ *(66–73)*

3. 57 — 3. ______________ *(66–73)*

4. 87 — 4. ______________ *(66–73)*

5. 97 — 5. ______________ *(66–73)*

Find the least common multiple (LCM) for each of the following sets of numbers.

6. 3 and 7 — 6. ______________ *(74–78)*

7. 6 and 14 — 7. ______________ *(74–78)*

8. 20 and 24 — 8. ______________ *(74–78)*

9. 6, 10, and 12 — 9. ______________ *(74–78)*

10. 12, 15, and 21 — 10. ______________ *(74–78)*

Name ______________________

Section ______________________

Date ______________________

Score ______________________

Post-Test 1

1. In 498,527, what is the value of the 5? 1. ______________ (*1.1*)

2. In 132,085, what digit is in the thousands' column? 2. ______________ (*1.1*)

3. Write 142,069 in words. 3. ______________________________ (*1.1*)

4. $\begin{array}{r} 579 \\ +862 \\ \hline \end{array}$ 4. ______________ (*1.2*)

5. $\begin{array}{r} 6812 \\ 9035 \\ +\ \ 744 \\ \hline \end{array}$ 5. ______________ (*1.2*)

6. $\begin{array}{r} 82 \\ 603 \\ 1492 \\ +8536 \\ \hline \end{array}$ 6. ______________ (*1.2*)

7. $\begin{array}{r} 644 \\ -328 \\ \hline \end{array}$ 7. ______________ (*1.3*)

8. $\begin{array}{r} 1125 \\ -\ \ 451 \\ \hline \end{array}$ 8. ______________ (*1.3*)

9. $\begin{array}{r} 20000 \\ -\ \ 8452 \\ \hline \end{array}$ 9. ______________ (*1.3*)

10. Find n: $n + 5 = 12$. 10. ______________ (*1.3*)

(*over*)

11. $\begin{array}{r} 64 \\ \times 83 \\ \hline \end{array}$

11. ____________ (*1.4*)

12. $\begin{array}{r} 2400 \\ \times \quad 69 \\ \hline \end{array}$

12. ____________ (*1.4*)

13. $\begin{array}{r} 659 \\ \times 372 \\ \hline \end{array}$

13. ____________ (*1.4*)

For problems 14 to 16, divide and show the quotient and remainder (if any).

14. $42\overline{)6297}$

14. ____________ (*1.5*)

15. $78\overline{)67392}$

15. ____________ (*1.5*)

16. $462\overline{)366505}$

16. ____________ (*1.5*)

17. Find x: $7 \cdot x = 42$.

17. ____________ (*1.5*)

For problems 18 and 19, factor composites into primes; if number is prime, write "Prime."

18. 140

18. ____________ (*1.6*)

19. 91

19. ____________ (*1.6*)

20. Find the least common multiple (LCM) of 15 and 24.

20. ____________ (*1.6*)

Name ____________

Section ____________

Date ____________

Score ____________

Post-Test 2

1. In 2,582,177, what is the value of the 8? 1. ____________ (*1.1*)

2. In 108,435, what digit is in the tens' column? 2. ____________ (*1.1*)

3. Write 263,015 in words. 3. ____________ (*1.1*)

4. $\begin{array}{r} 683 \\ +959 \\ \hline \end{array}$ 4. ____________ (*1.2*)

5. $\begin{array}{r} 7154 \\ 8587 \\ +\ 644 \\ \hline \end{array}$ 5. ____________ (*1.2*)

6. $\begin{array}{r} 91 \\ 436 \\ 6035 \\ +8568 \\ \hline \end{array}$ 6. ____________ (*1.2*)

7. $\begin{array}{r} 512 \\ -308 \\ \hline \end{array}$ 7. ____________ (*1.3*)

8. $\begin{array}{r} 2173 \\ -\ 885 \\ \hline \end{array}$ 8. ____________ (*1.3*)

9. $\begin{array}{r} 35000 \\ -18246 \\ \hline \end{array}$ 9. ____________ (*1.3*)

10. Find x: $x + 9 = 21$. 10. ____________ (*1.3*)

(*over*)

1

11. $\begin{array}{r} 73 \\ \times 56 \\ \hline \end{array}$

11. ______________ (*1.4*)

12. $\begin{array}{r} 45000 \\ \times \quad 17 \\ \hline \end{array}$

12. ______________ (*1.4*)

13. $\begin{array}{r} 812 \\ \times 563 \\ \hline \end{array}$

13. ______________ (*1.4*)

For problems 14 to 16, divide and show the quotient and remainder (if any).

14. $57\overline{)5094}$

14. ______________ (*1.5*)

15. $361\overline{)331037}$

15. ______________ (*1.5*)

16. $826\overline{)614057}$

16. ______________ (*1.5*)

17. Find n: $2 \cdot n = 20$.

17. ______________ (*1.5*)

For problems 18 and 19, factor composites into primes; if number is prime, write "Prime."

18. 210

18. ______________ (*1.6*)

19. 138

19. ______________ (*1.6*)

20. Find the least common multiple (LCM) of 15 and 21.

20. ______________ (*1.6*)

Name ____________________

Section ____________________

Date ____________________

Score ____________________

Pre-Test

2

1. Does $\frac{20}{42} = \frac{16}{36}$? — 1. ____________ (2.1)*

2. Reduce $\frac{45}{80}$ to simplest terms. — 2. ____________ (2.1)

3. Reduce $\frac{12}{90}$ to simplest terms. — 3. ____________ (2.1)

4. Write $6\frac{2}{5}$ as an improper fraction. — 4. ____________ (2.2)

5. Write $1\frac{1}{20}$ as an improper fraction. — 5. ____________ (2.2)

6. Write $\frac{86}{7}$ as a mixed number. — 6. ____________ (2.2)

7. Write $\frac{125}{16}$ as a mixed number. — 7. ____________ (2.2)

Perform the indicated operations.

8. $\frac{2}{5} \times \frac{3}{8}$ — 8. ____________ (2.3)

9. $\frac{1}{4} \times 6\frac{2}{3}$ — 9. ____________ (2.3)

10. $4\frac{1}{2} \times 3\frac{1}{3}$ — 10. ____________ (2.3)

(*over*)

*These numbers refer to the sections in the text.

2

11. $\frac{4}{9} \div \frac{16}{21}$ 11. ________________ (2.4)

12. $6\frac{1}{4} \div 7\frac{1}{2}$ 12. ________________ (2.4)

13. Divide $\frac{2}{3}$ by 18. 13. ________________ (2.4)

14. Add $\frac{1}{24} + \frac{2}{15}$. 14. ________________ (2.5)

15. Add $8\frac{7}{10} + 4\frac{3}{5}$. 15. ________________ (2.5)

16. Add $1\frac{1}{6} + \frac{13}{15}$. 16. ________________ (2.5)

17. Subtract $\frac{17}{20} - \frac{5}{12}$. 17. ________________ (2.6)

18. Subtract $17\frac{1}{12} - 14\frac{5}{8}$. 18. ________________ (2.6)

19. Subtract $18 - 4\frac{1}{3}$. 19. ________________ (2.6)

20. Find n: $n + \frac{1}{3} = 5$. 20. ________________ (2.6)

2.1 BASIC IDEAS OF FRACTIONS

Name ________________

Section ________________

Date ________________

Score ________________

Post-Test 1

1. What fraction of the square is shaded? — 1. __________ *(1–3)**

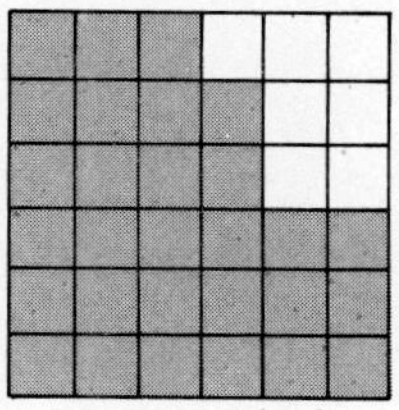

2. Shade the square at the right to show $\frac{5}{6}$. — 2. *(1–3)*

3. For $\frac{7}{10}$, identify the numerator and denominator. — 3. Numerator = ______ *(1–3)*
Denominator = ______

4. Does $\frac{14}{21} = \frac{26}{39}$? — 4. __________ *(4–8)*

5. Does $\frac{19}{25} = \frac{75}{100}$? — 5. __________ *(4–8)*

6. Does $\frac{51}{119} = \frac{33}{77}$? — 6. __________ *(4–8)*

Reduce the following fractions to simplest terms.

7. $\frac{36}{66}$ — 7. __________ *(9–15)*

8. $\frac{42}{48}$ — 8. __________ *(9–15)*

9. $\frac{39}{108}$ — 9. __________ *(9–15)*

10. $\frac{420}{540}$ — 10. __________ *(9–15)*

*These numbers refer to the examples in Chapter 2.

Name ______________________

Section ______________________

Date ______________________

Score ______________________

Post-Test 2

2

1. What fraction of the circle is shaded? 1. ______________ (1–3)

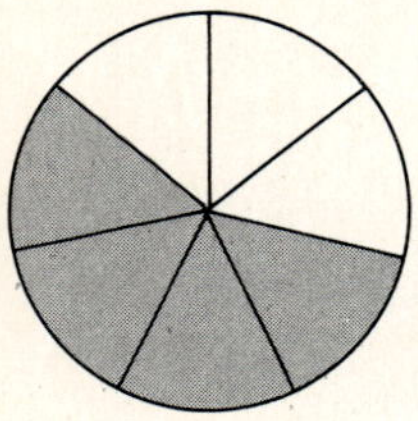

2. Shade the rectangle at the right to show $\frac{3}{10}$. 2. (1–3)

3. For $\frac{3}{8}$, identify the numerator and denominator. 3. Numerator = ________ (1–3)

Denominator = ______

4. Does $\frac{4}{14} = \frac{6}{21}$? 4. ______________ (4–8)

5. Does $\frac{6}{20} = \frac{10}{28}$? 5. ______________ (4–8)

6. Does $\frac{8}{28} = \frac{20}{64}$? 6. ______________ (4–8)

Reduce the following fractions to simplest terms.

7. $\frac{14}{48}$ 7. ______________ (9–15)

8. $\frac{60}{96}$ 8. ______________ (9–15)

9. $\frac{105}{180}$ 9. ______________ (9–15)

10. $\frac{54}{144}$ 10. ______________ (9–15)

Name ____________________

Section ____________________

Date ____________________

Score ____________________

Post-Test 1

Write the following mixed numbers as improper fractions.

1. $3\frac{1}{7}$ — 1. ____________ (16–20)

2. $10\frac{1}{2}$ — 2. ____________ (16–20)

3. $9\frac{3}{8}$ — 3. ____________ (16–20)

4. $21\frac{5}{12}$ — 4. ____________ (16–20)

5. $82\frac{10}{11}$ — 5. ____________ (16–20)

Write the following improper fractions as mixed numbers.

6. $\frac{9}{2}$ — 6. ____________ (21–24)

7. $\frac{25}{3}$ — 7. ____________ (21–24)

8. $\frac{75}{7}$ — 8. ____________ (21–24)

9. $\frac{163}{11}$ — 9. ____________ (21–24)

10. $\frac{386}{21}$ — 10. ____________ (21–24)

2.2 FRACTION/MIXED-NUMBER CONVERSIONS

Name ______________________

Section ______________________

Date ______________________

Score ______________________

Post-Test 2

2

Write the following mixed numbers as fractions.

1. $4\frac{7}{8}$ — 1. ______________ (16–20)

2. $5\frac{1}{13}$ — 2. ______________ (16–20)

3. $10\frac{3}{16}$ — 3. ______________ (16–20)

4. $37\frac{2}{9}$ — 4. ______________ (16–20)

5. $99\frac{44}{100}$ — 5. ______________ (16–20)

Write the following improper fractions as mixed numbers.

6. $\frac{10}{7}$ — 6. ______________ (21–24)

7. $\frac{21}{4}$ — 7. ______________ (21–24)

8. $\frac{37}{10}$ — 8. ______________ (21–24)

9. $\frac{201}{13}$ — 9. ______________ (21–24)

10. $\frac{447}{31}$ — 10. ______________ (21–24)

2.3 MULTIPLICATION OF FRACTIONS AND MIXED NUMBERS

Post-Test 1

Name ______________________

Section ______________________

Date ______________________

Score ______________________

Multiply the following fractions and mixed numbers.

1. $7 \times \frac{1}{4}$ — 1. ______________ (25, 26)

2. $\frac{2}{7} \times \frac{9}{10}$ — 2. ______________ (27–30)

3. $\frac{10}{21} \times \frac{7}{15}$ — 3. ______________ (27–30)

4. $\frac{1}{5} \times \frac{15}{19} \times \frac{7}{12}$ — 4. ______________ (27–30)

5. $2\frac{1}{2} \times 1\frac{1}{5}$ — 5. ______________ (31–35)

6. $4\frac{1}{5} \times 2\frac{1}{7}$ — 6. ______________ (31–35)

7. $\frac{2}{3} \times 6\frac{2}{5} \times 1\frac{7}{8}$ — 7. ______________ (31–35)

8. Find n: $n \div \frac{2}{3} = \frac{4}{7}$. — 8. ______________ (36–38)

9. Find the area of the rectangle. — 9. ______________ (30, 35)

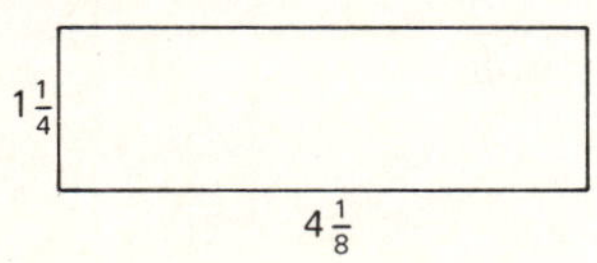

10. The chance of flipping seven straight heads with a fair coin is given by — 10. ______________ (30, 35)

$$\frac{1}{2} \times \frac{1}{2} \times \frac{1}{2} \times \frac{1}{2} \times \frac{1}{2} \times \frac{1}{2} \times \frac{1}{2}$$

Compute this.

2.3 MULTIPLICATION OF FRACTIONS AND MIXED NUMBERS

Name ______________

Section ______________

Date ______________

Score ______________

Post-Test 2

2

Multiply the following fractions and mixed numbers.

1. $8 \times \frac{4}{7}$ — 1. ______________ (25, 26)

2. $\frac{3}{14} \times \frac{2}{11}$ — 2. ______________ (27–30)

3. $\frac{15}{28} \times \frac{4}{21}$ — 3. ______________ (27–30)

4. $\frac{1}{12} \times \frac{3}{14} \times \frac{4}{9}$ — 4. ______________ (27–30)

5. $5\frac{3}{5} \times 2\frac{6}{7}$ — 5. ______________ (31–35)

6. $12\frac{1}{2} \times 8\frac{1}{10}$ — 6. ______________ (31–35)

7. $\frac{7}{10} \times 6\frac{2}{3} \times 2\frac{1}{10}$ — 7. ______________ (31–35)

8. Find n: $n \div \frac{4}{7} = \frac{2}{9}$. — 8. ______________ (36–38)

9. It takes Jake $2\frac{1}{3}$ hours to paint a room. How long will it take him to paint 8 rooms? — 9. ______________ (30, 35)

10. A turtle can move at $\frac{1}{20}$ mile per hour. How far can it move in $\frac{1}{6}$ hour (10 minutes)? — 10. ______________ (30, 35)

2.4 DIVISION OF FRACTIONS AND MIXED NUMBERS

Name ______________________

Section ______________________

Date ______________________

Score ______________________

Post-Test 1

Divide the following mixed numbers and fractions.

1. $\frac{2}{9} \div \frac{3}{5}$ 1. ______________ (39–43)

2. $\frac{8}{9} \div \frac{2}{3}$ 2. ______________ (39–43)

3. $\frac{18}{11} \div \frac{15}{22}$ 3. ______________ (39–43)

4. $\frac{7}{60} \div \frac{14}{15}$ 4. ______________ (39–43)

5. $3\frac{1}{4} \div 2\frac{1}{5}$ 5. ______________ (44–48)

6. $4\frac{3}{4} \div 6\frac{3}{7}$ 6. ______________ (44–48)

7. $21 \div \frac{3}{10}$ 7. ______________ (44–48)

8. Find n: $\frac{2}{3} \cdot n = \frac{7}{10}$.   8. ______________ (49–51)

9. The area of the rectangle shown is $12\frac{1}{2}$ square feet. Find the missing length.   9. ______________ (43, 48)

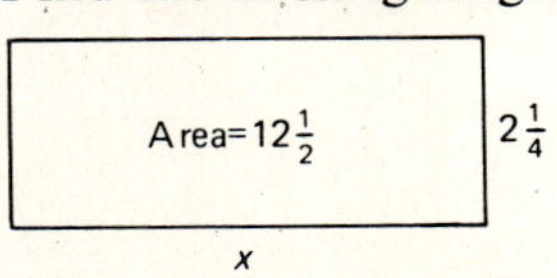

10. A town allows 18 acres of land set aside for families to plant. If each family gets $\frac{2}{75}$ acre, how many families can plant?   10. ______________ (43, 48)

2.4 DIVISION OF FRACTIONS AND MIXED NUMBERS

Name ____________

Section ____________

Date ____________

Score ____________

Post-Test 2

2

Divide the following fractions and mixed numbers.

1. $\frac{3}{11} \div \frac{2}{5}$ — 1. ____________ (39–43)

2. $\frac{6}{35} \div \frac{15}{14}$ — 2. ____________ (39–43)

3. $\frac{5}{18} \div \frac{25}{28}$ — 3. ____________ (39–43)

4. $\frac{11}{120} \div \frac{77}{20}$ — 4. ____________ (39–43)

5. $10\frac{1}{2} \div 1\frac{1}{13}$ — 5. ____________ (44–48)

6. $2\frac{2}{9} \div 6\frac{1}{4}$ — 6. ____________ (44–48)

7. $12\frac{3}{5} \div \frac{7}{10}$ — 7. ____________ (44–48)

8. Find n: $6 \cdot n = 2\frac{1}{4}$. — 8. ____________ (49–51)

9. A store has a bolt of cloth $45\frac{1}{2}$ yards long. How many $1\frac{1}{3}$-yard pieces can this be cut into? — 9. ____________ (43, 48)

10. Roger can run $\frac{1}{2}$ mile in $\frac{1}{24}$ hour. How fast is this is miles per hour? — 10. ____________ (43, 48)

2.5 ADDITION OF FRACTIONS AND MIXED NUMBERS

Name ____________

Section ____________

Date ____________

Score ____________

Post-Test 1

2

Add the following fractions and mixed numbers.

1. $\frac{4}{17} + \frac{6}{17}$ — 1. ____________ (52, 53)

2. $\frac{2}{5} + \frac{3}{10}$ — 2. ____________ (54–58)

3. $\frac{5}{12} + \frac{3}{16}$ — 3. ____________ (54–58)

4. $\frac{7}{18} + \frac{16}{27}$ — 4. ____________ (54–58)

5. $4\frac{1}{9} + 17\frac{7}{15}$ — 5. ____________ (59–63)

6. $121\frac{19}{20} + 93\frac{7}{18}$ — 6. ____________ (59–63)

7. $9\frac{1}{10} + 2\frac{4}{5} + 8\frac{9}{20}$ — 7. ____________ (59–63)

8. Find n: $n - \frac{1}{2} = \frac{3}{4}$. — 8. ____________ (64–66)

9. Ralph has three pieces of stereo wire: $14\frac{1}{4}$, $6\frac{1}{6}$, and $24\frac{1}{12}$ feet. How long a wire can he have if he splices them together? — 9. ____________ (58, 63)

10. Find the perimeter of the triangle. — 10. ____________ (58, 63)

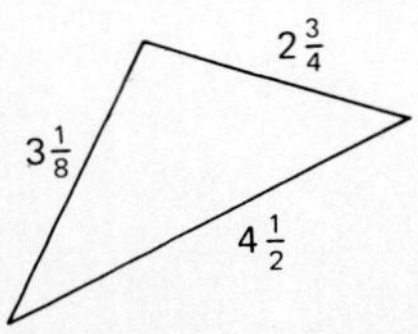

2.5 ADDITION OF FRACTIONS AND MIXED NUMBERS

Name ______________

Section ______________

Date ______________

Score ______________

Post-Test 2

2

Add the following fractions and mixed numbers.

1. $\frac{5}{11} + \frac{2}{11}$ — 1. ______________ (52, 53)

2. $\frac{5}{16} + \frac{3}{32}$ — 2. ______________ (54–58)

3. $\frac{7}{18} + \frac{2}{21}$ — 3. ______________ (54–58)

4. $\frac{4}{15} + \frac{3}{20}$ — 4. ______________ (54–58)

5. $6\frac{1}{5} + 11\frac{1}{4}$ — 5. ______________ (59–63)

6. $142\frac{13}{15} + 89\frac{7}{20}$ — 6. ______________ (59–63)

7. $16\frac{4}{25} + 21\frac{17}{20} + \frac{37}{50}$ — 7. ______________ (59–63)

8. Find n: $n - \frac{1}{3} = 2\frac{1}{2}$. — 8. ______________ (64–66)

9. Ellen owns $\frac{4}{21}$ of ABC Lamp store. She buys another $\frac{1}{15}$ of it. How much of the store does she now own? — 9. ______________ (58, 63)

10. Mitch makes four short films: $6\frac{1}{2}$, $8\frac{2}{3}$, $10\frac{1}{4}$, and $5\frac{7}{12}$ minutes. How long a showing would the four films make? — 10. ______________ (58, 63)

2.6 SUBTRACTION OF FRACTIONS AND MIXED NUMBERS

Name ______________

Section ______________

Date ______________

Score ______________

Post-Test 1

2

Subtract the following fractions and mixed numbers.

1. $\frac{6}{7} - \frac{4}{7}$ — 1. ______________ (67)

2. $\frac{7}{8} - \frac{1}{4}$ — 2. ______________ (68–71)

3. $\frac{3}{10} - \frac{2}{15}$ — 3. ______________ (68–71)

4. $\frac{7}{20} - \frac{5}{24}$ — 4. ______________ (68–71)

5. $10\frac{1}{2} - 6\frac{1}{5}$ — 5. ______________ (72–76)

6. $28\frac{1}{6} - 12\frac{5}{9}$ — 6. ______________ (72–76)

7. $4 - \frac{1}{40}$ — 7. ______________ (72–76)

8. Find n: $n + \frac{1}{3} = \frac{1}{2}$. — 8. ______________ (77–79)

9. Which fraction is larger, $\frac{7}{12}$ or $\frac{11}{18}$? — 9. ______________ (80, 81)

10. One year ago, Richard weighed $261\frac{1}{8}$ pounds. Now he weighs $179\frac{3}{4}$ pounds. How much weight did he lose? — 10. ______________ (76)

2.6 SUBTRACTION OF FRACTIONS AND MIXED NUMBERS

Name ______________________

Section ______________________

Date ______________________

Score ______________________

Post-Test 2

2

Subtract the following fractions and mixed numbers.

1. $\frac{29}{47} - \frac{17}{47}$ — 1. ______________ *(67)*

2. $\frac{1}{4} - \frac{1}{7}$ — 2. ______________ *(68–71)*

3. $\frac{3}{20} - \frac{1}{15}$ — 3. ______________ *(68–71)*

4. $\frac{11}{24} - \frac{5}{18}$ — 4. ______________ *(68–71)*

5. $19\frac{1}{6} - 18\frac{1}{8}$ — 5. ______________ *(72–76)*

6. $45\frac{1}{9} - 32\frac{4}{15}$ — 6. ______________ *(72–76)*

7. $1000 - 651\frac{1}{3}$ — 7. ______________ *(72–76)*

8. Find n: $n + \frac{2}{3} = 1\frac{1}{2}$. — 8. ______________ *(77–79)*

9. Which fraction is larger, $\frac{5}{8}$ or $\frac{7}{12}$? — 9. ______________ *(80, 81)*

10. Kathy is trying to jog 300 miles this year. So far, she has jogged $173\frac{2}{5}$ miles. How much farther does she need to go? — 10. ______________ *(76)*

Name ______________________
Section ______________________
Date ______________________
Score ______________________

Post-Test 1

1. Does $\frac{8}{9} = \frac{7}{8}$? 1. ______________ (2.1)

2. Reduce $\frac{48}{72}$ to simplest terms. 2. ______________ (2.1)

3. Reduce $\frac{98}{140}$ to simplest terms. 3. ______________ (2.1)

4. Write $10\frac{1}{5}$ as an improper fraction. 4. ______________ (2.2)

5. Write $6\frac{6}{7}$ as an improper fraction. 5. ______________ (2.2)

6. Write $\frac{41}{3}$ as a mixed number. 6. ______________ (2.2)

7. Write $\frac{301}{11}$ as a mixed number. 7. ______________ (2.2)

Perform the indicated operations.

8. $\frac{3}{4} \times \frac{2}{9}$ 8. ______________ (2.3)

9. $\frac{1}{5} \times 7\frac{1}{7}$ 9. ______________ (2.3)

10. Multiply $5\frac{2}{5}$ by $6\frac{2}{3}$. 10. ______________ (2.3)

(*over*)

11. $\frac{3}{5} \div \frac{21}{25}$ — 11. ____________ (2.4)

12. $4\frac{1}{5} \div 2\frac{7}{10}$ — 12. ____________ (2.4)

13. Divide 16 by $\frac{4}{5}$. — 13. ____________ (2.4)

14. $\frac{5}{18} + \frac{1}{24}$ — 14. ____________ (2.5)

15. $75\frac{4}{5} + 22\frac{7}{10}$ — 15. ____________ (2.5)

16. Add $\frac{7}{20} + 9\frac{4}{15}$. — 16. ____________ (2.5)

17. $\frac{4}{15} - \frac{1}{12}$ — 17. ____________ (2.6)

18. $21\frac{3}{8} - 15\frac{3}{4}$ — 18. ____________ (2.6)

19. Subtract $100 - 7\frac{1}{9}$. — 19. ____________ (2.6)

20. Find n: $n + \frac{1}{5} = 2$. — 20. ____________ (2.6)

CHAPTER 2 FRACTIONS AND MIXED NUMBERS

Name ______________________

Section ______________________

Date ______________________

Score ______________________

Post-Test 2

2

1. Does $\frac{20}{25} = \frac{28}{35}$? — 1. ______________ (2.1)

2. Reduce $\frac{15}{21}$ to simplest terms. — 2. ______________ (2.1)

3. Reduce $\frac{84}{120}$ to simplest terms. — 3. ______________ (2.1)

4. Write $8\frac{4}{7}$ as an improper fraction. — 4. ______________ (2.2)

5. Write $4\frac{15}{16}$ as an improper fraction. — 5. ______________ (2.2)

6. Write $\frac{39}{5}$ as a mixed number. — 6. ______________ (2.2)

7. Write $\frac{175}{8}$ as a mixed number. — 7. ______________ (2.2)

Perform the indicated operations.

8. $\frac{4}{7} \times \frac{21}{26}$ — 8. ______________ (2.3)

9. $\frac{1}{10} \times 6\frac{2}{3}$ — 9. ______________ (2.3)

10. Multiply $4\frac{2}{7} \times 4\frac{2}{3}$. — 10. ______________ (2.3)

(*over*)

2

11. $\frac{4}{9} \div \frac{16}{27}$ 11. ________________ (*2.4*)

12. $6\frac{1}{4} \div 5\frac{5}{6}$ 12. ________________ (*2.4*)

13. Divide $\frac{2}{15}$ by 10. 13. ________________ (*2.4*)

14. $\frac{5}{12} + \frac{11}{15}$ 14. ________________ (*2.5*)

15. $61\frac{4}{5} + 96\frac{13}{15}$ 15. ________________ (*2.5*)

16. $\frac{17}{20} + 5\frac{11}{24}$ 16. ________________ (*2.5*)

17. $\frac{9}{10} - \frac{7}{12}$ 17. ________________ (*2.6*)

18. $24\frac{1}{6} - 19\frac{8}{9}$ 18. ________________ (*2.6*)

19. Subtract $\frac{2}{9}$ from 200. 19. ________________ (*2.6*)

20. Find n: $n + \frac{1}{3} = 7$. 20. ________________ (*2.6*)

CHAPTER 3 DECIMALS

Name ____________________

Section ____________________

Pre-Test

Date ____________________

Score ____________________

---

1. In 8.97352, what is the value of the 3? 1. ______________ *(3.1)**

2. Round 6.78241 to the nearest tenth. 2. ______________ *(3.1)*

3. Round 1958.2 to the nearest hundred. 3. ______________ *(3.1)*

4. Determine the larger decimal: 0.7 or 0.666. 4. ______________ *(3.1)*

5. $\begin{array}{r} 6.01 \\ 8.523 \\ +9.66 \\ \hline \end{array}$ 5. ______________ *(3.2)*

6. $\begin{array}{r} 292.7 \\ -184.823 \\ \hline \end{array}$ 6. ______________ *(3.2)*

7. Add 8.3 + 9.06 + 12.84 + 5. 7. ______________ *(3.2)*

8. Subtract 450 - 382.73. 8. ______________ *(3.2)*

9. Find n: $n - 1.2 = 8.3$. 9. ______________ *(3.2)*

10. $\begin{array}{r} 5.2 \\ \times 4.8 \\ \hline \end{array}$ 10. ______________ *(3.3)*

11. $\begin{array}{r} 48.2 \\ \times 1.07 \\ \hline \end{array}$ 11. ______________ *(3.3)*

12. Multiply 0.0003 by 0.005. 12. ______________ *(3.3)*

(*over*)

*These numbers refer to the sections in the text.

3

Round answers to problems 13 and 14 to two decimal places.

13. $0.17\overline{)2.192}$ 13. ______ (3.4)

14. $0.07\overline{)8.5}$ 14. ______ (3.4)

15. Divide 602.1 by 18.4 and round your answer to the nearest tenth. 15. ______ (3.4)

16. Find n: $(0.6) \cdot n = 2.4$. 16. ______ (3.4)

17. Write $\frac{4}{7}$ as a two-place decimal. 17. ______ (3.5)

18. Write $7\frac{6}{13}$ as a three-place decimal. 18. ______ (3.5)

19. Write 0.85 as a reduced fraction. 19. ______ (3.5)

20. Write 3.14 as a reduced mixed number. 20. ______ (3.5)

3.1 DECIMAL PLACE VALUE AND ROUNDING OFF

Name ____________________

Section ____________________

Date ____________________

Score ____________________

Post-Test 1

1.	In 23.7489, what is the value of the 8?	1. ________	*(1–3)**
2.	In 390.03821, what is the value of the 2?	2. ________	*(1–3)*
3.	In 6.123456, what is the value of the 1?	3. ________	*(1–3)*
4.	Round 656.5115 to the nearest tenth.	4. ________	*(4–6)*
5.	Round 61.9616 to the nearest thousandth.	5. ________	*(4–6)*
6.	Round 5.667487 to the nearest ten-thousandth.	6. ________	*(4–6)*
7.	Round 0.2985 to the nearest hundredth.	7. ________	*(4–6)*
8.	Round 94.482 to the nearest ten.	8. ________	*(4–6)*
9.	Determine the larger decimal: 0.567 or 0.63.	9. ________	*(7–9)*
10.	Determine the largest decimal: 0.04, 0.004, 0.444, 0.44.	10. ________	*(7–9)*

3

*These numbers refer to the examples in Chapter 3.

3.1 DECIMAL PLACE VALUE AND ROUNDING OFF

Name ____________

Section ____________

Date ____________

Score ____________

Post-Test 2

---

1. In 48.09274, what is the value of the 9? 1. ____________ *(1–3)*

2. In 5.937264, what is the value of the 2? 2. ____________ *(1–3)*

3

3. In 90.837462, what is the value of the 6? 3. ____________ *(1–3)*

4. Round 84.519 to the nearest hundredth. 4. ____________ *(4–6)*

5. Round 7.028126 to the nearest ten-thousandth. 5. ____________ *(4–6)*

6. Round 981.07 to one decimal place. 6. ____________ *(4–6)*

7. Round 246.094 to the nearest tenth. 7. ____________ *(4–6)*

8. Round 123.761 to the nearest one. 8. ____________ *(4–6)*

9. Determine the larger decimal: 0.22 or 0.022. 9. ____________ *(7–9)*

10. Determine the largest decimal: 0.89, 0.98, 0.889, 0.8988, 0.9898. 10. ____________ *(7–9)*

3.2 ADDITION AND SUBTRACTION OF DECIMALS

Post-Test 1

Name ____________

Section ____________

Date ____________

Score ____________

Perform the indicated operations.

1.	$1.07 + 36.859$	1. ____________	*(10–12)*
2.	$8.37 + 3.952 + 27.6$	2. ____________	*(10–12)*
3.	$3.68 + 5.971 + 98.0231$	3. ____________	*(10–12)*
4.	$199.3 - 75.92$	4. ____________	*(13–16)*
5.	$595.5 - 38.216$	5. ____________	*(13–16)*
6.	$60.48 - 51.796$	6. ____________	*(13–16)*
7.	Find n: $n - 21.4 = 54.89$.	7. ____________	*(17, 18)*
8.	Find n: $n + 32.6 = 78.43$.	8. ____________	*(19, 20)*
9.	Mary writes the following checks: rent, \$155.00; groceries, \$27.86; jeans, \$21.57; drugstore, \$7.11. Find the total of these checks.	9. ____________	*(10)*
10.	Jonathan wins \$100,000 in a lottery. He pays \$37,982.79 in income tax. How much does he have left?	10. ____________	*(16)*

3

3.2 ADDITION AND SUBTRACTION OF DECIMALS

Name ______________

Section ______________

Date ______________

Score ______________

Post-Test 2

3

Perform the indicated operations.

1.	$74.92 + 23.673$	1. ______________	(*10–12*)
2.	$2.586 + 14.95 + 9.0065$	2. ______________	(*10–12*)
3.	$37.1 + 39.57 + 864.093$	3. ______________	(*10–12*)
4.	$345.527 - 57.98$	4. ______________	(*13–16*)
5.	$748.48 - 458.531$	5. ______________	(*13–16*)
6.	$489.02 - 421.458$	6. ______________	(*13–16*)
7.	Find n: $n - 32.4 = 12.86$.	7. ______________	(*17, 18*)
8.	Find n: $n + 14.7 = 20$.	8. ______________	(*19, 20*)
9.	Find the perimeter of the triangle.	9. ______________	(*12*)

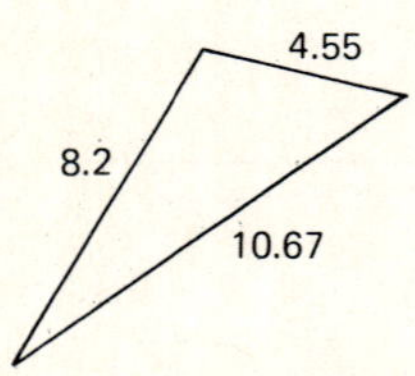

10.	Roger buys a \$400 stereo set, but he gets a \$37.75 discount. What does he pay?	10. ______________	(*16*)

3.3 MULTIPLICATION OF DECIMALS

Name ____________________

Section ____________________

Post-Test 1

Date ____________________

Score ____________________

Multiply the following decimals.

1. 0.1×9.2 — 1. ____________________ (*21, 22*)

2. 0.001×0.87 — 2. ____________________ (*21, 22*)

3. 4.04×24.04 — 3. ____________________ (*23–28*)

4. 7.4×0.0949 — 4. ____________________ (*23–28*)

5. 0.014×0.00012 — 5. ____________________ (*23–28*)

6. $0.15 \times 66{,}000$ — 6. ____________________ (*23–28*)

7. 35.5×40.44 — 7. ____________________ (*23–28*)

8. Find n: $n \div 1.5 = 40$. — 8. ____________________ (*29–31*)

9. Multiply 0.241×0.33 and round your answer to three decimal places. — 9. ____________________ (*25, 4*)

10. Tracy's tank holds 22.4 gallons of gasoline. Her car gets 19.2 miles per gallon. How far can she drive on one tank? — 10. ____________________ (*28*)

3

Name ______________

Section ______________

Date ______________

Score ______________

Post-Test 2

Multiply the following decimals.

1. 0.04 × 0.01 — 1. ______________ *(21, 22)*

2. 650 × 0.001 — 2. ______________ *(21, 22)*

3. 8.8 × 9.63 — 3. ______________ *(23–28)*

4. 5.17 × 7.002 — 4. ______________ *(23–28)*

5. 0.016 × 0.064 — 5. ______________ *(23–28)*

6. 0.035 × 54,000 — 6. ______________ *(23–28)*

7. 0.051 × 3.24 — 7. ______________ *(23–28)*

8. Find n: $n \div 2.3 = 500$. — 8. ______________ *(29–31)*

9. Multiply 21,584 × 0.35 and round your answer to the nearest tenth. — 9. ______________ *(24, 4)*

10. Karyn and Dave buy 41.3 square yards of carpeting that costs \$8.95 per square yard. How much does this cost? (Round your answer to the nearest penny, or two decimal places.) — 10. ______________ *(28)*

3.4 DIVISION OF DECIMALS

Name ______________

Section ______________

Date ______________

Score ______________

Post-Test 1

Divide the following decimals. Round your answers in problems 5 and 6 to three-place accuracy, in problems 7 and 8 to the nearest hundredth.

1. $0.1645 \div 7$ — 1. ______________ *(32, 33)*

2. $\dfrac{91.2}{38}$ — 2. ______________ *(32, 33)*

3. $300 \div 0.15$ — 3. ______________ *(34–36)*

4. $\dfrac{0.0002485}{0.035}$ — 4. ______________ *(34–36)*

5. $17 \div 2400$ — 5. ______________ *(37–39)*

6. $700.6 \div 84.7$ — 6. ______________ *(37–39)*

7. $\dfrac{41}{0.006}$ — 7. ______________ *(37–39)*

8. $\dfrac{3.9}{9}$ — 8. ______________ *(37–39)*

9. Find n: $2.5 \cdot n = 1500$. — 9. ______________ *(40–42)*

10. Rich drives 301.5 miles on 14.3 gallons of gasoline. How many miles per gallon is this, to the nearest tenth? — 10. ______________ *(38)*

3

3.4 DIVISION OF DECIMALS

Name ____________

Section ____________

Date ____________

Score ____________

Post-Test 2

Divide the following decimals. Round your answers to problems 5 and 6 to two-place accuracy, in problems 7 and 8 to the nearest thousandth.

1. $1.032 \div 8$ — 1. ____________ (32, 33)

2. $\dfrac{1.708}{61}$ — 2. ____________ (32, 33)

3. $0.0035 \div 0.014$ — 3. ____________ (34–36)

4. $\dfrac{1512}{0.42}$ — 4. ____________ (34–36)

5. $92 \div 97$ — 5. ____________ (37–39)

6. $\dfrac{12}{7.8}$ — 6. ____________ (37–39)

7. $63 \div 0.038$ — 7. ____________ (37–39)

8. $\dfrac{0.0079}{0.045}$ — 8. ____________ (37–39)

9. Find n: $1.6 \cdot n = 0.464$. — 9. ____________ (40–42)

10. A certain book has 622 pages (311 sheets). The total thickness of the pages is 3.25 cm. How thick is each sheet (to the nearest thousandth)? — 10. ____________ (39)

3

3.5 FRACTION-DECIMAL CONVERSIONS

Name ______________

Section ______________

Date ______________

Score ______________

Post-Test 1

---

1. Write $\frac{7}{12}$ as a two-place decimal. 1. ______________ (*43–46*)

2. Write $\frac{5}{11}$ as a three-place decimal. 2. ______________ (*43–46*)

3. Write $2\frac{1}{7}$ as a decimal accurate to the nearest thousandth. 3. ______________ (*43–46*)

4. Write $\frac{79}{150}$ as a decimal accurate to the nearest hundredth. 4. ______________ (*43–46*)

5. Two-sevenths of a group are Republicans. Write this as a two-place decimal. 5. ______________ (*43–46*)

6. Write 0.23 as a reduced fraction. 6. ______________ (*47–51*)

7. Write 0.065 as a reduced fraction. 7. ______________ (*47–51*)

8. Write 7.15 as a reduced mixed number. 8. ______________ (*47–51*)

9. Write 4.25 as a reduced improper fraction. 9. ______________ (*47–51*)

10. Chuck gets 21.35 miles per gallon. Write this as a reduced mixed number. 10. ______________ (*47–51*)

Name ______________________

Section ______________________

Date ______________________

Score ______________________

Post-Test 2

---

1. Write $\frac{1}{13}$ as a three-place decimal. 1. ______________ *(43–46)*

2. Write $\frac{4}{50}$ as a two-place decimal. 2. ______________ *(43–46)*

3. Write $4\frac{5}{9}$ as a decimal accurate to the nearest thousandth. 3. ______________ *(43–46)*

4. Write $\frac{177}{270}$ as a decimal accurate to the nearest hundredth. 4. ______________ *(43–46)*

5. Christy and Matt own $\frac{1}{54}$ of a townhouse co-op. Write this as a decimal accurate to the nearest ten-thousandth. 5. ______________ *(43–46)*

6. Write 0.83 as a reduced fraction. 6. ______________ *(47–51)*

7. Write 0.0025 as a reduced fraction. 7. ______________ *(47–51)*

8. Write 8.75 as a reduced mixed number. 8. ______________ *(47–51)*

9. Write 1.55 as a reduced improper fraction. 9. ______________ *(47–51)*

10. The Ritziks' driveway is 17.6 meters long. Write this as a reduced mixed number. 10. ______________ *(47–51)*

3

Name ____________

Section ____________

Date ____________

Score ____________

Post-Test 1

1. In 45.0892, what is the value of the 9? 1. ____________ (*3.1*)*

2. Round 81.02198 to the nearest hundredth. 2. ____________ (*3.1*)

3. Round 683462.1 to the nearest ten thousand. 3. ____________ (*3.1*)

4. Determine the larger decimal: 0.98 or 0.899. 4. ____________ (*3.1*)

5. $\begin{array}{r} 17.2 \\ 8.93 \\ +23.609 \\ \hline \end{array}$ 5. ____________ (*3.2*)

6. $\begin{array}{r} 667.25 \\ -185.6 \\ \hline \end{array}$ 6. ____________ (*3.2*)

7. Add 0.03 + 0.7 + 1.08 + 0.821. 7. ____________ (*3.2*)

8. Subtract 300 - 245.63. 8. ____________ (*3.2*)

9. Find n: $n - 2.5 = 4.1$. 9. ____________ (*3.2*)

10. $\begin{array}{r} 6.9 \\ \times 0.7 \\ \hline \end{array}$ 10. ____________ (*3.3*)

11. $\begin{array}{r} 6.15 \\ \times 1.08 \\ \hline \end{array}$ 11. ____________ (*3.3*)

12. Multiply 0.0015 by 4.05. 12. ____________ (*3.3*)

(*over*)

*These numbers refer to the sections in the text.

Round your answers to problems 13 and 14 to the nearest hundredth.

13. $0.201\overline{\smash{)}1.65}$ 13. ________________ *(3.4)*

14. $6.3\overline{\smash{)}0.397}$ 14. ________________ *(3.4)*

15. Divide 172.4 by 8.5 and round your answer to the nearest tenth.   15. ________________ *(3.4)*

16. Find n: $(0.4) \cdot n = 0.08$.   16. ________________ *(3.4)*

17. Write $\frac{11}{12}$ as a three-place decimal.   17. ________________ *(3.5)*

3

18. Write $14\frac{1}{6}$ as a two-place decimal.   18. ________________ *(3.5)*

19. Write 0.28 as a reduced fraction.   19. ________________ *(3.5)*

20. Write 12.8 as a reduced mixed number.   20. ________________ *(3.5)*

Name ________________

Section ________________

Date ________________

Score ________________

Post-Test 2

1. In 34.269, what is the value of the 6? 1. ________ (*3.1*)

2. Round 18.0983 to the nearest hundredth. 2. ________ (*3.1*)

3. Round 283,456.7813 to the nearest hundred thousand. 3. ________ (*3.1*)

4. Determine the larger decimal: 2.01 or 2.099. 4. ________ (*3.1*)

5. $$\begin{array}{r} 7.73 \\ 18.652 \\ +\ 9.3 \\ \hline \end{array}$$ 5. ________ (*3.2*)

6. $$\begin{array}{r} 482.1 \\ -\ 193.86 \\ \hline \end{array}$$ 6. ________ (*3.2*)

7. Add 23 + 8.5 + 17.03 + 4.241. 7. ________ (*3.2*)

8. Subtract 4000 - 385.17. 8. ________ (*3.2*)

9. Find n: $n - 0.3 = 2.9$. 9. ________ (*3.2*)

10. $$\begin{array}{r} 7.15 \\ \times\ 0.3 \\ \hline \end{array}$$ 10. ________ (*3.3*)

11. $$\begin{array}{r} 0.473 \\ \times\ 2.03 \\ \hline \end{array}$$ 11. ________ (*3.3*)

12. Multiply 0.00035 by 0.0062. 12. ________ (*3.3*)

(*over*)

3

Round your answers to problems 13 and 14 to the nearest thousandth.

13. $0.63\overline{)0.715}$ — 13. __________ (3.4)

14. $14.1\overline{)0.842}$ — 14. __________ (3.4)

15. Divide 621.6 by 36.5 and round your answer to two-decimal-place accuracy. — 15. __________ (3.4)

16. Find n: $(2.5) \cdot n = 0.5$. — 16. __________ (3.4)

3

17. Write $\frac{13}{15}$ as a three-place decimal. — 17. __________ (3.5)

18. Write $17\frac{1}{4}$ as a two-place decimal. — 18. __________ (3.5)

19. Write 0.96 as a reduced fraction. — 19. __________ (3.5)

20. Write 35.2 as a reduced mixed number. — 20. __________ (3.5)

Name ______________

Section ______________

Date ______________

Score ______________

Pre-Test

---

1. Write 42.7% as a decimal.	1.	______________	*(4.1)*
2. Write 0.01% as a decimal.	2.	______________	*(4.1)*
3. Write 24% as a reduced fraction.	3.	______________	*(4.1)*
4. Write 550% as a reduced mixed number.	4.	______________	*(4.1)*
5. Write 0.072 as a percent.	5.	______________	*(4.2)*
6. Write 6.08 as a percent.	6.	______________	*(4.2)*
7. Write $\frac{7}{31}$ as a percent with no decimal places.	7.	______________	*(4.2)*
8. Write $\frac{47}{49}$ as a percent with one decimal place.	8.	______________	*(4.2)*
9. Translate the following into an equation: What is 25% of 300?	9.	______________	*(4.3)*

(over)

Solve the following.

10. What is 9% of 1500? 10. ______________ (*4.3*)

11. Kevin Kampaign wins an election, 423 to 377. What percent of the vote did he get? 11. ______________ (*4.3*)

12. The Smiths clear $35,000 on their house *after* an 8% commission. What was the selling price? 12. ______________ (*4.3*)

13. What is to 8 as 12 is to 25? 13. ______________ (*4.4*)

14. Lori drove 175 miles on 9.0 gallons of gas. How much gas does she need to drive 3000 miles? 14. ______________ (*4.4*)

15. For their store, Joe put up $6000 and Jane put up $5000. In two years, this store is worth $25,000. What is Jane's share? 15. ______________ (*4.4*)

4.1 CHANGING PERCENTS TO FRACTIONS AND DECIMALS

Name ____________________

Section ____________________

Date ____________________

Score ____________________

Post-Test 1

Write the following percents as reduced fractions.

1. 41% 1. ____________ *(1–6)*

2. 21.2% 2. ____________ *(1–6)*

3. 0.5% 3. ____________ *(1–6)*

4. 125% 4. ____________ *(1–6)*

5. The O'Conners are paying $9\frac{1}{2}$% on their mortgage. Write this as a reduced fraction. 5. ____________ *(1–6)*

4

Write the following percents as decimals.

6. 6.2% 6. ____________ *(7–12)*

7. 0.06% 7. ____________ *(7–12)*

8. 305% 8. ____________ *(7–12)*

9. Jeni got a $6\frac{1}{2}$% raise. Write this as a decimal. 9. ____________ *(7–12)*

10. In a recent poll, 55.3% of those surveyed said that they would vote for Senator Blah. Write this as a decimal. 10. ____________ *(7–12)*

4.1 CHANGING PERCENTS TO FRACTIONS AND DECIMALS

Name ____________

Section ____________

Date ____________

Score ____________

Post-Test 2

Write the following percents as reduced fractions.

1. 56% — 1. ____________ *(1–6)*

2. 30.8% — 2. ____________ *(1–6)*

3. 0.04% — 3. ____________ *(1–6)*

4. 330% — 4. ____________ *(1–6)*

5. A nurse is mixing a 15% solution. Write this as a reduced fraction. — 5. ____________ *(1–6)*

4

Write the following percents as decimals.

6. $4\frac{1}{4}\%$ — 6. ____________ *(7–12)*

7. 0.1% — 7. ____________ *(7–12)*

8. 2000% — 8. ____________ *(7–12)*

9. 98.3% — 9. ____________ *(7–12)*

10. In Illinois, the state income tax is $2\frac{1}{2}\%$. Write this as a decimal. — 10. ____________ *(7–12)*

Name ______________

Section ______________

Date ______________

Score ______________

Post-Test 1

Write the following decimals and fractions as percents.

1. 0.61 — 1. ______________ (*13–18*)

2. 0.053 — 2. ______________ (*13–18*)

3. 3.5 — 3. ______________ (*13–18*)

4. 0.0002 — 4. ______________ (*13–18*)

5. Fred computes his salary increase as 0.064. Write this as a percent. — 5. ______________ (*13–18*)

6. $\frac{5}{7}$ (as a percent with a fraction) — 6. ______________ (*19–25*)

7. $\frac{17}{80}$ (as a percent with a fraction) — 7. ______________ (*19–25*)

8. $\frac{8}{11}$ (as a percent with one decimal place) — 8. ______________ (*19–25*)

9. $\frac{87}{472}$ (as a percent with one decimal place) — 9. ______________ (*19–25*)

10. The chance of Kathy winning a contest is $\frac{1}{60}$. Write this as a percent (with a fraction). — 10. ______________ (*19–25*)

4

4.2 CHANGING FRACTIONS AND DECIMALS TO PERCENTS

Name ____________

Section ____________

Date ____________

Score ____________

Post-Test 2

Write the following decimals and fractions as percents.

1.	0.723	1.	____________	(13–18)
2.	0.011	2.	____________	(13–18)
3.	7	3.	____________	(13–18)
4.	0.00005	4.	____________	(13–18)
5.	The proportion of women in a certain group is 0.279. Write this as a percent.	5.	____________	(13–18)
6.	$\frac{7}{9}$ (as a percent with a fraction)	6.	____________	(19–25)
7.	$\frac{113}{160}$ (as a percent with a fraction)	7.	____________	(19–25)
8.	$\frac{5}{19}$ (as a percent with one decimal place)	8.	____________	(19–25)
9.	$\frac{87}{102}$ (as a percent with one decimal place)	9.	____________	(19–25)
10.	In a group, $\frac{12}{67}$ of the group wore gym shoes. Write this as a percent (with one decimal place).	10.	____________	(19–25)

4

Name ______________
Section ______________
Date ______________
Score ______________

Post-Test 1

1. Translate into an equation: Twelve is 28% of what? 1. ______________ *(26)*

Solve the following.

2. Twenty-one is what percent of 60? 2. ______________ *(29–33)*

3. One hundred is 40% of what? 3. ______________ *(29–33)*

4. What is 70% of 6900? 4. ______________ *(29–33)*

5. Forest State College is 46% women. There are 1890 men. What is the total size of the school? 5. ______________ *(35–42)*

6. Of all pregnancies, 1.1% result in twins. If a certain country has 1,600,000 pregnancies per year, how many pairs of twins can it expect? 6. ______________ *(35–42)*

7. One year, a townhouse group budgeted \$6500 for insurance, \$11,600 for management, \$14,200 for maintenance, and \$13,700 for reserves. What percent of the budget is for maintenance? 7. ______________ *(35–42)*

8. Tom makes \$8300 on the sale of land after a 10% commission. What was the actual selling price? 8. ______________ *(35–42)*

9. The Conways put \$2200 in the bank at $6\frac{1}{2}\%$ interest. How much interest do they earn in 1 year? 9. ______________ *(35–42)*

10. A 46-ounce can of fruit drink claims to be about 12% actual fruit juice. How much fruit juice is this? 10. ______________ *(35–42)*

4

4.3 PERCENT PROBLEMS

Name ______________________

Section ______________________

Date ______________________

Score ______________________

Post-Test 2

1. Translate into an equation: Fifty is what percent of 270? — 1. ______________ (26)

Solve the following.

2. Eighty-five is 17% of what? — 2. ______________ (29–33)

3. Four hundred is 4% of what? — 3. ______________ (29–33)

4. What is 13% of 700? — 4. ______________ (29–33)

4

5. Denise buys a \$420 stereo set. She is told that this is 30% off list price. What was the list price? — 5. ______________ (35–42)

6. Mickey earns \$1200 per month. He spends about \$96 a month on dates. What percent is this? — 6. ______________ (35–42)

7. In a recent election, Adams got 4500 votes, Clay got 4100 votes, and Jones got 400 votes. What percent of the vote did Adams get? — 7. ______________ (35–42)

8. Al and Nancy have saved \$3200. In a new housing development, they need a 5% down payment. How expensive a house can they buy with the \$3200? — 8. ______________ (35–42)

9. In one year the price of oil was \$3.50 a barrel. Then it went up 250%. What was the new price? — 9. ______________ (35–42)

10. Fred earns \$13,250 per year. He pays \$1106 federal tax, \$310 state tax, \$737 social security tax, \$512 real estate tax, and \$410 in other taxes. What percent of his income goes to pay taxes? — 10. ______________ (35–42)

Name ____________

Section ____________

Date ____________

Score ____________

Post-Test 1

1. What is to 26 as 8 is to 13? 1. ____________ *(49–51)*

2. Ten is to what as 24 is to 45? 2. ____________ *(49–51)*

3. Nineteen is to 6 as 95 is to what? 3. ____________ *(49–51)*

4. Mike drove 180 miles on 12 gallons of gas. How much gas does he need to drive 500 miles? 4. ____________ *(52–59)*

5. The Smiths were able to sod a 450-square-foot yard for $250. How much will it cost the Bakers to sod 800 square feet? 5. ____________ *(52–59)*

6. One kilogram is 2.2 pounds. How many kilograms is a 250-pound football player? 6. ____________ *(52–59)*

7. In a group of 80, there are 23 smokers. How many smokers would you expect in a group of 500? 7. ____________ *(52–59)*

8. A 3.5-foot stick casts a 2.5-foot shadow. How tall is a building that casts a 40-foot shadow? 8. ____________ *(52–59)*

9. If one pregnancy in 90 results in twins, what is the twins' birthrate per 100,000 pregnancies? 9. ____________ *(52–59)*

10. Messrs. X, Y, and Z make a joint investment. X puts up $800, Y puts up $1300, and Z puts up $600. After a year, their investments are worth $1800. What is X's share? 10. ____________ *(52–59)*

Name ____________

Section ____________

Date ____________

Score ____________

Post-Test 2

1. What is to 12 as 11 is to 4? — 1. ____________ (49–51)

2. Five is to 13 as 16 is to what? — 2. ____________ (49–51)

3. Twelve is to what as 7 is to 12? — 3. ____________ (49–51)

4. Carol drove 400 miles on 21 gallons of gas. How far can she drive on 75 gallons of gas? — 4. ____________ (52–59)

5. One foot is 0.305 meter. How many meters is a 5-foot 6-inch (5.5-foot) woman? — 5. ____________ (52–59)

6. It costs Tony $240 to carpet 18 square yards of floor. How much will it cost Gina to carpet 32 square yards with the same carpeting? — 6. ____________ (52–59)

7. In a group of 160 people, 90 are Democrats. At this rate, how large a group would be needed to find 10,000 Democrats? — 7. ____________ (52–59)

8. A 1.5-meter stick casts a 2.0-meter shadow. How tall is a building that casts a 22-meter shadow? — 8. ____________ (52–59)

9. The birthrate in America is about 14.8 per 1000. If there are 220,000,000 Americans, about how many births can one expect in one year? — 9. ____________ (52–59)

10. Mr. Jones leaves $120,000 to his children, Ann, Bob, and Chuck, to be split 4 to 5 to 6, respectively. What is Ann's share? — 10. ____________ (52–59)

Name ______________________

Section ______________________

Date ______________________

Score ______________________

Post-Test 1

1. Write 16.5% as a decimal. 1. ______________ *(4.1)*

2. Write 0.7% as a decimal. 2. ______________ *(4.1)*

3. Write 65% as a reduced fraction. 3. ______________ *(4.1)*

4. Write 225% as a reduced mixed number. 4. ______________ *(4.1)*

5. Write 2.1 as a percent. 5. ______________ *(4.2)*

6. Write 0.732 as a percent. 6. ______________ *(4.2)*

7. Write $\frac{8}{29}$ as a percent with no decimal places. 7. ______________ *(4.2)*

8. Write $\frac{99}{103}$ as a percent with one decimal place. 8. ______________ *(4.2)*

9. Translate into an equation:
   Fourteen is 8% of what number? 9. ______________ *(4.3)*

(over)

Solve the following.

10. Twenty is what percent of 85?	**10.**	________________	(*4.3*)
11. The Conrads put \$620 in the bank at 5.25% interest. How much interest do they earn?	**11.**	________________	(*4.3*)
12. A person should spend about 25% of his or her income on rent. How much income is needed to afford a \$350-per-month apartment?	**12.**	________________	(*4.3*)
13. Five is to what as 3 is to 8?	**13.**	________________	(*4.4*)
14. A 1-meter stick casts a 1.3-meter shadow. How tall is a building that casts a 26-meter shadow?	**14.**	________________	(*4.4*)
15. There are 11 joggers in a group of 60 people. At this rate, how many joggers would be in a group of 1000?	**15.**	________________	(*4.4*)

Name ____________

Section ____________

Date ____________

Score ____________

Post-Test 2

1. Write 5.75% as a decimal. 1. ____________ (*4.1*)

2. Write $\frac{1}{2}$% as a decimal. 2. ____________ (*4.1*)

3. Write 98% as a reduced fraction. 3. ____________ (*4.1*)

4. Write 0.02% as a reduced fraction. 4. ____________ (*4.1*)

5. Write 0.632 as a percent. 5. ____________ (*4.2*)

6. Write 0.0003 as a percent. 6. ____________ (*4.2*)

7. Write $\frac{16}{43}$ as a percent with one decimal place. 7. ____________ (*4.2*)

8. Write $\frac{152}{250}$ as a percent with no decimal places. 8. ____________ (*4.2*)

9. Translate the following into an equation: Thirty is what percent of 900? 9. ____________ (*4.3*)

(*over*)

Solve the following.

10. Thirty is 18% of what? 10. ________ (4.3)

11. There are 650 people in a club, of whom 22% are women. How many men are in the club? 11. ________ (4.3)

12. In one year, Betty's monthly rent jumped from \$280 to \$310. What percent increase is this? 12. ________ (4.3)

13. Twelve is to what as 5 is to 7? 13. ________ (4.4)

14. One foot is 0.305 meter. How many feet is a 1.8-meter man? 14. ________ (4.4)

15. If 18 square yards of carpeting cost \$250, how much carpeting will \$700 buy? 15. ________ (4.4)

4

CHAPTER 5 MEASUREMENT AND METRIC GEOMETRY

Name ______________________

Section ______________________

Date ______________________

Score ______________________

Pre-Test

1. The approximate weight of a medium-sized apple is: 1. ______________ (5.1)
   (a) 1 gram
   (b) 7 grams
   (c) 50 grams
   (d) 130 grams

2. Write 24 ounces in pounds. 2. ______________ (5.1)

3. Write 10 miles in kilometers. 3. ______________ (5.1)

4. How many degrees are in one-third of a complete revolution? 4. ______________ (5.2)

5. What is the approximate measure of angle A? 5. ______________ (5.2)
   (a) 19°
   (b) 78°
   (c) 126°
   (d) 159°

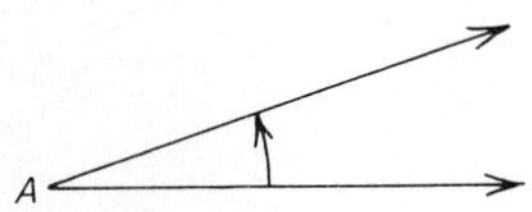

Find the perimeter of each of the following figures.

6. 6. ______________ (5.2)

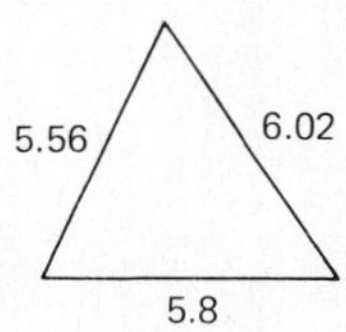

7. A rectangle: $10\frac{1}{5}$ by $14\frac{3}{10}$. 7. ______________ (5.2)

8. A circle: radius = 25.7. 8. ______________ (5.2)

(*over*)

Find the area of each of the following figures.

9. A triangle: base = 4.2, height = 3.1. 9. ______ (5.3)

10. A trapezoid: big base = $2\frac{1}{3}$, little base = $1\frac{1}{3}$, and height = $2\frac{1}{2}$. 10. ______ (5.3)

11. A circle: radius = 8.2. 11. ______ (5.3)

12. 12. ______ (5.3)

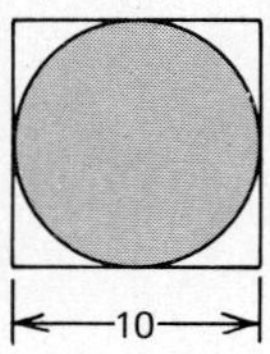

Find the volume of each of the following figures.

13. A rectangular box: $4\frac{1}{2}$ by $2\frac{2}{3}$ by $3\frac{1}{3}$. 13. ______ (5.4)

14. A cylinder: height = 6, diameter = 4. 14. ______ (5.4)

15. A sphere: radius = 6. 15. ______ (5.4)

5

5.1 MEASUREMENT AND THE METRIC SYSTEM

Name ______________________

Section ______________________

Date ______________________

Score ______________________

Post-Test 1

Choose the closest approximation for the following items in the units indicated.

1. Length of a dollar bill (centimeters): 1. __________ *(8–10)*
   (a) 4
   (b) 15
   (c) 40
   (d) 150
2. Weight of a pair of hiking boots (kilograms): 2. __________ *(8–10)*
   (a) 1
   (b) 8
   (c) 70
   (d) 600
3. Capacity of a stein of beer (liters): 3. __________ *(8–10)*
   (a) $\frac{1}{50}$
   (b) $\frac{1}{2}$
   (c) 5
   (d) 25

Perform the following unit conversions.

4. Express 48 quarts in gallons. 4. __________ *(4–7)*
5. Express 4 miles in feet. 5. __________ *(4–7)*
6. Express 3.4 meters in centimeters. 6. __________ *(11–16)*
7. Express 500 milliliters in liters. 7. __________ *(11–16)*
8. Express 75 feet in meters. 8. __________ *(11–16)*
9. Express 40 kilograms in pounds. 9. __________ *(11–16)*
10. Express 10 gallons in liters. 10. __________ *(11–16)*

5

5.1 MEASUREMENT AND THE METRIC SYSTEM

Name ____________

Section ____________

Date ____________

Score ____________

Post-Test 2

Choose the closest approximation for the following items in the units indicated.

1. Height of a tennis net (meters): 1. ____________ (8–10)
   (a) 1
   (b) 5
   (c) 30
   (d) 180

2. Weight of one medium-sized apple (grams): 2. ____________ (8–10)
   (a) 20
   (b) 130
   (c) 550
   (d) 3200

3. Capacity of a punch bowl (liters): 3. ____________ (8–10)
   (a) $\frac{1}{6}$
   (b) 1
   (c) 8
   (d) 50

5

Perform the following unit conversions.

4. Express $2\frac{1}{2}$ pounds in ounces. 4. ____________ (4–7)

5. Express 80 inches in feet. 5. ____________ (4–7)

6. Express 2.1 kilograms in grams. 6. ____________ (11–16)

7. Express 800 meters in kilometers. 7. ____________ (11–16)

8. Express 320 kilometers in miles. 8. ____________ (11–16)

9. Express 30 ounces in grams. 9. ____________ (11–16)

10. Express 6 liters in quarts. 10. ____________ (11–16)

5.2 ANGLES AND LENGTHS

Name ____________________

Section ____________________

Date ____________________

Score ____________________

Post-Test 1

---

1. What fraction of a whole circle is 120°? — 1. ____________ (17–19)

2. What angle is $\frac{1}{4}$ of a circle? — 2. ____________ (17–19)

3. Measure the angle. — 3. ____________ (17–19)

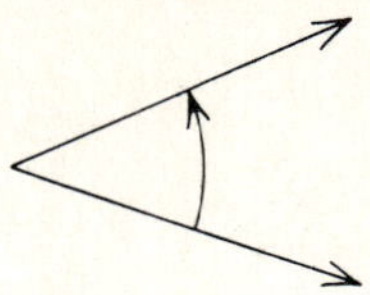

Compute the following perimeters.

4. 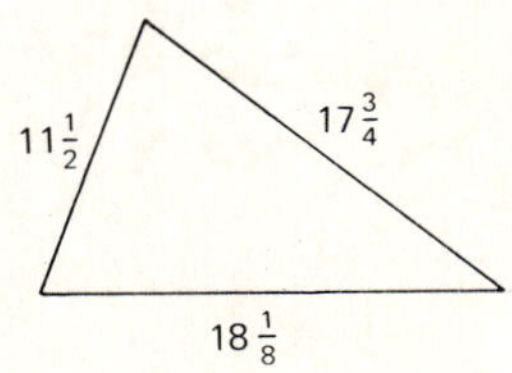

5. 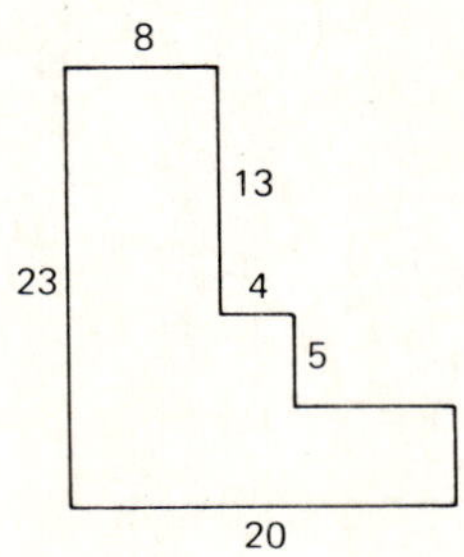

4. ____________ (20–22)

5. ____________ (20–22)

6. A rectangle: 10.6 by 7.85. — 6. ____________ (23)

7. A circle: radius = 20. — 7. ____________ (25–27)

8. — 8. ____________ (28–29)

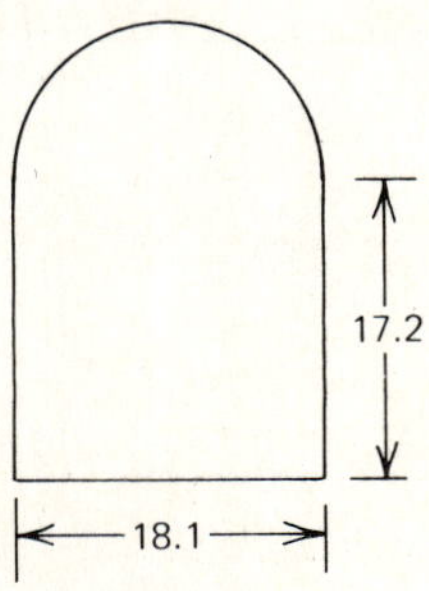

9. A sketch of Mike's den is shown below on the left. How many feet of floor molding will he need to go all the way around the room? — 9. ____________ (28–29)

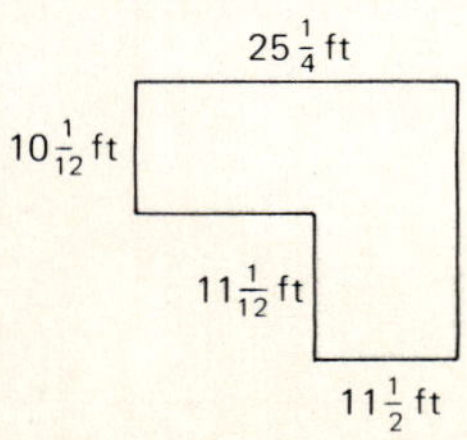

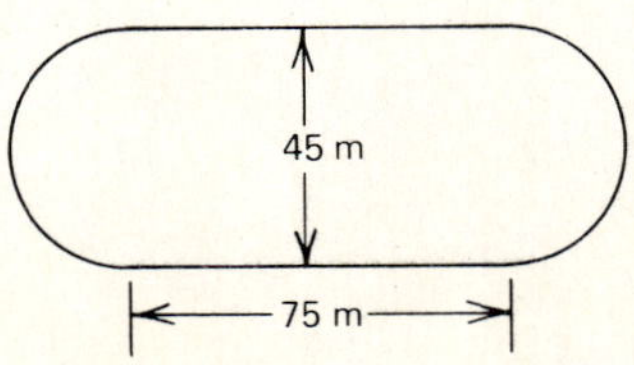

10. Shown above on the right is a sketch of an indoor track at Odell College. How far around is one lap? — 10. ____________ (28–29)

5

5.2 ANGLES AND LENGTHS

Name ______________

Section ______________

Date ______________

Score ______________

Post-Test 2

---

1. What fraction of a circle is 30°? — 1. ______________ *(17–19)*

2. What angle is $\frac{1}{6}$ of a circle? — 2. ______________ *(17–19)*

3. Measure the angle. — 3. ______________ *(17–19)*

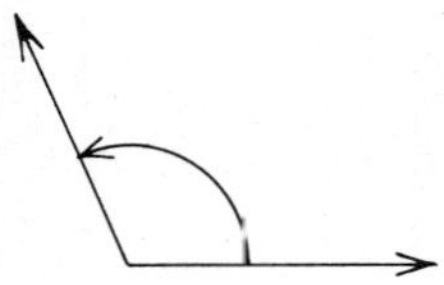

Compute the following perimeters.

4. — 4. ______________ *(20–22)*

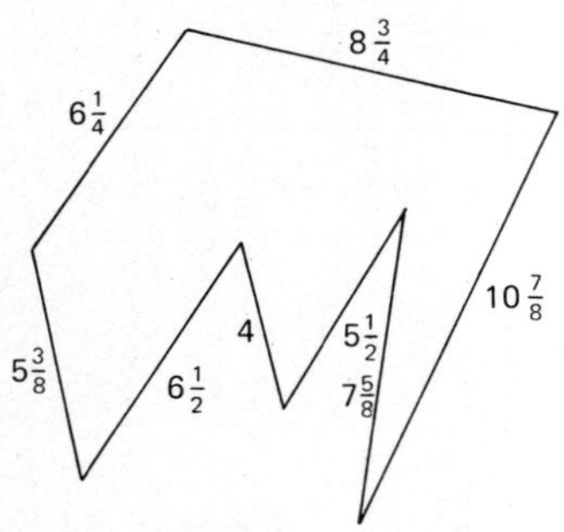

5. — 5. ______________ *(20–22)*

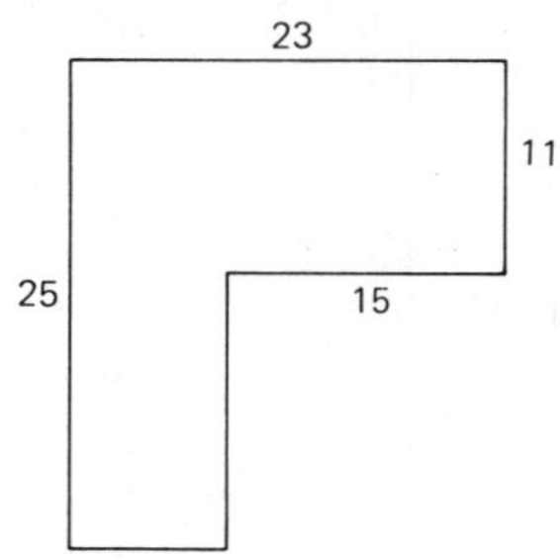

6. A rectangle: 38.9 by 10.3. — 6. ______________ *(23)*

7. A circle: radius = 50. — 7. ______________ *(25–27)*

8. — 8. ______________ *(28–29)*

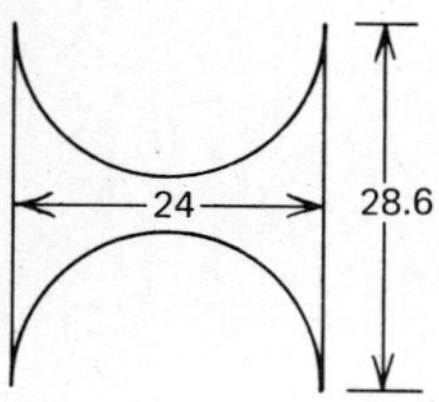

9. Below is a sketch of the Peters' patio. If they wish to put a trim around the edge, how much will they need? — 9. ______________ *(28–29)*

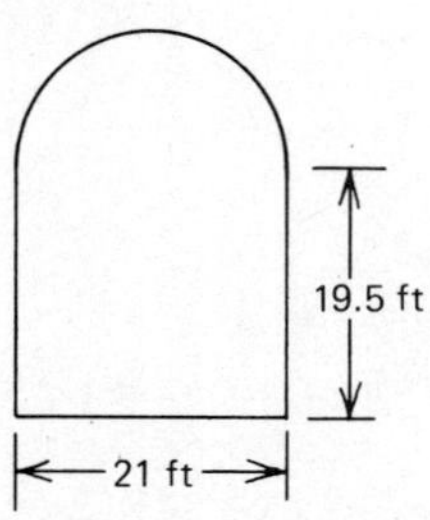

10. A bicycle wheel is 26 inches in diameter. How far does a bicycle go in one revolution of the wheel (what is the circumference)? — 10. ______________ *(28–29)*

5

5.3 AREA

Name ____________________

Section ____________________

Date ____________________

Score ____________________

Post-Test 1

Compute the following areas.

1. A rectangle: 5.4 by 2.5. — 1. ____________ (*30–32*)

2. A square: side = $2\frac{1}{4}$. — 2. ____________ (*30–32*)

3. A triangle: base = 150, height = 140. — 3. ____________ (*33–35*)

4. A trapezoid: big base = 9, little base = 5, and height = 4. — 4. ____________ (*36, 37*)

5. A circle: radius = 15. — 5. ____________ (*38–40*)

6. A semicircle: diameter = 12.2. — 6. ____________ (*38–40*)

7. — 7. ____________ (*41–43*)

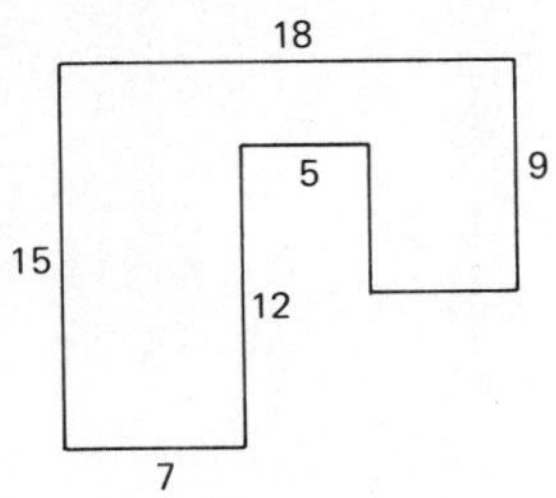

8. — 8. ____________ (*41–43*)

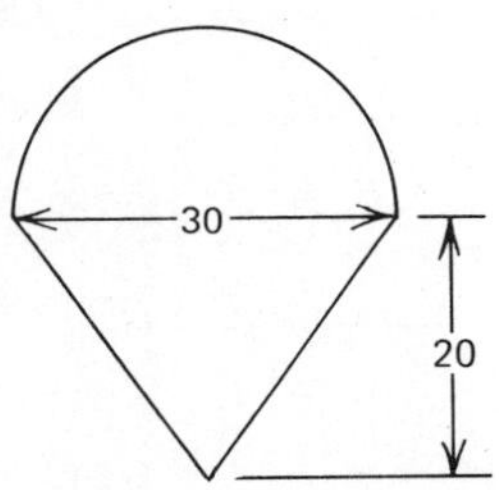

9. — 9. ____________ (*41–43*)

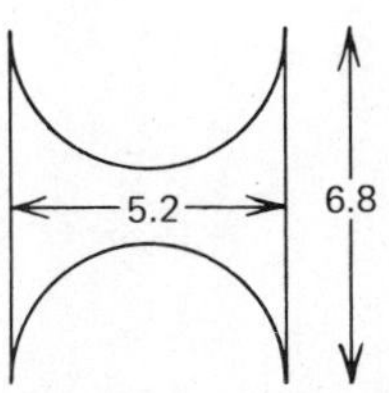

10. Nick and Ruth are going to carpet their living room/dining room (as sketched below). What is the area? — 10. ____________ (*41–43*)

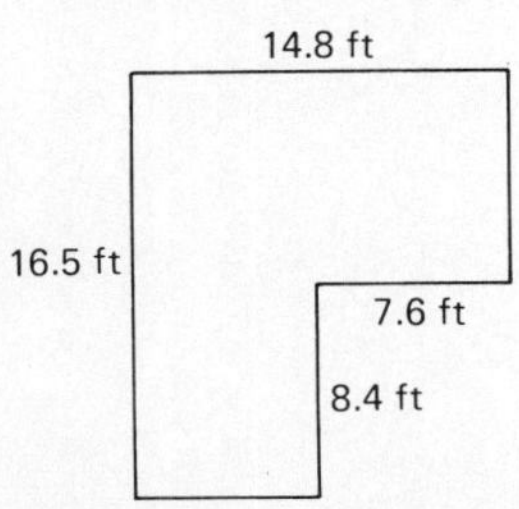

5.3 AREA

Post-Test 2

Name ______________________

Section ______________________

Date ______________________

Score ______________________

Compute the following areas.

1. A rectangle: $4\frac{1}{5}$ by $3\frac{1}{3}$. 1. ______________ (30–32)
2. A square: side = 18. 2. ______________ (30–32)
3. A triangle: base = 18.2, height = 26.4. 3. ______________ (33–35)
4. A trapezoid: big base = 140, little base = 60, and height = 80. 4. ______________ (36, 37)
5. A circle: diameter = 40. 5. ______________ (38–40)
6. A semicircle: radius = 1.2. 6. ______________ (38–40)
7. 7. ______________ (41–43)

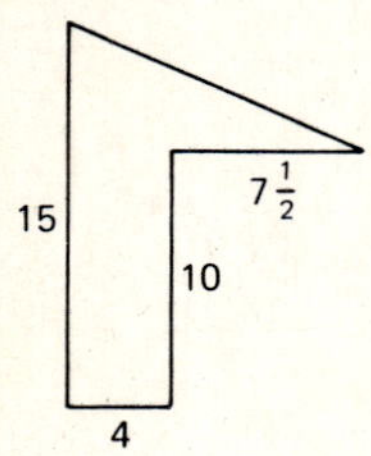

8. 8. ______________ (41–43)

9. Karen and Dennis are carpeting the rooms as shown. What is the area? 9. ______________ (41–43)

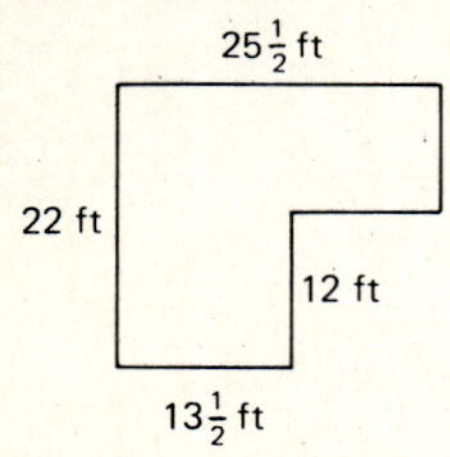

10. Bonzo State College is paving a U-shaped walkway as shown. What is the area? 10. ______________ (41–43)

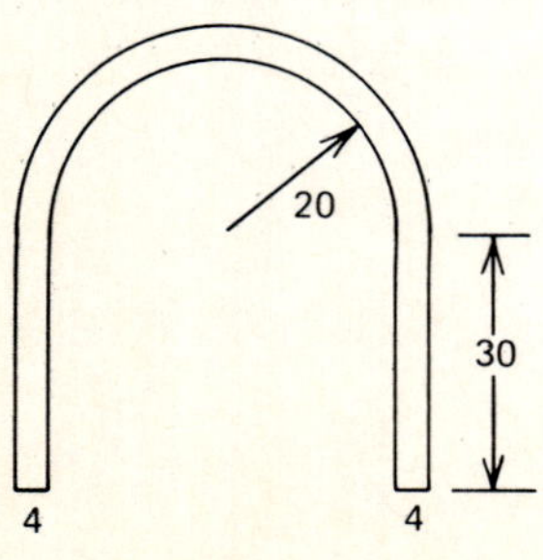

Name ____________
Section ____________
Date ____________
Score ____________

Post-Test 1

Compute the following volumes.

1. A rectangular box: $1\frac{1}{2}$ by $\frac{3}{4}$ by $\frac{2}{3}$. 1. ____________ (44, 45)

2. A rectangular box: 20 by 5.6 by 2.7. 2. ____________ (44, 45)

3. A cube: side = 8. 3. ____________ (46)

4. A cube: side = $\frac{2}{3}$. 4. ____________ (46)

5. A cylinder: height = 19, diameter = 24. 5. ____________ (47, 48)

6. A cylinder: height = 39, radius = 9. 6. ____________ (47, 48)

7. A sphere: radius = 40. 7. ____________ (49–51)

8. A sphere: diameter = 6. 8. ____________ (49–51)

9. 9. ____________ (44–51)

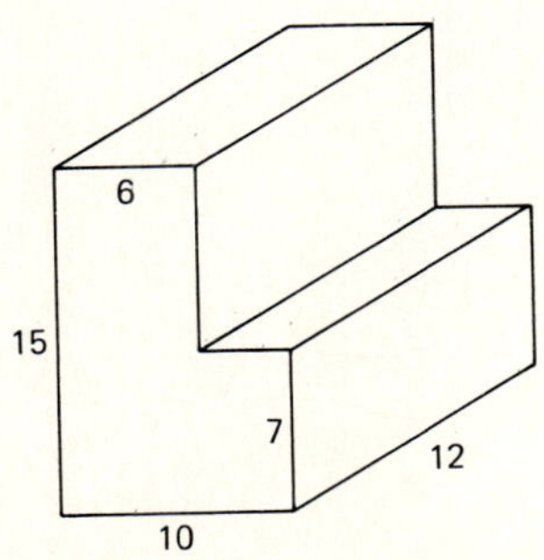

10. 10. ____________ (44–51)

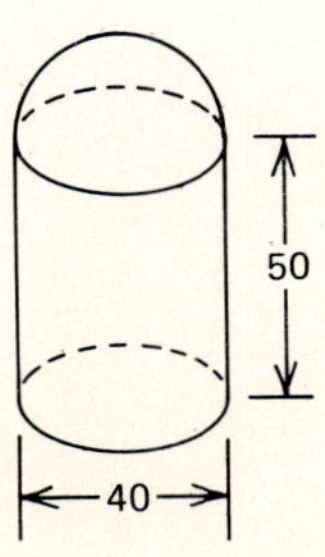

5

Name ____________

Section ____________

Post-Test 2

Date ____________

Score ____________

Compute the following volumes.

1. A rectangular box: 14 by 11 by 9. 1. ____________ (44, 45)

2. A rectangular box: 5 by $2\frac{1}{2}$ by $1\frac{1}{5}$. 2. ____________ (44, 45)

3. A cube: side = 11. 3. ____________ (46)

4. A cube: side = 0.2. 4. ____________ (46)

5. A cylinder: height = 11.3, diameter = 6.2. 5. ____________ (47, 48)

6. Jackie has a large cylindrical drum for storing things with height 1.6 meters and diameter 1.2 meters. What is its volume? 6. ____________ (47, 48)

7. A sphere: diameter = 20. 7. ____________ (49–51)

8. A hemisphere: radius = 18. 8. ____________ (49–51)

9. 9. ____________ (44–51)

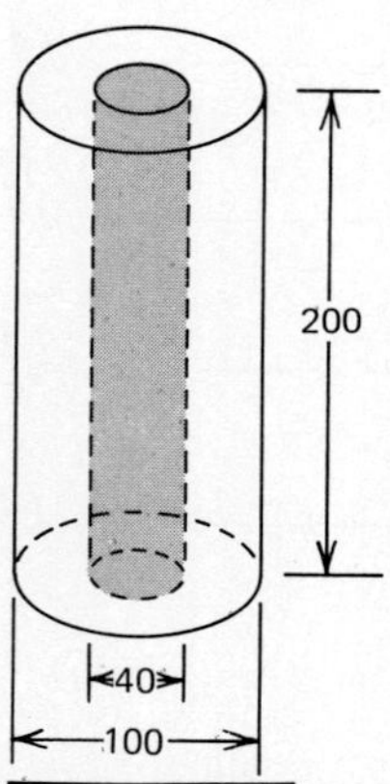

10. The Arbuckles' living room/dining room has the dimensions shown below. Find the volume. 10. ____________ (44–51)

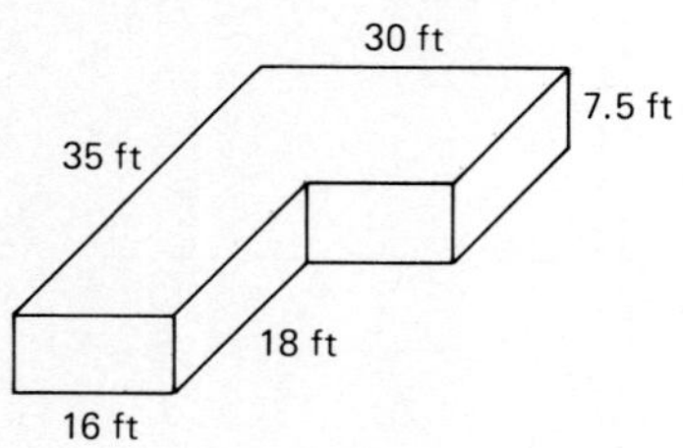

CHAPTER 5 MEASUREMENT AND METRIC GEOMETRY

Name ______________

Section ______________

Date ______________

Score ______________

Post-Test 1

1. The approximate height of a bottle of soda pop is: 1. ______________ (5.1)
   (a) 4 cm
   (b) 28 cm
   (c) 98 cm
   (d) 310 cm

2. Write 96 inches in feet. 2. ______________ (5.1)

3. Write 6 inches in centimeters. 3. ______________ (5.1)

4. Sixty degrees is what fraction of one complete revolution? 4. ______________ (5.2)

5. What is the approximate measure of angle A? 5. ______________ (5.2)
   (a) 21°
   (b) 69°
   (c) 108°
   (d) 155°

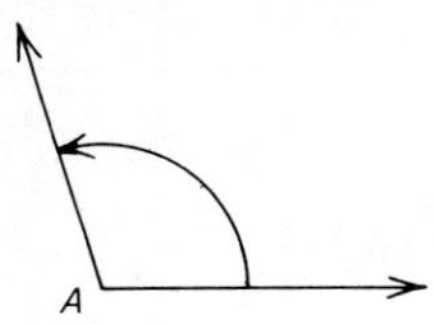

Find the perimeter of each of the following figures.

6. 6. ______________ (5.2)

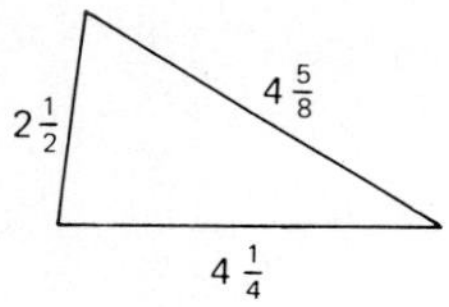

7. 7. ______________ (5.2)

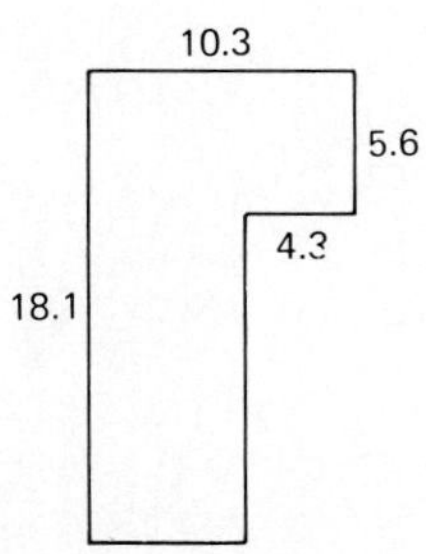

8. A circle: diameter = 12.9. 8. ______________ (5.2)

(*over*)

Find the area of each of the following figures.

9. A triangle: base = $2\frac{4}{5}$, height = $2\frac{1}{7}$. 9. ____________ (5.3)

10. A trapezoid: big base = 14.7, little base = 10.5, and height = 9.8. 10. ____________ (5.3)

11. 

11. ____________ (5.3)

12. 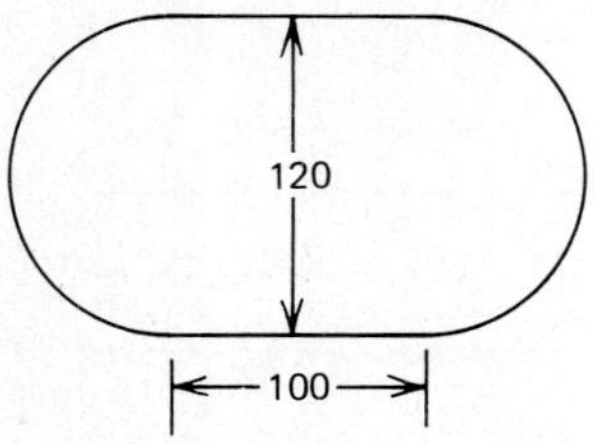

12. ____________ (5.3)

Find the volume of each of the following figures.

13. A rectangular box: $3\frac{3}{4}$ by $1\frac{3}{5}$ by $1\frac{1}{2}$. 13. ____________ (5.4)

14. A cylinder: height = 7, diameter = 4. 14. ____________ (5.4)

15. A sphere: radius = 6. 15. ____________ (5.4)

5

CHAPTER 5 MEASUREMENT AND METRIC GEOMETRY

Name ______________

Section ______________

Date ______________

Score ______________

Post-Test 2

1. The approximate weight of a big-city telephone book is: 1. ______________ (*5.1*)
   (a) 1.5 kg
   (b) 8 kg
   (c) 30 kg
   (d) 160 kg

2. Write 15 gallons in quarts. 2. ______________ (*5.1*)

3. Write 150 centimeters in inches. 3. ______________ (*5.1*)

4. What angle is one-tenth of a complete revolution? 4. ______________ (*5.2*)

5. What is the approximate measure of angle A? 5. ______________ (*5.2*)
   (a) 15°
   (b) 48°
   (c) 97°
   (d) 153°

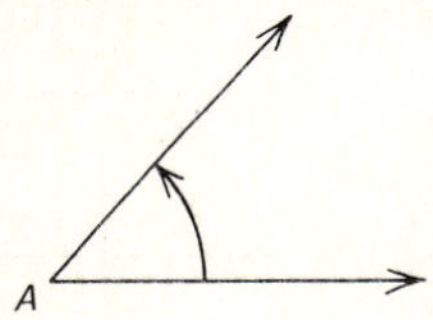

Find the perimeter of each of the following figures.

6. 6. ______________ (*5.2*)

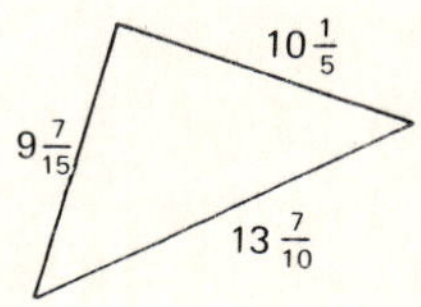

7. 7. ______________ (*5.2*)

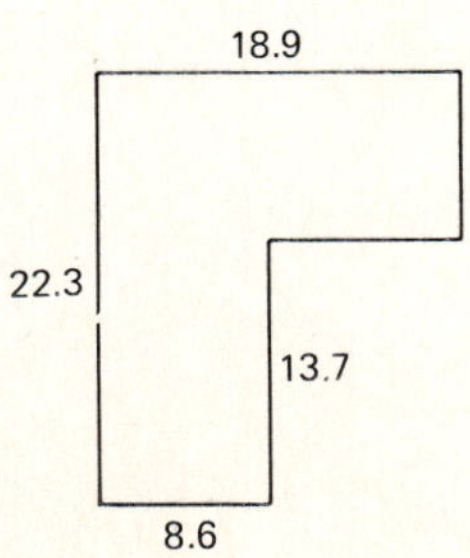

8. A circle: diameter = 98.6. 8. ______________ (*5.2*)

(*over*)

5

Find the area of each of the following figures.

9. A triangle: base = 7.9, height = 6.4. 9. ____________ (5.3)

10. A trapezoid: big base = $5\frac{3}{4}$, little base = $4\frac{3}{4}$, and height = $3\frac{1}{3}$. 10. ____________ (5.3)

11. A circle: radius = 4.2. 11. ____________ (5.3)

12. 12. ____________ (5.3)

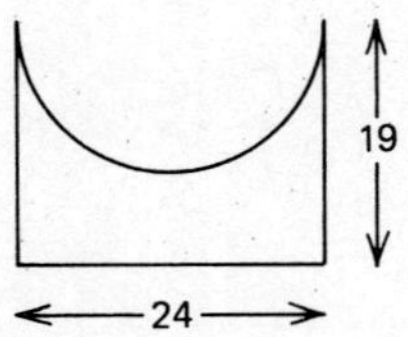

Find the volume of each of the following figures.

13. A rectangular box: $4\frac{1}{2}$ by $2\frac{1}{7}$ by $2\frac{4}{5}$. 13. ____________ (5.4)

14. A cylinder: height = 12, diameter = 10. 14. ____________ (5.4)

15. A sphere: radius = 9. 15. ____________ (5.4)

5

Pre-Test

Name ________________

Section ________________

Date ________________

Score ________________

---

1. Simplify $\dfrac{2 + 14}{5 - 1}$. 1. ________ (6.1)

2. Give the coordinate of point A. 2. ________ (6.1)

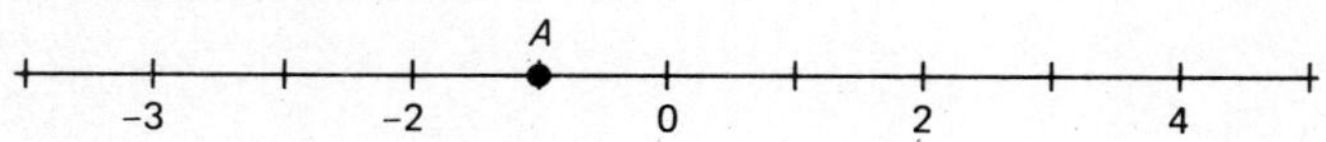

3. What is $-(-8)$? 3. ________ (6.1)

4. Simplify $-3 + (-8)$. 4. ________ (6.2)

5. Simplify $2 + (-3 + 4) + (-11)$. 5. ________ (6.2)

6. Simplify $-3 - (-4)$. 6. ________ (6.2)

7. Simplify $-6 - (-4) - 5 + (-7) + 3$. 7. ________ (6.2)

8. Simplify $(-7)(5)$. 8. ________ (6.3)

9. Simplify $4 + (-2)(-3)$. 9. ________ (6.3)

10. Simplify $\dfrac{-40}{-8}$. 10. ________ (6.3)

11. Simplify $\dfrac{(-1)(-2)(-3)(-4)}{(-6)}$. 11. ________ (6.3)

(*over*)

Simplify the following.

12. $5(3 - x)$ | 12. ____________ (6.4)

13. $6a - 4 - 3a + 7$ | 13. ____________ (6.4)

14. $4(a - 2) - 6(a + 1) - 4$ | 14. ____________ (6.4)

Solve the following equations and problems.

15. $7x = -28$ | 15. ____________ (6.5)

16. $3t + 2t + 2 = 17$ | 16. ____________ (6.5)

17. $5(2 - a) = 6(a + 9)$ | 17. ____________ (6.5)

18. One number is 13 more than another number. Their sum is 71. Find the numbers. | 18. ____________ (6.6)

19. In an election Brown defeats Green by 150 votes. If 1100 total votes were cast for them, how many did Brown get? | 19. ____________ (6.6)

20. The perimeter of a rectangle is 26. The length is 1 more than 3 times the width. Find the length. | 20. ____________ (6.6)

6

6.1 ORDER OF OPERATIONS AND THE REAL NUMBERS

Name ______________________

Section ______________________

Date ______________________

Score ______________________

Post-Test 1

Simplify the following.

1. 2^3 1. ______________ (1–4)

2. $5 + 2(3)$ 2. ______________ (1–4)

3. $4(5 - 2^2)$ 3. ______________ (1–4)

4. $\dfrac{4 \cdot 9}{3 + 3}$ 4. ______________ (1–4)

5. Give the coordinate of point *A*.   5. ______________ (5)

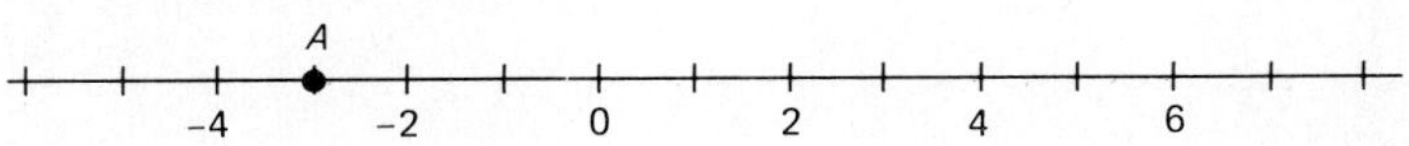

Locate the following points on the number line.

6. $B = -1.5$

6., 7. 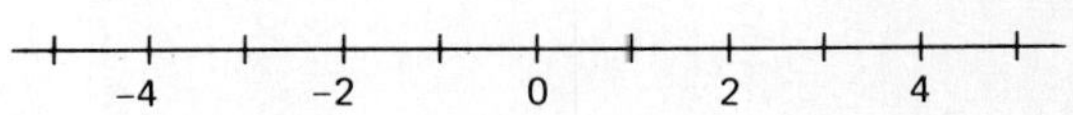

(5)

7. $C = 4$

Find the following numbers.

8. The opposite of 9.   8. ______________ (6, 7)

9. The absolute value of -10.   9. ______________ (6, 7)

10. $-(-3)$ 10. ______________ (6, 7)

6.1 ORDER OF OPERATIONS AND THE REAL NUMBERS

Name ______________________

Section ______________________

Date ______________________

Score ______________________

Post-Test 2

Simplify the following.

1. 5^2 — 1. ______________ *(1–4)*

2. $4 + 5(2)$ — 2. ______________ *(1–4)*

3. $4(2^2 + 1)$ — 3. ______________ *(1–4)*

4. $\dfrac{20 + 8}{2 \cdot 7}$ — 4. ______________ *(1–4)*

5. Give the coordinate of point A. — 5. ______________ *(5)*

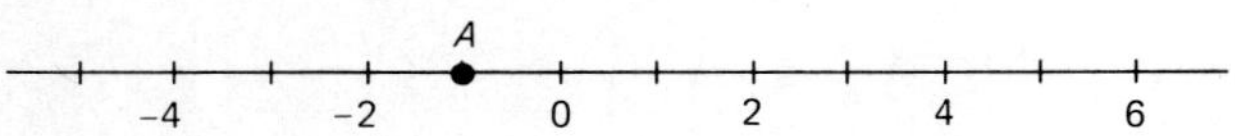

Locate the following points on the number line.

6. $B = -3$

7. $C = 2\frac{1}{2}$

6., 7.

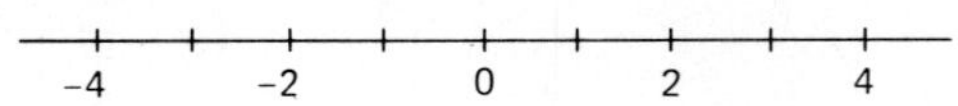

(5)

6

Find the following numbers.

8. The opposite of -7. — 8. ______________ *(6, 7)*

9. The absolute value of -8. — 9. ______________ *(6, 7)*

10. $-(-6)$ — 10. ______________ *(6, 7)*

6.2 ADDITION AND SUBTRACTION OF REAL NUMBERS

Post-Test 1

Name ______________

Section ______________

Date ______________

Score ______________

Simplify the following.

1. $-3 + (-8)$	1.	______________	*(9–15)*
2. $10 + (-3)$	2.	______________	*(9–15)*
3. $-5 + 9$	3.	______________	*(9–15)*
4. $2 + 3 + (-6)$	4.	______________	*(9–15)*
5. $-2 + (-4) + (-6)$	5.	______________	*(9–15)*
6. $-4 - 7$	6.	______________	*(16–20)*
7. $-9 - (-5)$	7.	______________	*(16–20)*
8. $10 - (-3)$	8.	______________	*(16–20)*
9. $4 - (7 - 2)$	9.	______________	*(16–20)*
10. $-3 - [-6 - (-5)]$	10.	______________	*(16–20)*

6

6.2 ADDITION AND SUBTRACTION OF REAL NUMBERS

Name ______________________

Section ______________________

Date ______________________

Score ______________________

Post-Test 2

Simplify the following.

1.	$-4 + 10$	1.	______________	(9–15)
2.	$-2 + (-11)$	2.	______________	(9–15)
3.	$5 + (-2)$	3.	______________	(9–15)
4.	$4 + (-5) + 7$	4.	______________	(9–15)
5.	$-4 + (-3) + 8$	5.	______________	(9–15)
6.	$-3 - 9$	6.	______________	(16–20)
7.	$-4 - (-10)$	7.	______________	(16–20)
8.	$8 - (-2)$	8.	______________	(16–20)
9.	$3 - (8 - 10)$	9.	______________	(16–20)
10.	$-2 - [-8 - (-3)]$	10.	______________	(16–20)

6

6.3 MULTIPLICATION AND DIVISION OF REAL NUMBERS

Name ______________________

Section ______________________

Date ______________________

Score ______________________

Post-Test 1

Simplify the following.

1. $(-3)(10)$ — 1. ________ *(21–23)*

2. $(-4)(-7)$ — 2. ________ *(21–23)*

3. $(2)(-11)$ — 3. ________ *(21–23)*

4. $(3)(6)$ — 4. ________ *(21–23)*

5. $\frac{-15}{-5}$ — 5. ________ *(24, 25)*

6. $\frac{-25}{5}$ — 6. ________ *(24, 25)*

7. $\frac{35}{-7}$ — 7. ________ *(24, 25)*

8. $\frac{0}{0}$ — 8. ________ *(24, 25)*

9. $\frac{(-1)(-3)(-10)}{(-2)(-5)}$ — 9. ________ *(26, 27)*

10. $-5 + (-2)(-3)$ — 10. ________ *(26, 27)*

6

Name ____________

Section ____________

Date ____________

Score ____________

Post-Test 2

Simplify the following.

1. $(-2)(7)$ — 1. ____________ (21–23)

2. $(-5)(-3)$ — 2. ____________ (21–23)

3. $(6)(-4)$ — 3. ____________ (21–23)

4. $(5)(2)$ — 4. ____________ (21–23)

5. $\frac{-10}{2}$ — 5. ____________ (24, 25)

6. $\frac{20}{-5}$ — 6. ____________ (24, 25)

7. $\frac{-30}{-6}$ — 7. ____________ (24, 25)

8. $\frac{10}{0}$ — 8. ____________ (24, 25)

9. $\frac{(-2)(-8)}{4(-4)}$ — 9. ____________ (26, 27)

10. $5 - (-2)(-3)$ — 10. ____________ (26, 27)

6.4 THE DISTRIBUTIVE LAW AND SIMPLIFYING EXPRESSIONS

Post-Test 1

Name ____________________

Section ____________________

Date ____________________

Score ____________________

Simplify the following expressions.

1. $4(a - 3)$ — 1. ____________ *(28)*

2. $-8(2 - 5y)$ — 2. ____________ *(28)*

3. $3x + 11x$ — 3. ____________ *(29)*

4. $10a + 2a - 5a$ — 4. ____________ *(29)*

5. $5t - 2 - 4 + 5$ — 5. ____________ *(30)*

6. $6x + 4 - 7 - 3$ — 6. ____________ *(30)*

7. $2u + 3u - 4u + 1$ — 7. ____________ *(31)*

8. $4x - 10x - 3x + 7$ — 8. ____________ *(31)*

9. $7k - 3 + 4k + 2 - 8$ — 9. ____________ *(32, 33)*

10. $5w + 4 + 3w - 2w - 5$ — 10. ____________ *(32, 33)*

Name ______________

Section ______________

Date ______________

Score ______________

Post-Test 2

Simplify the following expressions.

1.	$3(x - 7)$	1.	______________	(28)
2.	$-5(2a + 7)$	2.	______________	(28)
3.	$5r + 7r$	3.	______________	(29)
4.	$6y - 2y + 11y$	4.	______________	(29)
5.	$4a - 3 + 4 + 8$	5.	______________	(30)
6.	$5t + 5 - 2 + 7$	6.	______________	(30)
7.	$6u - 5u + u + 8$	7.	______________	(31)
8.	$4k - 7k - 2k - 3$	8.	______________	(31)
9.	$10m - 9m - 3 + 5 - 2m$	9.	______________	(32, 33)
10.	$4x - 3 + 8x + 5 - 10$	10.	______________	(32, 33)

6.5 SOLVING EQUATIONS

Name ____________

Section ____________

Date ____________

Score ____________

Post-Test 1

Solve the following equations.

1. $a - 4 = -1$ 1. ____________ (34, 35)

2. $r + 9 = 2$ 2. ____________ (34, 35)

3. $3x - 2x - 5 = 7$ 3. ____________ (36, 37)

4. $10m - 2 = 9m + 3$ 4. ____________ (36, 37)

5. $4x = 36$ 5. ____________ (38–41)

6. $-6k = -36$ 6. ____________ (38–41)

7. $\frac{n}{-2} = 8$ 7. ____________ (38–41)

8. $-u = 7$ 8. ____________ (38–41)

9. $10z + 3 = 4z + 15$ 9. ____________ (42–44)

10. $5(a + 2) = 2a + 1$ 10. ____________ (42–44)

Name ____________________

Section ____________________

Post-Test 2

Date ____________________

Score ____________________

Solve the following equations.

1. $x - 8 = 2$ 1. ____________ (34, 35)

2. $k + 10 = 3$ 2. ____________ (34, 35)

3. $7a - 6a - 2 = 9$ 3. ____________ (36, 37)

4. $5z - 4 = 4z + 7$ 4. ____________ (36, 37)

5. $5x = 35$ 5. ____________ (38–41)

6. $-8m = 40$ 6. ____________ (38–41)

7. $\dfrac{x}{-3} = 8$ 7. ____________ (38–41)

8. $-a = -5$ 8. ____________ (38–41)

9. $3u - 8 = u + 12$ 9. ____________ (42–44)

10. $4(n - 5) = n + 1$ 10. ____________ (42–44)

6.6 SOLVING PROBLEMS

Name ______________________

Section ______________________

Date ______________________

Score ______________________

Post-Test 1

Work the following problems.

1. One number is 8 more than another number. Their sum is 50. Find the numbers. — 1. ____________ (45, 46)

2. One number is 6 times another number. Their sum is 49. Find the numbers. — 2. ____________ (45, 46)

3. The sum of two consecutive odd integers is 60. Find the numbers. — 3. ____________ (45, 46)

4. The difference between two numbers is 9. Their sum is 39. Find the numbers. — 4. ____________ (45, 46)

5. The sum of two consecutive integers is 83. Find the numbers. — 5. ____________ (45, 46)

6. Peter weighs 25 pounds more than Paula. Their total weight is 271. Find their weights. — 6. ____________ (47)

7. Jones defeats Smith in an election by 15 votes. If there were 465 total votes for both men, how many votes did Jones get? — 7. ____________ (47)

8. In a room, there are 16 more women than men. If there are 138 total people, how many men are there in the room? — 8. ____________ (47)

9. The perimeter of a square is 20. What is the length of each side? — 9. ____________ (48)

10. The length of a certain rectangle is 4 more than the width. If the perimeter is 32, find the length and width. — 10. ____________ (48)

6

Name ____________________

Section ____________________

Post-Test 2

Date ____________________

Score ____________________

Work the following problems.

1. One number is 15 more than another number. Their sum is 61. Find the numbers. — 1. ____________ *(45, 46)*

2. One number is 3 times another number. Their sum is 36. Find the numbers. — 2. ____________ *(45, 46)*

3. The sum of two consecutive even integers is 54. Find the numbers. — 3. ____________ *(45, 46)*

4. The sum of two consecutive integers is 55. Find the numbers. — 4. ____________ *(45, 46)*

5. The difference between two numbers is 11. Their sum is 61. Find the numbers. — 5. ____________ *(45, 46)*

6. Ned is 3 years older than Shirley. Their total age is 121. Find each of their ages. — 6. ____________ *(47)*

7. A wrestling team, the Ugly Brothers, weigh 575 pounds together. If Lou weighs 15 pounds more than Bruno, how much does each of them weigh? — 7. ____________ *(47)*

8. In an apartment, there are 25 more one-bedroom units than two-bedroom units If there are 185 total units, how many of each are there? — 8. ____________ *(47)*

6

9. The perimeter of a square is 28. What is the length of each side? — 9. ____________ *(48)*

10. The length of a rectangle is 8 more than the width. If the perimeter is 36, find the length and width. — 10. ____________ *(48)*

CHAPTER 6 INTRODUCTION TO ALGEBRA

Name ____________

Section ____________

Date ____________

Score ____________

Post-Test 1

1. Simplify $\frac{17-1}{6+2}$. 1. ____________ (6.1)

2. Give the coordinate of point B. 2. ____________ (6.1)

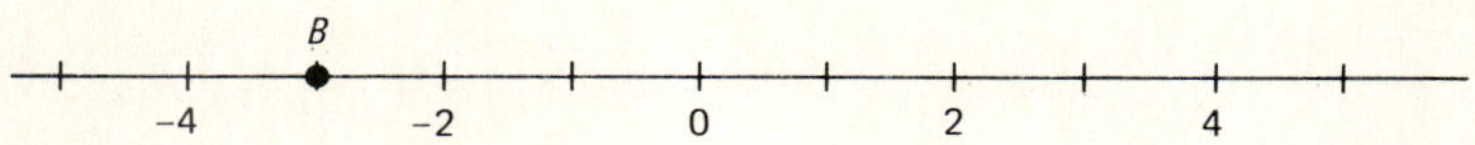

3. What is $-(-2)$? 3. ____________ (6.1)

Simplify the following.

4. $-4+(-10)$ 4. ____________ (6.2)

5. $4+(-5+6)+(-12)$ 5. ____________ (6.2)

6. $-9-(-2)$ 6. ____________ (6.2)

7. $-7-5+(-2)+4+(-1)-(-3)$ 7. ____________ (6.2)

8. $(-3)(-8)$ 8. ____________ (6.3)

9. $6+(2)(-3)$ 9. ____________ (6.3)

10. $\frac{32}{-8}$ 10. ____________ (6.3)

11. $\frac{(-2)(-3)(-4)}{(-1)(-12)}$ 11. ____________ (6.3)

(over)

6

12. $6(2 + x)$ 12. ______ (6.4)

13. $9x - 3 - 2x + 5$ 13. ______ (6.4)

14. $8(a - 1) - 5(a + 3) - 3$ 14. ______ (6.4)

Solve the following equations and problems.

15. $-2x = 14$ 15. ______ (6.5)

16. $5z + 2z - 2 = 19$ 16. ______ (6.5)

17. $6(a - 3) = a + 2$ 17. ______ (6.5)

18. One number is 11 more than another number. Their sum is 79. Find the numbers. 18. ______ (6.6)

19. Judy and Johnny earn $25,900 together. Judy earns $500 more than Johnny. How much does each earn? 19. ______ (6.6)

20. The length of one side of a triangle is 10 more than the second, which is 8 more than the third. If the perimeter is 86, find the three sides. 20. ______ (6.6)

6

Name ____________________

Section ____________________

Date ____________________

Score ____________________

Post-Test 2

1. Simplify $\frac{15 - 1}{3 + 4}$.   1. ____________ (6.1)

2. Give the coordinate of point C.   2. ____________ (6.1)

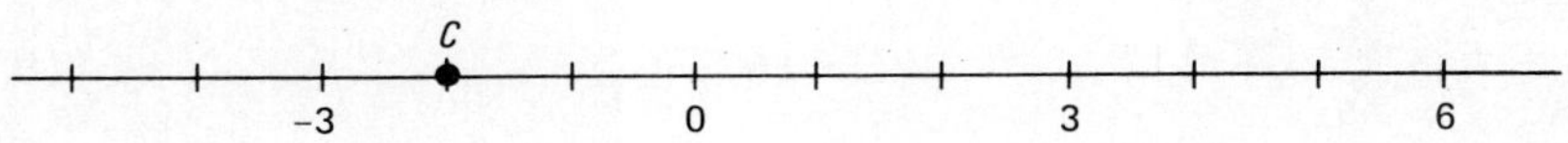

3. What is $-(-16)$?   3. ____________ (6.1)

Simplify the following.

4. $-8 + 13$ 4. ____________ (6.2)

5. $-4 + (-5 + 9) + (-8)$ 5. ____________ (6.2)

6. $-4 - (-3)$ 6. ____________ (6.2)

7. $-6 - 8 - (-5) + (-6) + 11$ 7. ____________ (6.2)

8. $(-7)(-8)$ 8. ____________ (6.3)

9. $5 + (-2)(8)$ 9. ____________ (6.3)

10. $\frac{-50}{-10}$ 10. ____________ (6.3)

11. $\frac{(-1)(-3)(-5)(-2)}{(-15)}$ 11. ____________ (6.3)

(over)

6

12. $4(x - 7)$ 12. ______________ (6.4)

13. $7k - 3 + 6k + 8$ 13. ______________ (6.4)

14. $3(r - 5) - 5(2r + 1) + 5$ 14. ______________ (6.4)

Solve the following equations and problems.

15. $r + 5 = -8$ 15. ______________ (6.5)

16. $5t + 2t + 9 = 2$ 16. ______________ (6.5)

17. $4(8 - a) = 5(a + 1)$ 17. ______________ (6.5)

18. The sum of two consecutive odd integers is 48. Find the numbers. 18. ______________ (6.6)

19. Al and Ted weigh 465 pounds together. Al weighs 25 pounds more than Ted. How much does each weigh? 19. ______________ (6.6)

20. The perimeter of a square is 440. What is the length of a side? 20. ______________ (6.6)

6

Name ____________

Section ____________

Date ____________

Score ____________

Pre-Test

Perform the following operations on a hand calculator.

1. $48.3 + 68.03 + 104.843$ — 1. ____________ (7.1)

2. $194.06 - 126.9674$ — 2. ____________ (7.1)

3. $(0.456)(0.467)$ — 3. ____________ (7.1)

4. $6.03 \div 0.749$ — 4. ____________ (7.1)

5. $\dfrac{(6.043)(4.32)}{5.04}$ — 5. ____________ (7.1)

6. $(4.1)(5.9) - (3.2)(1.3) + (6.1)(5)$ — 6. ____________ (7.2)

7. $\dfrac{2}{3} + \dfrac{5}{7} + \dfrac{10}{19}$ — 7. ____________ (7.2)

8. $\dfrac{6200}{4.93} + \dfrac{(7.2)(4.2)}{0.67} - \dfrac{5.39}{(4.2)}$ — 8. ____________ (7.2)

9. 21^5 — 9. ____________ (7.2)

10. $400(0.8)^4$ — 10. ____________ (7.2)

7

7.1 ELEMENTARY HAND CALCULATOR OPERATIONS

Name ____________________

Section ____________________

Date ____________________

Score ____________________

Post-Test 1

Perform the following operations and problems on a hand calculator.

1. $\begin{array}{r} 2375 \\ 6948 \\ +23486 \\ \hline \end{array}$ — 1. ____________ *(1–6)*

2. $\begin{array}{r} 208781 \\ -\ 37829 \\ \hline \end{array}$ — 2. ____________ *(1–6)*

3. $2.71 + 10.4 - 8.003$ — 3. ____________ *(1–6)*

4. 21.06×5.26 — 4. ____________ *(7–9)*

5. $80.17 \div 0.372$ — 5. ____________ *(7–9)*

6. 3213×198 — 6. ____________ *(7–9)*

7. $\dfrac{127}{453}$ — 7. ____________ *(10–11)*

8. $\dfrac{(1.4)\,(6.51)}{(8.3)\,(0.79)}$ — 8. ____________ *(10–11)*

9. Fifteen is what percent of 41? — 9. ____________ *(12–14)*

10. Find the area of a circle with radius = 1.7. — 10. ____________ *(12–14)*

7.1 ELEMENTARY HAND CALCULATOR OPERATIONS

Name ________________

Section ________________

Date ________________

Score ________________

Post-Test 2

Perform the following operations and problems on a hand calculator.

1. $\begin{array}{r} 4738 \\ 5782 \\ +68347 \\ \hline \end{array}$ 1. ____________ *(1–6)*

2. $\begin{array}{r} 463821 \\ -\ 38926 \\ \hline \end{array}$ 2. ____________ *(1–6)*

3. 3.96 + 45.2 - 32.065   3. ____________ *(1–6)*

4. 32.95 X 32.05   4. ____________ *(7–9)*

5. 48.39 ÷ 31.3   5. ____________ *(7–9)*

6. 4739 X 481   6. ____________ *(7–9)*

7. $\dfrac{235}{183}$ 7. ____________ *(10–11)*

8. $\dfrac{(4.3)\,(3.06)}{(0.31)\,(54.1)}$ 8. ____________ *(10–11)*

9. Fifty is 23.4% of what number?   9. ____________ *(12–14)*

10. Find the area of a triangle with base = 14.8 and height = 10.7.   10. ____________ *(12–14)*

Name ______________

Section ______________

Date ______________

Score ______________

Post-Test 1

Perform the following operations on a hand calculator.

1.	$3(12) + 5(17)$	1. ______________	(15–17)
2.	$8(7.2) - 3.7(4.2)$	2. ______________	(15–17)
3.	$(0.07)(3.14) + (1.9)(0.98) - 21(0.75)$	3. ______________	(15–17)
4.	$\frac{1}{2} + \frac{1}{3} + \frac{1}{4}$	4. ______________	(18–20)
5.	$\frac{21}{46} - \frac{5}{17} + \frac{2}{27}$	5. ______________	(18–20)
6.	$\frac{24}{29} + \frac{6.2}{8.5} - \frac{1.3}{6}$	6. ______________	(18–20)
7.	$\frac{(4.5)(6.2)}{3.7} - \frac{(61.7)(0.73)}{0.85}$	7. ______________	(18–20)
8.	3^6	8. ______________	(21–23)
9.	$(1.084)^{10}$	9. ______________	(21–23)
10.	$2000(1.09)^8$	10. ______________	(21–23)

7

7.2 THE MEMORY AND EXPONENTS

Name ______________

Section ______________

Date ______________

Score ______________

Post-Test 2

Perform the following operations on a hand calculator.

1. $4(11) + 6(19)$ — 1. ______________ *(15–17)*

2. $6(4.3) - 9(0.45)$ — 2. ______________ *(15–17)*

3. $(0.03)(4.36) + 92(2.1) - 43(3.2)$ — 3. ______________ *(15–17)*

4. $\frac{2}{7} + \frac{3}{8} + \frac{4}{9}$ — 4. ______________ *(18–20)*

5. $\frac{21}{57} - \frac{7}{17} + \frac{19}{32}$ — 5. ______________ *(18–20)*

6. $\frac{36}{41} - \frac{6.43}{8.02} + \frac{2.45}{5}$ — 6. ______________ *(18–20)*

7. $\frac{(5.34)(4.1)}{7.45} - \frac{5.483}{(2.048)(0.57)}$ — 7. ______________ *(18–20)*

8. 5^7 — 8. ______________ *(21–23)*

9. $(1.079)^{11}$ — 9. ______________ *(21–23)*

10. $3200(1.085)^{12}$ — 10. ______________ *(21–23)*

7

Name ____________

Section ____________

Date ____________

Score ____________

Post-Test 1

Perform the following operations on a hand calcualtor.

1. $5.38 + 94.032 + 34.062$ — 1. ____________ (7.1)

2. $294.05 - 189.382$ — 2. ____________ (7.1)

3. $(4.302)(20.69)$ — 3. ____________ (7.1)

4. $4.962 \div 0.83742$ — 4. ____________ (7.1)

5. $\dfrac{(2.049)(4157)}{84.038}$ — 5. ____________ (7.1)

6. $(3.01)(2.9) + (4.7)(56) - 23(0.68)$ — 6. ____________ (7.2)

7. $\dfrac{4}{7} + \dfrac{8}{9} + \dfrac{1}{6}$ — 7. ____________ (7.2)

8. $\dfrac{5301}{3.27} + \dfrac{(5.39)(86)}{0.86} - \dfrac{403}{1.6}$ — 8. ____________ (7.2)

9. 19^6 — 9. ____________ (7.2)

10. $560(1.05)^7$ — 10. ____________ (7.2)

Name ____________

Section ____________

Date ____________

Score ____________

Post-Test 2

Perform the following operations on a hand calculator.

1. $593 + 49.30 + 302.059$ — 1. ____________ *(7.1)*

2. $473.954 - 382.0594$ — 2. ____________ *(7.1)*

3. $(3.095)(23.054)$ — 3. ____________ *(7.1)*

4. $56.932 - 9.0493$ — 4. ____________ *(7.1)*

5. $\dfrac{(40.932)(2.394)}{37.049}$ — 5. ____________ *(7.1)*

6. $(2.109)(59) + (3.2)(7.98) - 23(5.2)$ — 6. ____________ *(7.2)*

7. $\dfrac{4}{9} + \dfrac{5}{13} + \dfrac{2}{11}$ — 7. ____________ *(7.2)*

8. $\dfrac{67.3}{1.79} + \dfrac{6.854}{(0.93)(24)} - \dfrac{502}{23.1}$ — 8. ____________ *(7.2)*

9. 17^7 — 9. ____________ *(7.2)*

10. $350(1.2)^8$ — 10. ____________ *(7.2)*

ANSWERS TO MISCELLANEOUS EXERCISES AND REVIEW TESTS

MISCELLANEOUS EXERCISES 1.1

1. 40 + 5 **2.** 60 + 9 **3.** 10 + 8 **4.** 500 + 70 + 9 **5.** 200 + 10 **6.** 300 + 7
7. 5000 + 300 + 90 + 6 **8.** 40,000 + 30 + 1 **9.** 800,000 + 5000 + 700 + 3 **10.** 900 **11.** 2000
12. 80 **13.** 900,000 **14.** 5,000,000 **15.** 60,000 **16.** 8 **17.** 9 **18.** 7 **19.** 3
20. 0 **21.** 8 **22.** Three hundred sixty-nine **23.** Six hundred twelve
24. One hundred twenty-eight **25.** Nine hundred three **26.** Three thousand, five hundred eight-three
27. Sixty-one thousand, three hundred forty-six **28.** One hundred nine thousand, eight hundred fifty-four
29. Twenty-three million, five hundred forty-seven thousand, nine hundred one **30.** 4700 **31.** 20,000
32. 574,000 **33.** 486,000,000
34. Four thousand, seven hundred sixty-two; fifty-six; three hundred fifty-two

MISCELLANEOUS EXERCISES 1.2

1. 9 **2.** 15 **3.** 15 **4.** 85 **5.** 586 **6.** 3898 **7.** 102 **8.** 87 **9.** 1159
10. 16,060 **11.** 18,270 **12.** 109,740 **13.** 122 **14.** 551 **15.** 1024 **16.** 1163
17. 1459 **18.** 10,709
19. **(a)** Arbuckle: $314,000; Brown: $220,200; Munchkin: $244,000; Zimkin: $368,000
(b) Life: $764,000; auto: $209,700; home: $172,500
20. **(a)** Bruce: 2260; Susan: 1645; Nancy: 1795 **(b)** Breakfast: 1010; lunch: 1695; dinner: 2995
21. Smiths: $3721; O'Connors: $5513; LaRues: $2562 **22.** 1587 km

MISCELLANEOUS EXERCISES 1.3

1. 5 **2.** 4 **3.** 2 **4.** 15 **5.** 222 **6.** 3316 **7.** 25 **8.** 27 **9.** 67 **10.** 144
11. 223 **12.** 4062 **13.** 37 **14.** 31 **15.** 98 **16.** 293 **17.** 867 **18.** 6531
19. 7 **20.** 9 **21.** 25 **22.** 52 **23.** 9 **24.** 13 **25.** 58 **26.** 106
27. $140,000; $25,583; $237,998 **28.** 1825 **29.** 180° **30.** $396 **31.** $6329 **32.** $827
33. 118,550 **34.** 3,009,059,550 **35.** 19

MISCELLANEOUS EXERCISES 1.4

1. 30 **2.** 24 **3.** 63 **4.** 3700 **5.** 20,000 **6.** 4,500,000 **7.** 8000 **8.** 4,800,000
9. 450,000,000 **10.** 192 **11.** 161 **12.** 5058 **13.** 2680 **14.** 438,600 **15.** 5,877,900
16. 1058 **17.** 819 **18.** 2744 **19.** 3402 **20.** 31,144 **21.** 29,304 **22.** 91,630
23. 456,896 **24.** 730,191 **25.** 2,534,049 **26.** 2,187,320 **27.** 3,145,446 **28.** 30,000,000
29. 120,000,000 **30.** 1,500,000,000 **31.** 28,000,000,000 **32.** $106,400 **33.** $3150
34. $24,000,000 **35.** $7395 **36.** 55,480 **37.** 5400 **38.** $792 **39.** 120,000
40. 39,200 **41.** 240 **42.** 31,536,000 **43.** 4500

MISCELLANEOUS EXERCISES 1.5

1. 8 **2.** 6 **3.** 9 **4.** 29 **5.** 53 **6.** 123 **7.** 663 R1 **8.** 1579 R1 **9.** 1712 R2
10. 58 **11.** 82 **12.** 621 **13.** 244 R11 **14.** 491 R10 **15.** 1469 R132 **16.** 7
17. 9 **18.** 20 **19.** 12 **20.** 16 **21.** 180 **22.** $227 **23.** 13 **24.** 7 **25.** 21
26. 25,000 **27.** 65 **28.** 15 **29.** $35,130 **30.** $7 **31.** 6 **32.** 11,747 **33.** 400

MISCELLANEOUS EXERCISES 1.6

1. $2 \cdot 2 \cdot 2 \cdot 3$ **2.** $2 \cdot 2 \cdot 2 \cdot 7$ **3.** $3 \cdot 5 \cdot 5$ **4.** Prime **5.** $3 \cdot 3 \cdot 11$ **6.** $3 \cdot 43$ **7.** 48
8. 40 **9.** 225 **10.** 150 **11.** 240 **12.** 720

CHAPTER 1 REVIEW TEST

1. (b) **2.** (b) **3.** (a) **4.** (e) **5.** (e) **6.** (d) **7.** (e) **8.** (a) **9.** (e) **10.** (b)
11. (c) **12.** (d) **13.** (c) **14.** (b) **15.** (e) **16.** (c) **17.** (d) **18.** (d) **19.** (b)
20. (b)

MISCELLANEOUS EXERCISES 2.1

1. 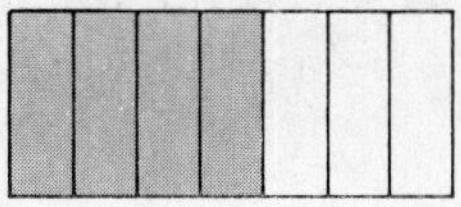 **2.** 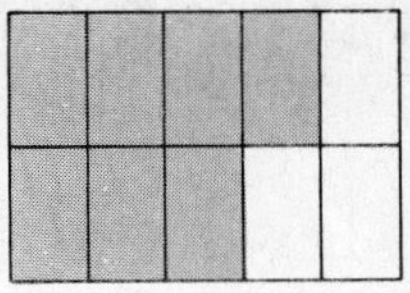 **3.** 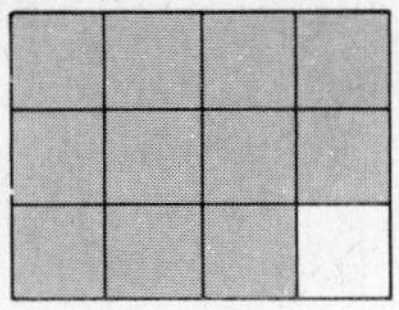 **4.** Yes **5.** No **6.** Yes

7. Yes **8.** $\frac{3}{5}$ **9.** $\frac{2}{5}$ **10.** $\frac{3}{5}$ **11.** $\frac{4}{5}$ **12.** $\frac{14}{27}$ **13.** $\frac{8}{7}$ **14.** $\frac{1}{125}$ **15.** $\frac{1}{8}$

16. $\frac{5}{16}$ **17.** No

MISCELLANEOUS EXERCISES 2.2

1. $\frac{16}{5}$ **2.** $\frac{37}{7}$ **3.** $\frac{9}{2}$ **4.** $\frac{31}{3}$ **5.** $\frac{31}{15}$ **6.** $\frac{142}{11}$ **7.** $2\frac{2}{5}$ **8.** $1\frac{1}{8}$ **9.** $2\frac{3}{7}$

10. $3\frac{5}{11}$ **11.** $11\frac{3}{5}$ **12.** $16\frac{11}{19}$ **13.** $\frac{13}{4}$ **14.** $6\frac{7}{15}$ **15.** $\frac{17}{4}$; 17 quarter cups **16.** $2\frac{17}{20}$

MISCELLANEOUS EXERCISES 2.3

1. $\frac{12}{5}$ **2.** 2 **3.** $\frac{25}{8}$ **4.** $\frac{15}{28}$ **5.** $\frac{8}{45}$ **6.** $\frac{10}{27}$ **7.** $\frac{5}{18}$ **8.** $\frac{35}{36}$ **9.** $\frac{1}{27}$

10. $10\frac{5}{6}$ **11.** $6\frac{6}{7}$ **12.** 20 **13.** $3\frac{3}{20}$ **14.** $22\frac{11}{12}$ **15.** 3 **16.** 1 **17.** $2\frac{23}{32}$

18. \$215,000 **19.** 90,000 **20.** \$$1666\frac{2}{3}$ **21.** $\frac{3}{8}$-grain **22.** 2 grains **23.** $\frac{9}{64}$ **24.** \$108

25. $280\frac{1}{3}$ **26.** \$10

MISCELLANEOUS EXERCISES 2.4

1. $\frac{14}{15}$ **2.** $\frac{21}{32}$ **3.** $\frac{10}{21}$ **4.** $\frac{3}{2}$ **5.** $\frac{7}{8}$ **6.** $\frac{14}{5}$ **7.** $2\frac{4}{9}$ **8.** $\frac{9}{20}$ **9.** $1\frac{3}{5}$ **10.** $\frac{7}{9}$

11. $1\frac{5}{9}$ **12.** $8\frac{2}{3}$ **13.** $\frac{1}{9}$ **14.** $3\frac{3}{8}$ **15.** $3\frac{1}{3}$ **16.** 80 **17.** $8\frac{5}{7}$; 9 (rounded up)

18. 450 **19.** 18 **20.** $9\frac{3}{10}$ **21.** 15 **22.** 20 **23.** \$1800 per year; \$150 per month

24. 20

MISCELLANEOUS EXERCISES 2.5

1. $\frac{5}{7}$ **2.** $\frac{2}{3}$ **3.** 1 **4.** $\frac{5}{4}$ **5.** $\frac{7}{18}$ **6.** $\frac{5}{6}$ **7.** $\frac{85}{72}$ **8.** $\frac{31}{40}$ **9.** $\frac{173}{90}$ **10.** $6\frac{3}{5}$

11. $9\frac{1}{2}$ **12.** $13\frac{19}{20}$ **13.** $6\frac{7}{10}$ **14.** $7\frac{11}{12}$ **15.** $72\frac{119}{120}$ **16.** $\frac{5}{6}$ **17.** $7\frac{7}{12}$

18. 22; $7\frac{1}{3}$ **19.** $41\frac{3}{4}$; $8\frac{7}{20}$ **20.** $57\frac{5}{8}$ **21.** $\frac{7}{100}$ **22.** $123\frac{5}{16}$ **23.** $L = 10\frac{3}{4}$; $W = 7\frac{7}{12}$

24. 4; 1; $\frac{4}{7}$ **25.** $\frac{9}{16}$ **26.** $3\frac{3}{8}$

MISCELLANEOUS EXERCISES 2.6

1. $\frac{1}{3}$ **2.** $\frac{2}{5}$ **3.** $\frac{2}{5}$ **4.** $\frac{1}{6}$ **5.** $\frac{5}{8}$ **6.** $\frac{7}{30}$ **7.** $\frac{13}{36}$ **8.** $\frac{11}{24}$ **9.** $\frac{19}{120}$ **10.** $4\frac{4}{7}$

11. $1\frac{1}{2}$ **12.** $7\frac{11}{24}$ **13.** $3\frac{1}{2}$ **14.** $5\frac{31}{45}$ **15.** $10\frac{11}{30}$ **16.** $\frac{1}{7}$ **17.** $\frac{5}{8}$ **18.** $1\frac{3}{4}$
19. $\frac{4}{9}$ **20.** $\frac{5}{8}$ **21.** $2\frac{3}{8}$ **22.** $\frac{11}{15}$ **23.** $2\frac{1}{4}$ **24.** $\frac{1}{50}$ **25.** $\frac{3}{8}$ **26.** $24\frac{5}{8}$ **27.** $3\frac{1}{4}$
28. $x = 3\frac{3}{4}; y = 4\frac{3}{4}$ **29.** $\frac{3}{8}$

CHAPTER 2 REVIEW TEST

1. (e) **2.** (e) **3.** (c) **4.** (b) **5.** (c) **6.** (b) **7.** (e) **8.** (b) **9.** (e) **10.** (d)
11. (b) **12.** (b) **13.** (a) **14.** (e) **15.** (e) **16.** (d) **17.** (e) **18.** (a) **19.** (e)
20. (a)

MISCELLANEOUS EXERCISES 3.1

1. $\frac{2}{100}$ **2.** $\frac{2}{1000}$ **3.** $\frac{1}{10}$ **4.** $\frac{5}{100{,}000}$ **5.** 4.9 **6.** 76.04 **7.** 1.386 **8.** 47.9
9. 5.029 **10.** 579.08 **11.** 4700 **12.** 20,000 **13.** 0.67 **14.** 3.15 **15.** 0.2 **16.** 1
17. $12,700,000 **18.** 3.1 **19.** 0.56
20. One hundred twelve and $\frac{37}{100}$; twenty-one and $\frac{72}{100}$; eighty-seven and $\frac{42}{100}$.

MISCELLANEOUS EXERCISES 3.2

1. 6.99 **2.** 53.012 **3.** 160.21 **4.** 197.854 **5.** 627.139 **6.** 39.2802 **7.** 14.972
8. 15.152 **9.** 205.77 **10.** 432.527 **11.** 44.419 **12.** 0.00012 **13.** 94.4 **14.** 0.015
15. 19.7 **16.** $103,987.79
17. **(a)** Kotowski: $1398.27; Dohner: $1715.69; Humbert: $1619.43
(b) Mon.: $824.18; Tue.: $713.83; Wed.: $953.35; Thur.: $1090.05; Fri.: $1151.98
18. **(a)** 1.262 **(b)** 1.238 **19.** 61.499 **20.** $129.16 **21.** 63.7 **22.** 16.25
23. **(a)** $x = 2.3$; $y = 6.4$ **(b)** 45

MISCELLANEOUS EXERCISES 3.3

1. 0.045789 **2.** 59.32 **3.** 59.438 **4.** 0.067285 **5.** 13.34 **6.** 18.27 **7.** 26.718
8. 113.6 **9.** 0.01196 **10.** 0.00000135 **11.** 5.4 **12.** 0.0036 **13.** 25.8 **14.** 28.98
15. 10.08 **16.** $22.50 **17.** $1610 **18.** $92,638 **19.** 0.0625 **20.** 15
21. $X = 0.255$; $Y = 1.11$ **22.** 1.555 **23.** 1.05 **24.** 530.4 **25.** **(a)** 46.556 **(b)** 22.644
26. $108.16 **27.** 494.91 **28.** $21.80 **29.** $89.31 **30.** 16.12 **31.** 3487.26
32. 59.928 **33.** 85.428 **34.** 587.16

MISCELLANEOUS EXERCISES 3.4

1. 54.8 **2.** 1.68 **3.** 4.98 **4.** 0.0259 **5.** 3.9 **6.** 0.123 **7.** 4.58 **8.** 78.56
9. 0.035 **10.** 0.0327 **11.** 5.3 **12.** 189.3 **13.** 0.64 **14.** 666.67 **15.** 2673.913
16. 6.333 **17.** 1.4 **18.** 17.44 **19.** 300 **20.** 0.009 **21.** $0.018 **22.** $250,000
23. **(a)** 0.9947 **(b)** 0.0053 **24.** 1.08 **25.** **(a)** 0.13 **(b)** 0.13 **26.** **(a)** 0.87 **(b)** 1.23
27. 1.43 **28.** **(a)** 3.75 **(b)** 80 **29.** 67.4 **30.** 6.4
31. **(a)** $1092.48 **(b)** $5644.48 **(c)** $156.79 **32.** **(a)** 1158.4 **(b)** 57.9 **(c)** 20.0

MISCELLANEOUS EXERCISES 3.5

1. 0.857 **2.** 0.556 **3.** 0.846 **4.** 1.103 **5.** 3.25 **6.** 23.364 **7.** $\frac{37}{50}$ **8.** $\frac{1}{500}$
9. $\frac{7}{20}$ **10.** $\frac{1}{2000}$ **11.** $1\frac{1}{5}$ **12.** $5\frac{1}{4}$ **13.** 0.327 **14.** $\frac{1}{400}$ **15.** $\frac{3}{8}$
16. **(a)** $\frac{1}{8}$ **(b)** 0.125 **17.** 0.375; 0.625; 0.5625 **18.** **(a)** $1\frac{4}{5}$ **(b)** 1.8 **19.** **(a)** 0.25 **(b)** $5.94
20. $\frac{7}{25}$ **21.** **(a)** $14\frac{3}{4}; 9\frac{2}{3}$ **(b)** 14.75; 9.67 **(c)** $142\frac{7}{12}$; 142.63

CHAPTER 3 REVIEW TEST

1. (b) **2.** (a) **3.** (c) **4.** (d) **5.** (a) **6.** (e) **7.** (c) **8.** (e) **9.** (a) **10.** (a)
11. (c) **12.** (d) **13.** (e) **14.** (a) **15.** (d) **16.** (d) **17.** (c) **18.** (c) **19.** (d)
20. (b)

MISCELLANEOUS EXERCISES 4.1

1. $\frac{37}{100}$ **2.** $\frac{49}{100}$ **3.** $\frac{3}{20}$ **4.** $\frac{11}{200}$ **5.** $3\frac{1}{2}$ **6.** $\frac{1}{2500}$ **7.** 0.29 **8.** 0.038
9. 0.025 **10.** 0.00007 **11.** 1.25 **12.** 0.3 **13.** $\frac{6}{25}$; 0.24 **14.** $\frac{1}{20,000}$; 0.00005
15. $4\frac{1}{2}$; 4.5 **16.** $\frac{1}{80}$; 0.0125 **17.** $\frac{1}{8}$ **18.** 0.27 **19.** 0.51 **20.** $\frac{3}{4}$ **21.** 0.009
22. 0.63 **23.** $1\frac{1}{4}$ **24.** $\frac{1}{10}$ **25.** 0.0625 **26.** 0.078 **27.** $\frac{1}{5}$ **28.** 0.31 **29.** $\frac{4}{5}$

MISCELLANEOUS EXERCISES 4.2

1. 43% **2.** 0.23% **3.** 430% **4.** 3.62% **5.** 40% **6.** 15% **7.** 55.6% **8.** 35.7%
9. 2% **10.** 0.075% **11.** 375% **12.** 28.05% **13.** $\frac{21}{50}$; 0.42 **14.** $\frac{1}{5000}$; 0.0002
15. 28.6%; 0.286 **16.** 0.18%; 0.0018 **17.** 60%; $\frac{3}{5}$ **18.** 220%; $2\frac{1}{5}$
19. **(a)** 2.04% **(b)** 32.65% **(c)** 65.31% **20.** 68.75% **21.** 9.835% **22.** 66.9% **23.** 12%
24. 0.25% **25.** 2.5% **26.** 33.3% **27.** 9.645% **28.** 8.3%

MISCELLANEOUS EXERCISES 4.3

1. $x = 0.54 \cdot 490$ **2.** $21 = x \cdot 460$ **3.** $50 = 0.03 \cdot x$ **4.** 8 **5.** $5\frac{2}{3}$ **6.** 192 **7.** 17,500
8. 750 **9.** 6% **10.** 2000 **11.** 25% **12.** 10 **13.** 1350 **14.** 300
15. **(a)** 4250 **(b)** 6200 **(c)** 4750 **(d)** 5300 **16.** 23,800,000 **17.** 8.75% **18.** 20%
19. 400 **20.** 42.24 **21.** 1.8% **22.** **(a)** 16% **(b)** 34%
23. 1 letter: 3.9%; 2 letters: 18.5%; 3 letters: 24.2%; 4 letters: 19.8%; 5 letters: 11.2%; 6 letters: 6.7%
7 letters: 5.7%; 8 letters: 4.5%; 9 letters: 2.4%; 10 letters: 1.4%; 11 letters: 0.7%; 12 letters: 0.4%
13 or more letters: 0.5%
24. **(a)** 58.5 **(b)** 39% **25.** 3000 **26.** 200 **27.** 27,500 **28.** 43.75% **29.** 9.75
30. 60%

MISCELLANEOUS EXERCISES 4.4

1. $\frac{312 \text{ mi}}{12.6 \text{ gal}}$ **2.** $\frac{1 \text{ kg}}{2.2 \text{ lb}}$ **3.** $\frac{x}{8} = \frac{3}{19}$ **4.** $\frac{4}{x} = \frac{5}{23}$ **5.** 20 **6.** 36 **7.** 5.6 **8.** 8
9. 4 **10.** 10.5 **11.** 7.2 **12.** 4.8 **13.** 26.7 **14.** 250,000 **15.** 13.3 **16.** 5
17. 74.1 **18.** 19.6 **19.** 1.875 in. **20.** $W = 880$; $X = 680$; $Y = 1320$; $Z = 1120$ **21.** 4200
22. 60 **23.** 1080 **24.** $2\frac{2}{3}$ **25.** 11.2 **26.** 56.5 **27.** 3.8 **28.** 186.3 **29.** 339.6

CHAPTER 4 REVIEW TEST

1. (e) **2.** (c) **3.** (d) **4.** (b) **5.** (a) **6.** (b) **7.** (d) **8.** (d) **9.** (c)
10. (b) **11.** (a) **12.** (a) **13.** (b) **14.** (b) **15.** (d)

MISCELLANEOUS EXERCISES 5.1

Note: There may be several correct answers to some of the conversion problems. (But all the answers should be close.)

1. 20 lb **2.** 30 in. **3.** 2 qt **4.** 100 g **5.** 90 m **6.** 2 L **7.** 45 kg **8.** 10 cm

9. 15 ml **10.** 1 m; 39.4 in.; 3.3 ft; 1.1 yd **11.** 300 cm; 3 m; 120 in.; $3\frac{1}{3}$ yd

12. 91.4 cm; 0.91 m; 3 ft; 1 yd **13.** 18.75 mi **14.** 320 km **15.** 56,000 mg; 0.056 kg; 2 oz; $\frac{1}{8}$ lb

16. 112,000 mg; 112 g; 0.112 kg; $\frac{1}{4}$ lb **17.** 3,000,000 mg; 3000 g; 107 oz; 6.6 lb

18. 1887 ml; 1.9 L; $\frac{1}{2}$ gal **19.** 5000 ml; 5.3 qt; 1.3 gal **20.** **(a)** $\frac{1}{3}$ grain; 333 mg **(b)** 0.5 g; 7.5 grains

21. **(a)** 1500 minims **(b)** $\frac{1}{2}$ t **22.** 1.27 cm **23.** 66 lb **24.** 114 L **25.** $117\frac{¢}{m}$

26. 5500 tons

27.

Animal	Brain Weight (oz)	Body Weight (lb)
Human	46	132
Elephant	214	15400
Gorilla	21	550
Deer	7.5	132
Beaver	1	13

28. 20 L; 21.2 qt; 84.8 cups **29.** $10.5\frac{\text{km}}{\text{L}}$ **30.** 1.65 yd **31.** 220 lb and $5\frac{1}{2}$ ft: overweight

32. $118\frac{\text{g P}}{\text{lb}}$ **33.** Give him 2.54 centimeters, and he will take 1.6 kilometers.

34. Twenty-eight grams of prevention is worth 0.45 kilogram of cure. **35.** Seats on the 45.45-meter line.

36. The cowboy wore a 38-liter hat. **37.** It hit me like 909 kilograms of bricks.

38. A miss is as good as 1.6 kilometers. **39.** 0.47 liters is 0.454 kilograms, the whole world round.

MISCELLANEOUS EXERCISES 5.2

1. 180° **2.** 36° **3.** $\frac{1}{6}$ **4.** $\frac{1}{4}$ **5.** 127° **6.** 43° **7.** 59° **8.** 30° **9.** 130°

10. 8.6 **11.** 13.5 **12.** 62 **13.** 28 **14.** 12 **15.** 21.98 **16.** 25.12 **17.** 9.42

18. 319.9 **19.** 594 **20.** **(a)** 90° **(b)** 60° **(c)** 120° **21.** **(a)** 15.7 in. **(b)** 12.7 times

22. 6000 miles (Chicago to Denver to Los Angeles to Dallas to New York to Chicago, or the reverse route)

23.

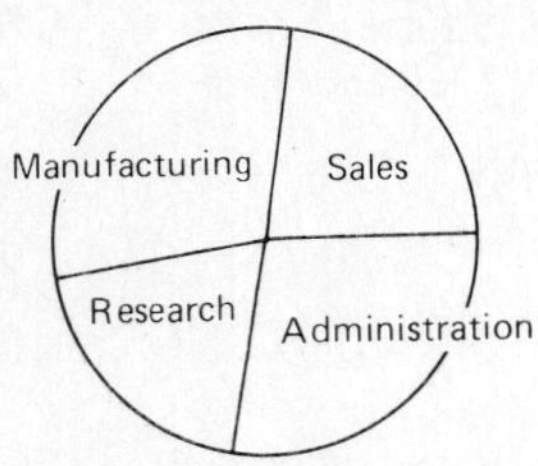

24. 179.23 **25.** 138° **26.** 25,120 **27.** 104° **28.** 54

29. 15.7 **30.** 17.792

31. **(a)** 9.6; 180°; **(b)** $1\frac{1}{24}$; 180°; **(c)** 129; 180°

MISCELLANEOUS EXERCISES 5.3

1. 21 **2.** 3.6 **3.** 25 **4.** 3.61 **5.** 14 **6.** $3\frac{1}{2}$ **7.** 1.92 **8.** $3\frac{25}{32}$ **9.** 8

10. 696 **11.** 200.96 **12.** 452.16 **13.** 0.7693 **14.** 62,800 **15.** 1392 **16.** 18.28

17. 942 **18.** 1026 **19.** **(a)** 9 mm^2 **(b)** 0.04 mm^2 **20.** 78.5 cm^2

21. **(a)** Top = bottom = 12.56 **(b)** Side = 87.92 **(c)** Total = 113.04 **22.** 664 **23.** 2826 mi^2

24. **(a)** 1500 **(b)** 1100 **(c)** 1650 **(d)** 2400 **(e)** 6650 **25.** 17850 **26.** $7\frac{1}{14}$ mi^2

27. *f*/2: 803.84; *f*/2.8: 379.94; *f*/4: 200.96; *f*/5.6: 98.4704; *f*/8: 50.24; *f*/11: 24.6176; *f*/16: 12.56
(Notice that each is about one-half of the one preceding.)

28. 16 **29.** **(a)** 137.5 **(b)** 15.3 **(c)** \$183.60 **30.** $88\frac{3}{4}$ ft^2 **31.** 18.7544 ft^2 **32.** 33.93

33. 7704

MISCELLANEOUS EXERCISES 5.4

1. 350 **2.** 2.64 **3.** 125 **4.** $2\frac{10}{27}$ **5.** 1004.8 **6.** 678.24 **7.** 4186.67 **8.** 19.386
9. **(a)** 180 **(b)** 200 **10.** 94.2 **11.** **(a)** 904.32 in^3 **(b)** 235.12 lb **12.** 523.33 cm^3
13. 423,900 ft^3 **14.** 192 **15.** 35.325 **16.** 261.67 **17.** 27,000
18. **(a)** 0.19625 **(b)** 1.9625 **(c)** 1 **19.** **(a)** 12.76 in.^{3} **(b)** 448.69 in.^{3} **(c)** 7.065 in.^{3} **(d)** 8.18 in.^{3}

CHAPTER 5 REVIEW TEST

1. (b) **2.** (e) **3.** (c) **4.** (a) **5.** (d) **6.** (e) **7.** (e) **8.** (c) **9.** (c) **10.** (b)
11. (e) **12.** (b) **13.** (e) **14.** (d) **15.** (b)

MISCELLANEOUS EXERCISES 6.1

1. 11 **2.** 2 **3.** 35 **4.** 6 **5.** 7 **6.** 10 **7.** 33 **8.** 1 **9.** 1 **10.** 3
11. 0 **12.** $-3\frac{1}{2}$ **13.** -5 **14.** $-\frac{1}{4}$

15–20.

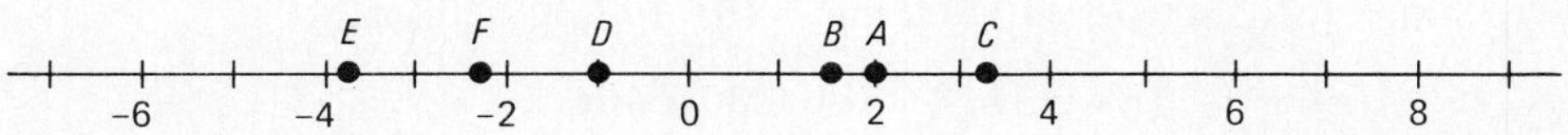

21. 55 **22.** -15 **23.** 20 **24.** 6 **25.** -4 **26.** $-2\frac{1}{4}$ **27.** -200 **28.** 20
29. 4000 **30.** -200 **31.** -18; 18 **32.** 3; 3 **33.** 11; 11 **34.** -14; 14 **35.** 2; 2
36. -10; 10 **37.** -9; 9 **38.** 7; 7

39.

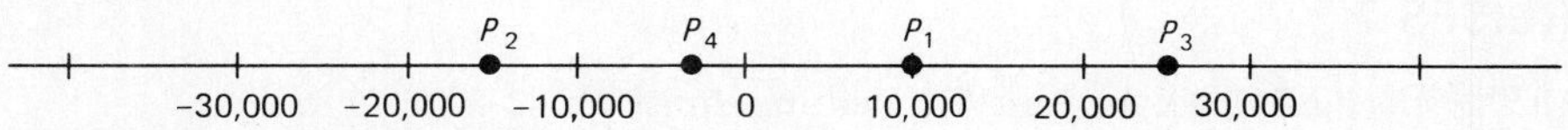

40. 410,000 **41.** \$5724.50 **42.** 0.488

43.

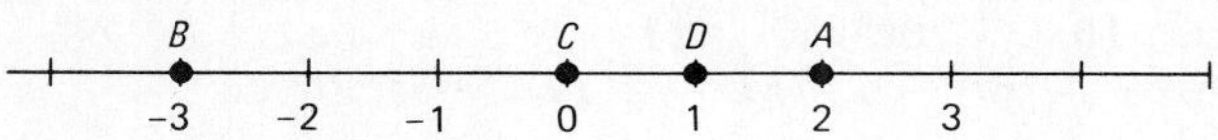

MISCELLANEOUS EXERCISES 6.2

1. 6 **2.** -17 **3.** -5 **4.** -23 **5.** 5 **6.** -8 **7.** -24 **8.** -37 **9.** -6
10. -11 **11.** 4 **12.** 21 **13.** -12 **14.** 10 **15.** -10 **16.** -11 **17.** 6
18. \$4000 **19.** 5000; -8000; 5000; $-10{,}000$ **20.** -3 **21.** A: -13; 13 B: -17; 16 C: -23; 11
22. 16; -41; 7; 18 **23.** -22 **24.** **(a)** -2 **(b)** 0

MISCELLANEOUS EXERCISES 6.3

1. -12 **2.** 48 **3.** -100 **4.** 84 **5.** -40 **6.** $\frac{1}{8}$ **7.** -41 **8.** 0 **9.** 10
10. -5 **11.** 3 **12.** 0 **13.** -50 **14.** 1 **15.** Not defined **16.** -17 **17.** 26
18. -3 **19.** 4 **20.** 3; -3; -6; 8 **21.** **(a)** 2 **(b)** -1 **(c)** -3 **(d)** 0
22. 5000; $-16{,}000$; -6000; 10,000 **23.** 2; $\frac{1}{2}$; $\frac{1}{10}$; 5 **24.** 10, -3; -4

MISCELLANEOUS EXERCISES 6.4

1. $7a + 7b$ **2.** $8x - 8y$ **3.** $5p + 5q + 20$ **4.** $10n - 10m - 30$ **5.** $11x$ **6.** $7a$ **7.** $17p$
8. $-6t$ **9.** $4m + 15$ **10.** $9t - 1$ **11.** $7y + 6$ **12.** $9r + 2$ **13.** $5t - 3$ **14.** $10x + 5$
15. $8z + 3$ **16.** $8u - 5$ **17.** $-10t + 3$ **18.** $14n + 7$

MISCELLANEOUS EXERCISES 6.5

1. 17 **2.** 24 **3.** 2 **4.** -6 **5.** 9 **6.** 2 **7.** 5 **8.** 7 **9.** 9 **10.** -10
11. -5 **12.** 6 **13.** 18 **14.** 8 **15.** -5 **16.** 9 **17.** 2 **18.** 3 **19.** 4
20. -3 **21.** 6 **22.** -2 **23.** 6 **24.** \$15,000 **25.** \$120,000 **26.** \$5000 **27.** 4
28. 150 **29.** 10,000 **30.** 20 **31.** 86

MISCELLANEOUS EXERCISES 6.6

1. 35 and 46 **2.** 34 and 41 **3.** 8 and 48 **4.** 52 and 53 **5.** 48 and 50 **6.** 28 and 31
7. 100 and 190 **8.** Ann: \$16,000; Bob: \$13,000; Chuck: \$9000 **9.** 27 **10.** Width = 7; length = 27

CHAPTER 6 REVIEW TEST

1. (b) **2.** (e) **3.** (a) **4.** (c) **5.** (d) **6.** (c) **7.** (a) **8.** (e) **9.** (b) **10.** (d)
11. (b) **12.** (e) **13.** (a) **14.** (c) **15.** (d) **16.** (c) **17.** (a) **18.** (e) **19.** (b)
20. (d)

MISCELLANEOUS EXERCISES 7.1

Note: Answers may differ in the last decimal places, because different machines round off differently.

1. 9 **2.** 9 **3.** 30,000 **4.** 116,158 **5.** 30.68 **6.** 21.066 **7.** 40 **8.** 12
9. 7,681,162 **10.** 3754.2995 **11.** 87.29151 **12.** 1769.8125 **13.** 0.368421
14. 0.1764705 **15.** 3.537219 **16.** 0.0003248 **17.** \$4,485,620 **18.** 160,244.28
19. 20.8% **20.** 27.8% **21.** **(a)** 14.13 in. **(b)** 353.25 in.
22. **(a)** 3963.5 mi **(b)** 24,890.78 mi **(c)** 260,678,340,000 mi^3 **(d)** 197,309,200 mi^2
23. 34.49 **24.** 0.01776 **25.** **(a)** $\frac{5}{16}$ **(b)** 0.3125 **(c)** 0.3066404
26. **(a)** 1243.6 **(b)** 65.9 **(c)** 18.9 **27.** 360.25
28. **(a)** 170.01 **(b)** 2515.01 **(c)** 182.34 **(d)** 2697.35 **29.** 223.29

MISCELLANEOUS EXERCISES 7.2

Note: Answers may differ in the last decimal places, because different machines round off differently.

1. 227 **2.** 18.02 **3.** 1072.33 **4.** 360,311.5 **5.** 0.3789682 **6.** 0.7595238
7. 1.5117969 **8.** 0.9960937 **9.** 8.479264 **10.** 1.700568 **11.** 65536 **12.** 32768
13. 118.58787 **14.** 4159.8579 **15.** 9653.618 **16.** 26,572.05 **17.** 663,204.36
18. 29,229.26 **19.** 53,325 **20.** 10.99 **21.** 3794

REVIEW TEST

1. (c) **2.** (a) **3.** (d) **4.** (e) **5.** (b) **6.** (a) **7.** (e) **8.** (b) **9.** (c) **10.** (d)

INDEX OF MATHEMATICAL TERMS

INDEX OF APPLICATIONS